Decade[illegible]ry-Scale Climate Variability and Change

A Science Strategy

Panel on Climate Variability on Decade-to-Century Time Scales
Board on Atmospheric Sciences and Climate
Commission on Geosciences, Environment, and Resources
National Research Council

NATIONAL ACADEMY PRESS
Washington, D.C. 1998

NOTICE: The project that is the subject of this report was approved by the Governing Board of the National Research Council, whose members are drawn from the councils of the National Academy of Sciences, the National Academy of Engineering, and the Institute of Medicine. The members of the committee responsible for the report were chosen for their special competencies and with regard for appropriate balance.

Support for this project was provided by the National Oceanic and Atmospheric Administration under Contract No. 50-DKNA-5-00015. Any opinions, findings, and conclusions or recommendations expressed in this publication are those of the author(s) and do not necessarily reflect the views of the above-mentioned agency.

Library of Congress Catalog Card Number 98-88439
International Standard Book Number 0-309-06098-2

Additional copies of this report are available from:

National Academy Press
2101 Constitution Avenue, NW
Box 285
Washington, DC 20055
800-624-6242
202-334-3313 (in the Washington Metropolitan Area)
http://www.nap.edu

COVER: *The Day It Happened*, the oil painting reproduced on the cover of this book, is the work of Ilana Cernat of Bat-Yam, Israel. Dr. Cernat is linked to the world of intermediate-scale climate change through her son Michael Ghil, a member of the panel on Climate Variability on Decade-to-Century Time Scales. *The Day It Happened* (1988) is one of several of her paintings that express her concern for the future, particularly what sort of world we will be leaving to the generations to come. A lawyer by training and profession, Dr. Cernat began studying painting in her teens. Her work has been exhibited in Romania, Hungary, Israel, and the United States, and hangs in collections in other countries as well. Her 1989 painting *The Eye of the Storm* appeared on the cover of the 1995 NRC report on natural climate variability on decade-to-century time scales.

Printed in the United States of America

PANEL ON CLIMATE VARIABILITY ON DECADE-TO-CENTURY TIME SCALES

DOUGLAS G. MARTINSON (*Chair*), Lamont-Doherty Earth Observatory of Columbia University, Palisades, New York
DAVID S. BATTISTI, University of Washington, Seattle
RAYMOND S. BRADLEY, University of Massachusetts, Amherst
JULIA E. COLE, University of Colorado, Boulder
RANA A. FINE, University of Miami, Florida
MICHAEL GHIL, University of California, Los Angeles
YOCHANAN KUSHNIR, Lamont-Doherty Earth Observatory of Columbia University, Palisades, New York
SYUKURO MANABE, Earth Frontier Research System, Tokyo, Japan
MICHAEL S. McCARTNEY, Woods Hole Oceanographic Institution, Massachusetts
M. PATRICK McCORMICK, Hampton University, Virginia
MICHAEL J. PRATHER, University of California, Irvine
EDWARD S. SARACHIK, University of Washington, Seattle
PIETER TANS, National Oceanic and Atmospheric Administration, Boulder, Colorado
LONNIE G. THOMPSON, Ohio State University, Columbus
MICHAEL WINTON, Princeton University, New Jersey

Staff

ELLEN F. RICE, Program Officer (ending September 1, 1998)
PETER A. SCHULTZ, Program Officer
DIANE L. GUSTAFSON, Administrative Assistant

CLIMATE RESEARCH COMMITTEE

Ex Officio Members

Staff

Preface

In 1990, the Intergovernmental Panel on Climate Change (IPCC) released its monumental scientific assessment on climate change. This document presented, for the first time, a broad international scientific perspective on the status of our understanding of global climate change, focusing predominantly on anthropogenic change. While first recognized as a scientific issue nearly 100 years ago and the subject of many reports, this first attempt at producing a comprehensive assessment of the problem was both timely and energizing. It helped focus our collective scientific attention on key issues, by identifying, among other things, critical gaps in our understanding of the fundamental physics, chemistry, and biology of global change.

One significant gap involved our meager understanding and documentation of natural variability in the Earth's climate system which provides a context for evaluating the significance of human-induced changes. The climate change and variability that we experience will be a commingling of the ever changing natural climate state with any anthropogenic change. While we are ultimately interested in understanding and predicting how climate will change, regardless of the cause, an ability to differentiate anthropogenic change from natural variability is fundamental to help guide policy decisions, treaty negotiations, and adaptation versus mitigation strategies. Without a clear understanding of how climate has changed naturally in the past, and the mechanisms involved, our ability to interpret any future change will be significantly confounded and our ability to predict future change severely curtailed.

Recognizing this gap, the Climate Research Committee of the National Research Council's Board on Atmospheric Sciences and Climate, organized a workshop in 1992 involving the world's most prominent climate researchers, to assess the state of understanding of natural climate variability. The workshop focused on natural climate change that occurs slowly, sometimes remaining almost imperceptible for many years, decades, or even a century. These "decade-to-century" (dec-cen) time scales are the same ones over which anthropogenic climate change is expected to manifest itself, and thus the ones most likely to confound our interpretation and prediction of observed climate change as it relates to anthropogenic change. The results of this workshop, elaborated and published in a peer-reviewed National Academy of Science volume in 1995, showed considerable progress in our understanding of dec-cen climate variability on a broad number of fronts.

At the same time the NRC workshop was being organized in the United States, the Joint Scientific Committee, an international scientific oversight body for guiding international climate research under the auspices of the World Climate Research Programme (WCRP), called on a group of experts to consider possible future directions for climate research. The results of their deliberations were published in 1992 in a report entitled *CLIVAR—A Study of Climate Variability and Predictability*. This document proposed the need for a new internationally-coordinated, interdisciplinary research program on climate variability and predictability, with decade-to-century time scale variability (natural and anthropogenic) playing a central role. As the science plan for the CLIVAR program was being developed, the United States Global Change Research Program (USGCRP), also active in the international process, began formalizing plans to advance the development of a U.S. national science plan for addressing climate variability and change on decade-to-century time scales. The manifestation of these plans would contribute to the international effort while clearly defining and articulating our own particular national scientific interests. This led to the formation of the NRC panel on Climate Variability on Decade-to-Century Time Scales (the Dec-Cen panel).

The NRC's Climate Research Committee (CRC) is the U.S. national committee to the WCRP. The Dec-Cen panel, as well as the complementary Global Ocean-Atmosphere-Land System (GOALS) and Global Energy and Water Cycle Experiment (GEWEX) panels (addressing shorter time scales and key processes), were established under the CRC to interface with the WCRP and CLIVAR organizational structures. However, consistent with the USGCRP's broad perspective regarding global change, the Dec-Cen panel not only addressed the physics-oriented CLIVAR objectives (the primary focus of WCRP projects), but also the biogeochemical aspects critical to a full understanding of the carbon cycle as well as paleoclimatology. The paleoclimate records are invaluable in their ability for providing a long record of natural climate change prior to the introduction of anthropogenic gases to the atmosphere, among other things.

In addition to its primary task of preparing an overall guiding document for a national research strategy, the Dec-Cen panel was tasked with providing scientific leadership and oversight for national programs on decade-to-century-scale climate variability; developing a strategy for monitoring, modeling, and assessing the forcing and state of the climatic system on decade-to-century time scales; reporting on our understanding of, and ability to predict, natural and anthropogenic climate variations on these time scales; serving as a formal two-way channel for apprising the scientific community and agencies of each other's needs and resources; and fostering increased communication and interaction across disciplines.

This report completes the primary task of the panel of preparing a national research strategy. To coordinate the national and international science plans, the Dec-Cen panel worked closely with the international community involved in developing the international science plan. Through an aggressive outreach program the panel worked to assure that the opinions of the U.S. climate community were accurately represented. Input was solicited by inviting prominent climate scientists to present their views to the panel, through discussions with federal agency and industrial representatives, and through broad-based e-mail solicitations. Panel representatives also participated in national workshops, major conferences, and discussion groups organized by a variety of specific disciplinary groups for the purpose of identifying and articulating the central dec-cen issues. The outcome of these efforts resulted in the development of the Science Strategy presented in this document for the advancement of our understanding of climate variability and change on decade-to-century time scales.

Understanding, predicting, and detecting future climate variability and change is an immense problem; it will not be solved by anything short of a well-focused coherent international scientific assault on the problem. This report represents those scientific issues and infrastructure considerations required to most effectively advance our understanding of climate variability and change, on decade-to-century time scales, with an emphasis toward more confidently predicting future climate conditions and detecting climate change. The report emphasizes U.S. national interests, while recognizing the global nature of the problem. Together with the international efforts, as well as other national efforts addressing phenomena dominating climate change on different time scales, or other particular aspects of the climate problem, these plans collectively represent an overall holistic perspective required to attack this problem.

Clearly, this document could not have been completed without the considerable input from members of the science community, industrial community, and federal agencies. In fact, the list of significant contributions is so long that it was deemed impractical to attempt to explicitly list all of those individuals who contributed to this document. Therefore, the Dec-Cen panel would like to extend a collective thanks and general acknowledgment to the entire community for its active involvement in this task.

Douglas G. Martinson
Chair, Dec-Cen Panel

Acknowledgment of Reviewers

This report has been reviewed in draft form by individuals chosen for their diverse perspectives and technical expertise, in accordance with procedures approved by the NRC's Report Review Committee. The purpose of this independent review is to provide candid and critical comments that will assist the institution in making the published report as sound as possible, and to ensure that the report meets institutional standards for objectivity, evidence, and responsiveness to the study charge. The review comments and draft manuscript remain confidential to protect the integrity of the deliberative process. We wish to thank the following individuals for their participation in the review of this report:

Russ E. Davis, University of California, San Diego
W. Lawrence Gates, Lawrence Livermore National Laboratory
George M. Hornberger, University of Virginia, Charlottesville
Upmanu Lall, Utah State University
Gerald A. Meehl, National Center for Atmospheric Research
Richard S. Stolarski, NASA Goddard Space Flight Center
John M. Wallace, University of Washington
John E. Walsh, University of Illinois at Urbana-Champaign
Warren A. Washington, National Center for Atmospheric Research

In addition, we appreciate the post-review material and comments from Russ Davis, Robert Dickinson, Upmanu Lall, and Peter Niiler that helped to provide a more balanced discussion in some key areas. While the individuals listed above have provided constructive comments and suggestions, it must be emphasized that responsibility for the final content of this report rests entirely with the authoring committee and the institution.

The National Academy of Sciences is a private, nonprofit, self-perpetuating society of distinguished scholars engaged in scientific and engineering research, dedicated to the furtherance of science and technology and to their use for the general welfare. Upon the authority of the charter granted to it by the Congress in 1863, the Academy has a mandate that requires it to advise the federal government on scientific and technical matters. Dr. Bruce M. Alberts is president of the National Academy of Sciences.

The National Academy of Engineering was established in 1964, under the charter of the National Academy of Sciences, as a parallel organization of outstanding engineers. It is autonomous in its administration and in the selection of its members, sharing with the National Academy of Sciences the responsibility for advising the federal government. The National Academy of Engineering also sponsors engineering programs aimed at meeting national needs, encourages education and research, and recognizes the superior achievements of engineers. Dr. William A. Wulf is president of the National Academy of Engineering.

The Institute of Medicine was established in 1970 by the National Academy of Sciences to secure the services of eminent members of appropriate professions in the examination of policy matters pertaining to the health of the public. The Institute acts under the responsibility given to the National Academy of Sciences by its congressional charter to be an adviser to the federal government and, upon its own initiative, to identify issues of medical care, research, and education. Dr. Kenneth I. Shine is president of the Institute of Medicine.

The National Research Council was organized by the National Academy of Sciences in 1916 to associate the broad community of science and technology with the Academy's purposes of furthering knowledge and advising the federal government. Functioning in accordance with general policies determined by the Academy, the Council has become the principal operating agency of both the National Academy of Sciences and the National Academy of Engineering in providing services to the government, the public, and the scientific and engineering communities. The Council is administered jointly by both Academies and the Institute of Medicine. Dr. Bruce M. Alberts and Dr. William A. Wulf are chairman and vice chairman, respectively, of the National Research Council.

Contents

Decade-to-Century-Scale Climate Variability and Change

A Science Strategy

Executive Summary

The evidence of natural variations in the climate system—which was once assumed to be relatively stable—clearly reveals that climate has changed, is changing, and will continue to do so with or without anthropogenic influences. Such variability influences society through a multitude of impacts while operating over a continuum of time scales, from seasonal through centennial and longer. Variability in climate (the time-averaged weather) manifests itself in a variety of forms, such as trends, cycles, and complex regional interactions.

Recent progress in the documentation of climate variability, and in our ability to predict certain climatic events and their regional impacts, has had a significant beneficial effect on society. Climate prediction, typically a season to a year or so in advance, has been successful to some extent for certain regions and climatic phenomena—for example, El Niño and its remote effects, tropical rainfall in Africa and Brazil, and precipitation in northwestern Europe and western North America. Some climatic properties, such as temperature or pressure, tend to preserve their spatial structure through time, or assume a limited number of related shapes, while their amplitude, phase, and sometimes geographic position change. Some of these patterns have also begun to be recognized and documented. The apparent persistence of such patterns, even allowing for their slow evolution, indicates that it may be possible to exploit these "signals" to help us understand and predict future climate variability and change.

The success of the short-term climate predictions mentioned above, together with the growing documentation of coherent climate patterns, provides some confidence that improved understanding might lead to broader and longer-range forecasting skills, with commensurate benefits to society. The extension of forecasts beyond seasonal-to-interannual time scales is not, however, a simple undertaking. This difficulty reflects the complexity of the Earth's climate system, which involves the atmosphere, ocean, land, biomass, and cryosphere, together with their various interactions. The complexity of this system is further enhanced by anthropogenic and natural changes in radiative forcing, as well as the anthropogenic changes that result from increased production of greenhouse gases. Furthermore, the slow evolution of climate associated with variability on decade-to-century scales may lead to gradual, but compounded changes in short-time-scale phenomena. Therefore, simply maintaining our present prediction capabilities may require continued investment in developing our understanding of the interaction of climate phenomena operating on different time and space scales.

The international scientific community and policymakers recognize the importance of the current successes and the potential for building on them, as well as the complexity of the problem. They have undertaken to organize global efforts to identify the most important problems yet to be resolved, and to target key research areas and regions. The present report is motivated by this international effort, as well as by our national interest in most efficiently using limited research dollars to improve the understanding of climate change, establish the means for detecting climate change, and ultimately realize broader and longer climate predictions.

DECADE-TO-CENTURY-SCALE CLIMATE VARIABILITY

This report focuses on decade-to-century-scale (dec-cen) climate variability and change. A separate report by the NRC's Global Ocean-Atmosphere-Land System Panel (GOALS) deals with seasonal-to-interannual prediction (NRC, 1998a). Certainly the physics of climate change extend across all these time scales. There are, however, fundamental differences in the nature of dec-cen and seasonal-to-interannual change: the dominant processes driving the change, the manner in which they affect society, and the manner in which society might use the predictive information. For example, on longer time scales, more of the slower components of the climate system can typically be expected to play a role in climate change. On short time scales, while the properties of the upper layer of the ocean play an important role in climate variations, as demonstrated by El Niño predictions, the deep ocean currents have little direct influence, given their slow rate of movement and change. However, on longer and longer time scales, these slow, vast deep-ocean currents may play a more important role, perhaps even a dominant one.

Similarly, the anthropogenic increase of greenhouse gases is not likely to produce a perceptible difference in climate from one year to the next, so it should have little impact on seasonal or interannual predictions for the next few years. Over longer periods, however, the effects can accumulate to produce a significant change. Likewise, climate processes that introduce very small alterations on a year-to-year basis, such as subtle yearly changes in the thickness of sea ice in polar regions (affecting the surface temperature of the ice and its albedo), might show long-term accumulative influences extending well beyond their immediate region. Therefore, although such processes might be ignored for climate prediction on seasonal-to-interannual time scales, they must be considered for longer-term climate prediction, so that the resulting "drift" of climate away from its previous state is taken into account.

In addition to emphasizing different physical components, short- and long-term climate predictions have different implications for society. Predictions involving short-time-scale events such as El Niño can be used to formulate short-term mitigating actions for any negative impacts, such as reinforcing sea walls, preparing for excessive rain, or increasing local disaster-relief funds. Anticipated variations in conditions from one season or year to the next also could be taken advantage of through short-term adaptive measures, such as altering the crops planted, adjusting water-resource management strategies, or reallocating energy resources.

On longer time scales, climate variations can lead to prolonged droughts, or can alter the frequency and distribution of severe weather events for many years. They also can influence the nature of short-term events, such as the frequency with which El Niño occurs, or their duration or severity. Such long-term changes have the potential for greatly surpassing shorter-scale variations in their societal, economic, and political impacts. Areas of the economy that could be significantly affected include agriculture, energy production and utilization, fisheries, forestry, insurance, recreation, and transportation. Water resources and quality, air quality, human health, and natural ecosystems also could suffer or improve. Consequently, responses to climate changes on decade-to-century time scales may involve investments in infrastructure and changes in policy.

For example, if it becomes clear that midwestern floods like those of 1993 and 1997 will occur more often in the next few decades, or that El Niño events will continue to be as frequent as in the 1990s, or that sea level will keep rising, then efforts to mitigate such effects will certainly involve both policy (e.g., modifications of building codes) and infrastructure changes (e.g., construction of protective and adaptive structures such as dikes and irrigation systems). Such predictions also might enable society to capitalize on opportunities, like improved planning for water-resource management, expansion or relocation of agricultural regions, prediction of prime fishing sites and times, or adjustment of insurance rates to better reflect the true risks over the lifetime of a policy.

The aforementioned differences, along with the practical need to subdivide the climate problem into more manageable units, at least for organizational purposes, have led to this initial division of climate-prediction efforts on the basis of time scale. This report represents the first stage in developing a coherent national effort for addressing decadal to centennial climate variability and change. It does this in the form of a science strategy that articulates the fundamental scientific issues that must be addressed, outlines the concepts underlying an observational and modeling program, and recommends the development of a formal, national dec-cen program that would ensure a balanced, effective approach to the scientific issues and the observational and modeling needs.

A DEC-CEN SCIENCE STRATEGY

The science strategy proposed in this volume focuses on six of the attributes of the Earth's environmental system that are considered to be most directly relevant to society and that have displayed variations on dec-cen time scales in the past, and on six components of the climate system that control these attributes. The attributes are precipitation and water availability, temperature, solar radiation, storms, sea level, and ecosystems. The controling climate components are the atmospheric composition and radiative forcing, atmospheric circulation, hydrologic cycle, ocean circulation, land and vegetation, and cryosphere.

This report describes each attribute and component, illustrates how they have varied in the past over decade-to-century time scales, explains the mechanisms involved and the interactions among them, describes their predictabilities, and presents the remaining issues and questions that surround them. The report also discusses the current state of knowledge of the climate patterns identified so far, and the outstanding issues related to such patterns. This approach maintains a scientifically sound yet socially relevant focus while naturally leading to an overall science strategy for future research, since the issues and questions serve to define and justify the science requirements presented in the plan.

DEC-CEN ISSUES

The outstanding scientific questions related to dec-cen

climate variability are fairly extensive and disparate. The panel's selection of principal issues was both guided and constrained by a few overarching questions:

- What are the patterns in both space and time of dec-cen variability, and what mechanisms give rise to them?
- What is the relationship between natural dec-cen variability and observed global warming? For example, what do we have to know about natural variability in order to detect anthropogenic change?
- How does variability of forcing (natural and anthropogenic) affect dec-cen variability?
- What is the role of the interactions among the climate components in generating and sustaining dec-cen variability?
- To what extent is dec-cen variability predictable, and what unresolved issues must be addressed to realize that predictability?

A few of the issues raised in the report are briefly described below.

Climate Patterns and Long-Term Climate Change

It has become clear that much of the accelerated global warming that has occurred since the mid-1970s is associated with a low-frequency relative phasing of two of the predominant patterns of the Northern Hemisphere: the Pacific-North American (PNA) teleconnection and the North Atlantic Oscillation (NAO). This phasing produces a spatial warming pattern that is similar to one of the "greenhouse fingerprints" predicted by models. This similarity raises several questions regarding the interaction between anthropogenic change and natural variability. For example, is the accelerated warming the result of natural variability caused by an unusually persistent coincidence of the NAO and PNA, or the result of the modification of natural modes (patterns) by anthropogenic changes in radiative forcing that alter the phasing, or some combination of both of these?

Likewise, there appears to have been a distinct change in the character (frequency and severity) of El Niño and La Niña events during this period of accelerated warming. Is this a consequence of the influence of anthropogenic change on the dominant natural modes of climate variability, or is it a natural, low-frequency (dec-cen) modulation of a high-frequency (interannual) mode? In addition to answering these specific questions, it is clear that we must determine the longevity of the current patterns, as well as their spatio-temporal variability; identify the best ways of characterizing the patterns; identify new patterns, especially in data-poor regions such as the Southern Hemisphere; determine which ones are statistical products of our analysis tools, rather than true dynamic modes; and determine the mechanisms and underlying physics that control the patterns, their sensitivities, and their evolution in space and time, including their interaction with anthropogenic and natural changes in radiative forcing, and their interaction amongst themselves.

Radiative Forcing

The redistribution of greenhouse gases between the ocean, atmosphere, and biosphere, and the manner in which these fluxes influence tropospheric greenhouse-gas concentrations on dec-cen time scales, are of central importance to dec-cen climate change. Accurate predictions of changes in greenhouse-gas concentrations and their effects on radiative forcing will require an improved understanding of the complex interactions between the physics (including radiation), biology, and chemistry (including photochemistry) of the climate system that are driving the fluxes, redistributions, and changes. For some of these gases, such as methane, N_2O, and extratropical ozone, it is not even clear what has caused the recent changes. Better understanding of the cloud/water-vapor/radiation processes and feedbacks is needed as well, because they will strongly influence climatic response to the increased radiative forcing associated with higher greenhouse-gas concentrations.

Understanding of the radiation, chemistry, and dynamic coupling in the upper troposphere and lower stratosphere must be improved in order to more accurately predict how stratospheric ozone will recover from depletion by anthropogenic halocarbons, and to determine how aerosols influence dec-cen climate variability. A number of issues must be addressed regarding the potential role of changes in solar radiation on climate. For instance, how representative are proxies for solar activity (e.g., sunspots) of actual total solar irradiance on dec-cen time scales? To what extent are dec-cen climate changes related to changes in the sun's output?

Other Issues

A variety of discipline-specific issues require attention in order to better understand dec-cen climate variability. Many of these issues require inter- and multi-disciplinary perspectives and efforts. A better understanding needs to be obtained of the nature of dec-cen changes and variations that have occurred in the portions of the climate system that are especially sensitive to climate changes. These sensitive components include cloud cover, sea-ice fields, terrestrial water, and the subsurface ocean. Improved understanding is also needed of how the various components of the climate system interact to produce dec-cen climate variability, as distinguished from the variability that is a product of variations internal to a single component. The atmosphere works on very fast time scales, so coherent decadal-to-centennial variability within the atmosphere is expected to be attributable, at least in part, to interactions with the other, slower components of the climate system, or else to be a response to changes in the radiative forcing. (Nonlinear internal feedbacks are a distinct possibility, though). On the other hand, modeling studies suggest that the ocean is susceptible to cycles of variability on dec-cen and longer scales that are caused by mechanisms internal to the ocean alone. Are these or other mechanisms responsible for driving dec-cen climate variability and its spatial patterns? A related set of questions is: How do processes and changes within one component influence other aspects of the climate system, how are changes in one region transmitted to other regions, and what components and time scales are involved in such telecommunication? For example, how do large-scale dec-cen changes in the atmospheric circulation influence the seasonal-to-interannual variability of severe storms? How do the details of the planetary boundary-layer physics and biogeochemistry, and the Earth's surface characteristics, influence the propagation of climatic variability or the transfer of greenhouse gases between various components of the climate system?

A U.S. DEC-CEN PROGRAM

In view of dec-cen climate variability's intrinsic scientific interest, its direct importance to society, and its involvement with variability on other time scales, the NRC Dec-Cen panel recommends the initiation of a national program designed to increase understanding of this topic. The initial design of this program would address the issues that are outlined above, while maintaining flexibility and adaptability so that new directions and opportunities can be pursued as our understanding is improved and research directions are refined.

In order to address the key issues, the U.S. Dec-Cen Program must include:

- a long-term, stable observing system;
- a hierarchical modeling program;
- appropriate process studies; and
- a means for producing and disseminating long-term proxy and instrumental data sets.

It is essential that all of these elements be present, because the paradigm developed for the study of climate variability on seasonal-to-interannual time scales cannot be applied to the study of dec-cen climate problems. Studies of short-time-scale climate problems have generally used a process of generating hypotheses and models that can be quickly evaluated and improved through analysis of existing historical records or observations of near-term climate variability. For dec-cen problems, the paleoclimate records are still too sparse and the historical records too short for this process to be applied; as for future records, multiple decades of observations are required before even a nominal comparison to model predictions can be made. Furthermore, the change in atmospheric composition as a consequence of anthropogenic emissions represents a forcing whose future trends can be estimated only with considerable uncertainty. As a result, making progress in dec-cen-scale prediction will require heavy reliance on improved and faster models, an expanded paleoclimate data base, and assumed scenarios for anthropogenic emission. Without the benefit of real-time observations for constant model validation and improvement, a substantial effort will be needed to validate models through alternate means, to improve understanding of the limits and implications of the proxy indicators constituting the paleoclimate records, and to monitor actual rates of emissions and atmospheric concentrations of radiatively active atmospheric constituents. As for future observations, we can only now begin collection of the data that will ultimately aid future generations of scientists in further understanding dec-cen climate variability and change.

The ultimate research objective of a U.S. Dec-Cen Program would be to define, understand, and model dec-cen climate variability and change (natural and anthropogenic), so that the extent to which they are predictable can be determined. If it can be shown that they are indeed predictable, the ultimate practical aims of a Dec-Cen Program would be to design and implement a complete prediction system, building on the emerging seasonal-to-interannual prediction systems now being constructed; to predict future decadal-to-centennial variations to the extent possible; and to learn to use these predictions for the benefit of all. The program also would provide a means for detecting climate change, which will be necessary for national and international policy decisions.

Climate variability and change on dec-cen time scales involve all of the elements of the U.S. Global Change Research Program: natural and anthropogenic variability and change; past, present, and future observational networks and data bases; modeling; and the physical, chemical, and biological sciences, with their implications for society. A U.S. Dec-Cen Program should encompass the dimensions of these elements pertaining to dec-cen climate variability and change, with particular attention to important aspects that have not received coordinated programmatic support, such as the effects of aerosols on climate, and coupled ocean-atmosphere modeling of long-term climate change, among others. The U.S. Dec-Cen Program would serve as the primary contribution of the United States to the DecCen and Anthropogenic Climate Change (ACC) components of the international Climate Variability and Predictability (CLIVAR) Programme of the World Climate Research Programme (WCRP) and would provide invaluable input to the Intergovernmental Panel on Climate Change (IPCC). It must also be well coordinated with CLIVAR's seasonal-to-interannual climate-variability component (GOALS) and other WCRP activities such as GEWEX, ACSYS, and WOCE, as well as with the components of the International Geosphere-Biosphere Programme, such as the PAGES (Past Global Changes) program. A successful Dec-Cen Program should lead to a deeper knowledge of the interconnections between seasonal-to-interannual variability and dec-cen variability, an improved ability to detect anthropogenic climate change, and an improved understanding of how natural dec-cen variability is anthropogenically modified.

Decadal-to-centennial climate variability is a subtle phenomenon. It proceeds too slowly to be perceptible to our senses, but its cumulative effects ultimately define the life prospects of future generations. Informed stewardship of the Earth's resources for the generations to come must draw on the insights afforded by model predictions and extrapolation of observations that can give us a glimpse of the future. A U.S. Dec-Cen Program will be the first step toward assuming this responsibility.

1

Introduction

SOCIETY AND A VARYING CLIMATE SYSTEM

As the glaciers of the last ice age receded and temperatures rose, humans moved into new territories and began to raise crops rather than seek them. The establishment of agriculture contributed to the gathering together of people in stable communities, and to the creation of early cities. In many regions, these agricultural communities were sensitive to the stability of the climate. While they might survive a singularly bad year or two, they were often vulnerable to prolonged or abrupt anomalies. Indeed, there is much circumstantial evidence to suggest that prolonged climatic variations contributed to the collapse of several well-established civilizations at certain times in the past (Weiss et al., 1993). For example, shifts in precipitation patterns in the early part of this millennium led to the demise of irrigation-based agriculture in Central America and the Peruvian highlands, causing starvation, population dispersal, and the end of once-prosperous civilizations.

Today we tend to think of society as well insulated from such catastrophes, yet with agriculture increasingly focused in certain regions of the globe, populations concentrated in large urban agglomerations, and world economic markets more responsive and competitive, our society's well-being and stability may be even more susceptible to global-scale climate change than were the societies of earlier civilizations. With the world's population nearing six billion, and continuing to increase at unprecedented rates, the security provided by a stable climate, and our potential vulnerability to its change, is becoming increasingly recognized.

With this increasing awareness comes greater evidence that climate has varied significantly in the past, and will continue to vary over time scales of decades to centuries. In 1992 and 1993, ice cores approximately 3 km long were extracted from the heart of the Greenland ice sheet, revealing changes in the Earth's climate system over the last 150,000 years or so (White et al., 1997a). One of the most remarkable revelations of these cores was the fact that the climate in the Holocene (the last 10,000 years)—a period that we might consider representative of our modern climate conditions—has undergone considerable natural variation. For instance, evidence from the tropics shows that large hydrologic changes have characterized much of the Holocene, with major impacts on biota and human societies. Prior to the Holocene, as the Earth warmed from the last glacial maximum (approximately 18,000 years ago), the climate system underwent large swings or cycles, and, even more surprisingly, abrupt temperature changes in decades or even shorter periods.

The long-held, implicit assumption that we live in a relatively stable climate system is thus no longer tenable. Furthermore, compounding the inevitable hazard of natural climate variations is the potential for long-term anthropogenic climatic alteration. The likelihood of changes arising from human influence adds another element of doubt to the possibility of predicting future climatic states and stability on these longer time scales; moreover, the uncertainty associated with the natural variability of the climate system precludes our ability to clearly assess human-induced climate change. Together, the evidence of natural variations and the potential for anthropogenic change have altered our way of viewing the climate system: Climate has changed and will continue to do so with or without anthropogenic influences, and a society that has been built around the perception of a stable climate system can only benefit by improving the understanding, assessment, prediction, and early detection of such change—both the natural variability and any possible anthropogenic changes.

Better understanding and prediction are particularly important for climate variability over long time scales, since such change has the potential for surpassing the significant social, economic, and political impacts of shorter-scale variations, which are often addressed through disaster relief. Over decades to centuries, the impacts of climate change can be considerable, and adaptation and mitigation (of both the forcing and the response) are likely to involve policy decisions and investments in infrastructure. Changes in frequency and intensity of extreme weather events may accompany such changes in climate (Karl et al., 1996), such as the devastating Midwestern floods that struck the United States in 1993 and again in 1997. The remarkable change in the flood frequency of the American River above Sacramento, California, is the subject of a current NRC study in the Water Science and Technology Board. The Folsom Dam was built in 1945 to provide flood protection for Sacramento. Eight floods greater than the largest flood in the 1905-1945 period have occurred since 1945. A similar situation exists for several of the other Sierra Nevada rivers in California. These high floods have led people to question the level of flood protection actually provided by the dam, and, more important, how flood risk should be analyzed.

Better information on likely climate change and the associated regional patterns—for example, the probability that such floods may occur in clusters, say six or seven times over a 20-year period—would not only permit the mitigation of negative impacts but afford the opportunity to exploit positive impacts. Governments and individuals alike would benefit from advance knowledge of any climate changes that would have a major impact on agriculture, energy production and utilization, water resources and quality, air quality, health, fisheries, forestry, insurance, recreation, and transportation—all fundamental to society's well-being, all vulnerable to any prolonged change or abrupt shift in our climate system. Not only would society benefit from increased climate-prediction skill by being better prepared to ward off adverse climatic consequences, but advance knowledge of climate variations would also enable society to capitalize on opportunities, such as increased geographical ranges for certain crops.

Unfortunately, the subtlety of slow changes over long time scales (relative to diurnal, seasonal, and interannual variations) tends to disguise their potential long-term severity, and thus limits society's willingness to address them in advance; this lack of urgency is exacerbated by the uncertainty in scientists' ability to forecast such change. Given the requisite understanding of climate variability, we hope to ultimately forecast and detect alterations in climate change (distinguishing natural variability from anthropogenic change), providing a rational basis for future policy and infrastructure-management decisions.

The limitations of the instrumental data on which our current state of understanding is based are readily exposed by evaluating their ability to help answer some of our most fundamental questions involving decadal or centennial change. For example, questions such as "Is the planet getting warmer? Is the hydrologic cycle changing? Are the atmospheric and oceanic circulations changing? Are the weather and climate becoming more extreme or variable? Is the radiative forcing of climate changing?" cannot yet be answered definitively. Each one of these apparently simple questions is actually quite complex, both because of its multivariate aspects and because global spatial and temporal sampling is required to address it adequately. The global observing systems needed to provide the answers are either inadequate or non-existent. For science to provide society with the information it needs, better data are essential. The models that will yield predictions require these data to improve our understanding of decade-to-century-scale climate change, its rate and range of variability, its likelihood and distribution of occurrence, and the sensitivity of the climate to changes in the forcing (natural and anthropogenic).

A U.S. DEC-CEN SCIENCE STRATEGY

The fundamental need to develop a good scientific understanding of climate variability and change over decade-to-century time scales, the inadequacy of our current understanding, and the limited resources available to increase this understanding all point to the need for a nationally recognized dec-cen science plan. The present report articulates the primary scientific issues that must be addressed in order to advance most efficiently toward the necessary understanding. In developing this plan, the members of the Dec-Cen panel have taken special care to recognize that research directed toward decade-to-century-scale change and variability will differ in two remarkable respects from research directed at shorter-time-scale variability.

First, research on these intermediate time scales is relatively new. As noted above, only recently have we obtained sufficiently long high-resolution paleoclimate records to allow the examination of past change on dec-cen time scales, and acquired faster computers and improved models that can perform the long simulations needed for studying such change. Consequently, we are on the steep slope of the learning curve, with new results and dramatic insights arising at an impressive rate. The fundamental scientific issues requiring our primary attention are evolving rapidly. Flexibility and adaptability in response to new opportunities and promising directions will be imperative if we are to optimally advance our understanding of medium- and long-range climate change and variability.

Second, the paradigm developed for the study of climate change on seasonal-to-interannual time scales cannot be applied to the study of climate problems on longer time scales. We have recently achieved considerable success in studying short-time-scale climate problems by generating hypotheses and models that are quickly evaluated and improved through analysis of the existing and rapidly expanding instrumental records. For longer-time-scale problems, the existing paleoclimate records are still too sparse and the historical records too short; as for future records, multiple decades will be required before even a nominal comparison with model predictions becomes possible. Furthermore, the change in atmospheric composition as a consequence of human actions represents a forcing whose future trends can be estimated only with considerable uncertainty. Making progress in dec-cen climate prediction will require heavy reliance on improved and faster models, an expanded paleoclimate database, and assumed anthropogenic and natural forcing scenarios. The inherent slowness of obtaining new dec-cen time-scale climate observations necessitates the use of additional climate-data sources (e.g., paleoclimate proxy data) to most efficiently validate and improve the models used to assess dec-cen climate variability and change. Considerable effort is required to use such alternative means, because of the steps that must be taken to understand the limits and implications of the proxy indicators constituting the paleoclimate records. Considerable effort is also needed to monitor actual rates of anthropogenic emissions, as well as natural concentrations of radiatively active atmospheric constituents that force climate on dec-cen time scales. We can only begin collection of those data that will ultimately aid future generations of scientists in understanding decade-to-century-scale climate variability and change.

This Dec-Cen report identifies the fundamental science issues that must be addressed in order to realize the following ultimate goals:

• *Characterize and assess natural climate variability.* Achieving this objective will require a solid statistical grasp of natural variability that will serve as a baseline for gauging anthropogenic change. This will help to reduce a vast, complex system to manageable components that encapsulate its key aspects and allow us to evaluate its mechanisms and determine the likelihood of future changes. Meeting this goal will depend on the availability of greatly expanded paleoclimate and historical databases, and on believable simulations by comprehensive climate models.

• *Design a comprehensive system to forecast change in the climatic mean and in climate variability.* Developing such a predictive capability demands a good understanding of the climate system, tested through controlled hindcasting experiments. A forecasting system is required in order to assess the likely response to changes in the forcing, which will then permit us to address important questions regarding adaptation versus mitigation measures, especially for anthropogenic climate change. Some reliable indication of future change can be realized in the interim through existing models or statistical formulations.

• *Develop a strategy for detecting climate change.* This strategy will provide the basis for testing and refining our ultimate predictive capabilities, while the relevant observations will provide the ground truth for such predictions. Reaching this goal will require identification of the sensitive components of the climate system that must be monitored to evaluate both natural and anthropogenic climate change. Understanding and characterization of the natural variability of the climate system on dec-cen time scales are crucial if the anthropogenic "signal" is to be distinguished from the natural climatic "noise." All statements about detection of anthropogenic climate change imply knowledge of the background variability, so we must achieve greater certainty about the latter.

• *Provide the physico-biogeochemical parameters or parameterizations required by social scientists for socioeconomic and environmental impact assessments and basic human-dimensions studies.* The societal consequences of climate variability on dec-cen scales—those of the human lifetime—are likely to be quite different from those of both shorter and longer time scales. Human-dimensions studies specific to the dec-cen time scale need to be performed, and scientists must be able to provide the necessary climate-related information.

Predicting and assessing the consequences of climate change and climate variability over dec-cen time scales will involve considerable scientific breadth: observing past, present, and future climate; understanding the processes of natural and anthropogenic change and variability; and modeling variability and change through a hierarchy of approaches. Potential consequences can be properly addressed only within the holistic perspective afforded by such breadth. This science strategy attempts to provide that perspective. Our strategy for achieving it is to include components that have already received considerable and widespread attention (e.g., those aspects of anthropogenic climate change highlighted in the recent Intergovernmental Panel on Climate Change document (IPCC, 1996a), while fleshing out the relevant issues of components that have received less institutional consideration (e.g., natural variability, and the interactions between natural and anthropogenic influences). Thus, the bulk of this report describes the latter, while including overviews of the former at the level needed to confer the necessary holistic dec-cen perspective.

2

Climate Attributes That Influence Society

Examining the ways people and their activities depend on or are controlled by climate reveals the climate variables that have the most influence on our lives and well-being. Our food production and agricultural systems depend highly on local temperature, precipitation, and the amount of sunlight during the growing season. Our forests, which provide wood for construction as well as habitat for our heritage of flora and fauna, depend similarly on particular temperature and precipitation regimes. Our water resources depend on precipitation and temperature (which controls evaporation and melting of ice and snow) and on the natural ecosystems and man-made features that affect water storage and control runoff. Extremes of precipitation (or other processes in the hydrologic system) can lead to drought or floods, often accompanied by fire or severe erosion, which endanger natural habitats, communities, and resources. Our coastal regions are especially sensitive to changes in sea level; long-term changes may involve only a slow rise in sea level, but water levels can be raised rapidly by storm surges, inundating coastal communities and wetlands. Our health depends directly on temperature and humidity (both of which contribute to heat stress) and on ultraviolet radiation (which contributes to skin cancer); indirectly, human health is affected by temperature, precipitation, land cover, and land use, which together contribute to determining the pathways for disease-bearing vectors. Ecosystems on land and in the oceans provide habitat for the biological diversity of flora and fauna that supply us with food, medicine, recreation, and other resources; all are dependent on a wide array of climatic variables. Finally, climate is a fundamental driver of our economic activities, contributing to the demands for energy, the maintenance of our food and fiber resources, the safety of our transportation systems, the appeal and lifestyles of different regions, the maintenance of our natural biological resources, the availability of outdoor recreation, and much more.

These complex and extensive interconnections point to the six attributes influencing the Earth's climate system that appear to be of most importance to society. These are:

- precipitation and water availability;
- temperature;
- solar radiation;
- storms;
- sea level; and
- ecosystems.

The value of their mean state or condition, how they vary over time (on scales of days, seasons, years, decades, and centuries), their character and extent geographically, and the frequency and persistence of extreme values all determine the availability of resources on which we depend, while influencing our ability to lead healthy, productive lives. Certainly other attributes are important to our environment and well-being, such as winds (which we treat below, under "Storms"), air quality, and even atmospheric composition in general. However, with regard to developing a science strategy for understanding future climate change and variability, these six attributes are of particular importance. They have been demonstrated to undergo significant variability over decade-to-century time scales in the past and are therefore likely to do so in the future, and they are intimately entwined within the climate system that is the focus of this science plan. Atmospheric composition also satisfies these criteria, but it is so essential to the Earth's radiative balance and atmospheric dynamics that we have opted to include it in our discussions of fundamental climate-system components in Chapter 5.

The research strategy outlined in this report focuses on the scientific issues that must be addressed to best advance our understanding of how these six attributes can be expected to change in the future. Being able to predict changes in their properties, and to recognize the conditions that indicate when significant changes are actually underway, will enable society to optimize its activities and prepare for the changes in the most cost-effective way possible. These capabilities will enhance our environmental security and sustain our continued economic success.

This chapter expands on some of the important, most societally relevant influences exerted by these attributes, documents our knowledge of how they have changed in the past, and explains why we have chosen to focus on them now.

PRECIPITATION AND WATER AVAILABILITY

Freshwater is the very basis of terrestrial life, and is arguably its most precious natural resource. Water influences nearly every aspect of society and day-to-day life. From the huge amounts of freshwater required for modern industrial and agricultural production to the bucket of clean water so highly prized in less developed countries, the uninterrupted supply of clean freshwater is necessary for the overall health and continuance of our societies and economies. In addition, freshwater distribution influences energy production and utilization, water quality, fisheries and land ecosystems, forestry, insurance, recreation, and transportation. The longevity of aquifers, the reliable flow of rivers, and the fall of rain determine where civilizations can grow and prosper. Significant investments in our infrastructure, such as the construction of dams and levees, and water-resource planning and management in general, are based on our current understanding of the supply, storage, and dispersal of freshwater. Any changes or disruptions in the freshwater cycle as we have come to know and rely on it can thus have widespread consequences, with implications for all levels of society and every individual in it.

Variations in the water supply will have more serious effects on some societies than on others. Less developed countries, particularly those with semi-arid climates, marginal agriculture, and rigid social structures, are clearly vulnerable to growing-season failures: The history of northeastern Brazil is replete with examples of major failures of growing-season rainfall (see Figure 2-1) that caused mass migrations of Nordestinos to other parts of Brazil (Magalhaes and Magee, 1994). More developed societies, through their economic prowess, are less vulnerable to the year-to-year variations of precipitation. For example, the record rainfalls over the midsection of the United States in June to August of 1993 led to record flooding (Kunkel et al., 1994; Bell and Janowiak, 1995); the Mississippi River was above flood stage for almost three months at St. Louis. This resulted in extraordinary damage (estimated at $15-20 billion—see Changnon, 1996), yet the flood, while causing considerable local hardship, produced only a blip in the U.S. economy. Similarly, the record summer drought of 1988 caused an estimated $30 billion in agricultural damage alone (Trenberth and Branstator, 1992), but the strength of the U.S. economy (if not the balance sheets of the people in the region) was easily able to withstand this climatic event.

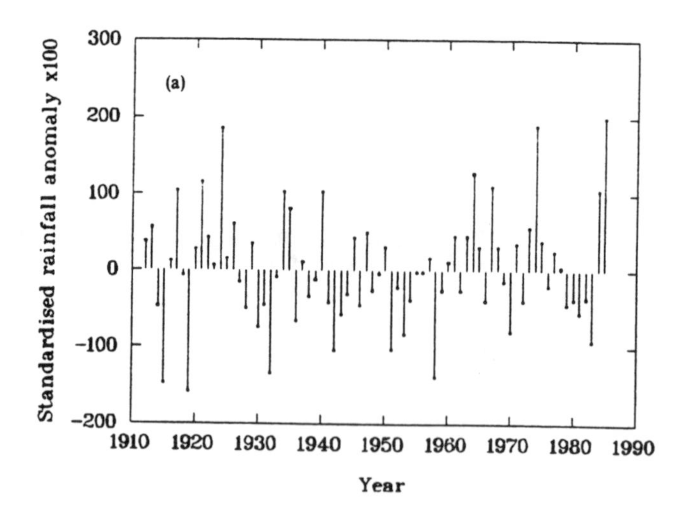

FIGURE 2-1 Northeast Brazil rainy-season (Feb-May) standardized precipitation anomalies. (From Ward and Folland, 1991; reprinted with permission of John Wiley and Sons, Ltd.)

As the time scale of precipitation variability increases, even the most developed countries become vulnerable. The United States enjoys an enviable agricultural sector that has become more efficient over the years, employing an ever-decreasing share of the population in the task of feeding its, and the world's, people, and showing great resilience in recovering from the random flood or drought, no matter how severe. But during the 1930s, when the economy was particularly fragile, an entire decade of low rainfall caused migrations and dislocations in the United States similar to those of northeastern Brazil. Indeed, recent paleoclimatic evidence from enclosed lakes (Laird et al., 1996) suggests that such droughts were considerably more severe and longer-lived in the past, relative to what we have experienced in the past few hundred years (Figure 2-2). Other parts of the paleoclimate record also suggest that such severe droughts are not unprecedented. Even today, when the economy of the United States is far more stable than during the Depression years, a decade of poor rainfall in the fertile agricultural regions would lead to economic dislocations and would place grave strains on the national and global economy. More frequent occurrence of floods like those in the Midwest in 1993 and 1997 would have similar types of effects.

The patterns of rainfall in the Sahel region of Africa also show decadal- and centennial-scale variability (Figure 2-3). The devastating impact of prolonged low rainfall on the mostly nomadic societies of sub-Saharan Africa has required massive and continuing infusions of world resources to avoid even greater disasters (Glantz, 1994). Such long periods of drought leave little room for adaptation by vulnerable populations, or for future economic development, and they affect the intellectual development of children in ways that will echo through generations.

An even longer drought may well have spelled the doom of the Classic Maya civilization (Hodell et al., 1995). The period AD 800-1000 was unusually dry, and corresponded with the decline of the Classic Maya. It is not difficult to imagine that a society well adapted to a given level of precipitation would find it difficult to adapt to a long period of drought, in this case one lasting 200 years.

The longest precipitation record comes from the Greenland ice cores (see, e.g., Grootes, 1995), which provide information on snow accumulation on the Greenland ice sheet for the last 100,000 years and more. Annual snow layers are discernible for the more recent millennia; Figure 2-4 shows snow accumulation during an 8,000-year period during which Greenland suddenly emerged from glacial conditions at about 15,000 BP, fell back into glacial conditions during the Younger Dryas period, and finally emerged into our current interglacial state. Superimposed on these upheavals was a background of precipitation variability on time scales ranging from interannual to centennial. We may safely assume that dec-cen variability in precipitation has always been with us, and thus can be expected in our future as well.

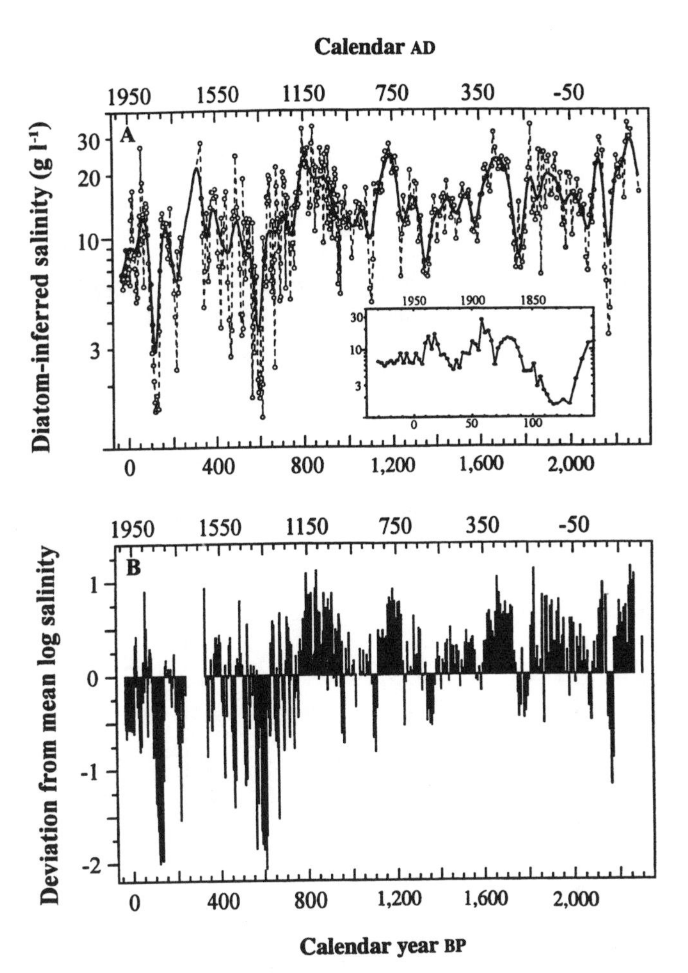

FIGURE 2-2 Reconstruction of the salinity of Moon Lake in the Northern Great Plains of the United States for the last 2,300 years. Upper panel: Logarithms of diatom-inferred salinity; smooth curve is filtered data. Bottom panel: Log deviation from 2300-year mean. Greater salinity implies greater desiccation in both curves. Note that the 1930s drought (see inset) is relatively minor relative to the more severe and long-lived droughts of the preceding millennium. (From Laird et al., 1996; reprinted with permission of Macmillan Magazines, Ltd.)

TEMPERATURE

Temperature probably influences our day-to-day comfort level more than any other environmental factor. Society is especially vulnerable to long-term changes in temperature because of the importance of temperature conditions to crop growth, heat stress, energy usage, and recreation. Also, the deleterious impacts of decade-to-century-long anomalies in temperature may be compounded by anomalous hydrologic conditions, with direct effects on agriculture, water supply, ecosystem stability, and so on. Agricultural vulnerability is higher in the less developed countries, where there are few safeguards to limit the consequences of temperature changes. However, as with freshwater, even highly developed countries are vulnerable to temperature changes as the duration of anomalies and magnitude of variability increase. Such changes may increase with a changing mean climate as noted by Karl et al. (1996), because even small shifts in the climatic mean may lead to significant increases in the number of extreme temperature events, such as the heat waves that sporadically strike during summer months in the midwestern and eastern United States, or the number of frost days. These directly influence agriculture through a variety of means.

Temperature changes are also directly related to changes in energy consumption for heating and cooling (Figure 2-5), which can have immediate impacts on consumer supplies and energy costs. From a broader perspective, because the global temperature distribution is the engine driving the Earth's climate system, temperature controls the primary circulation patterns, which in turn influence the precipitation and evaporation patterns, the tracks and intensities of storms, and other large-scale climate patterns as well. Consequently, while the direct impact of temperature change on society is considerable, its indirect impacts, through its influence on the other important climate attributes, are enormous.

There is considerable evidence documenting modern change in temperature over decades and centuries. Of particular relevance are those variations in the period leading up to the recent warming trends observed in instrumental records, because they give a clear indication of the nature and magnitude of the climate system's natural variability. Variations during the last 150 years or so of the modern instrumental records may reflect both natural and anthropogenic change. Over this last 150 years, temperatures have increased in most parts of the globe. However, this increase has not been uniform geographically and has not been steady over time (Figure 2-6). In the Southern Hemisphere, temperatures have increased more or less monotonically since the turn of the century, whereas in the Northern Hemisphere,

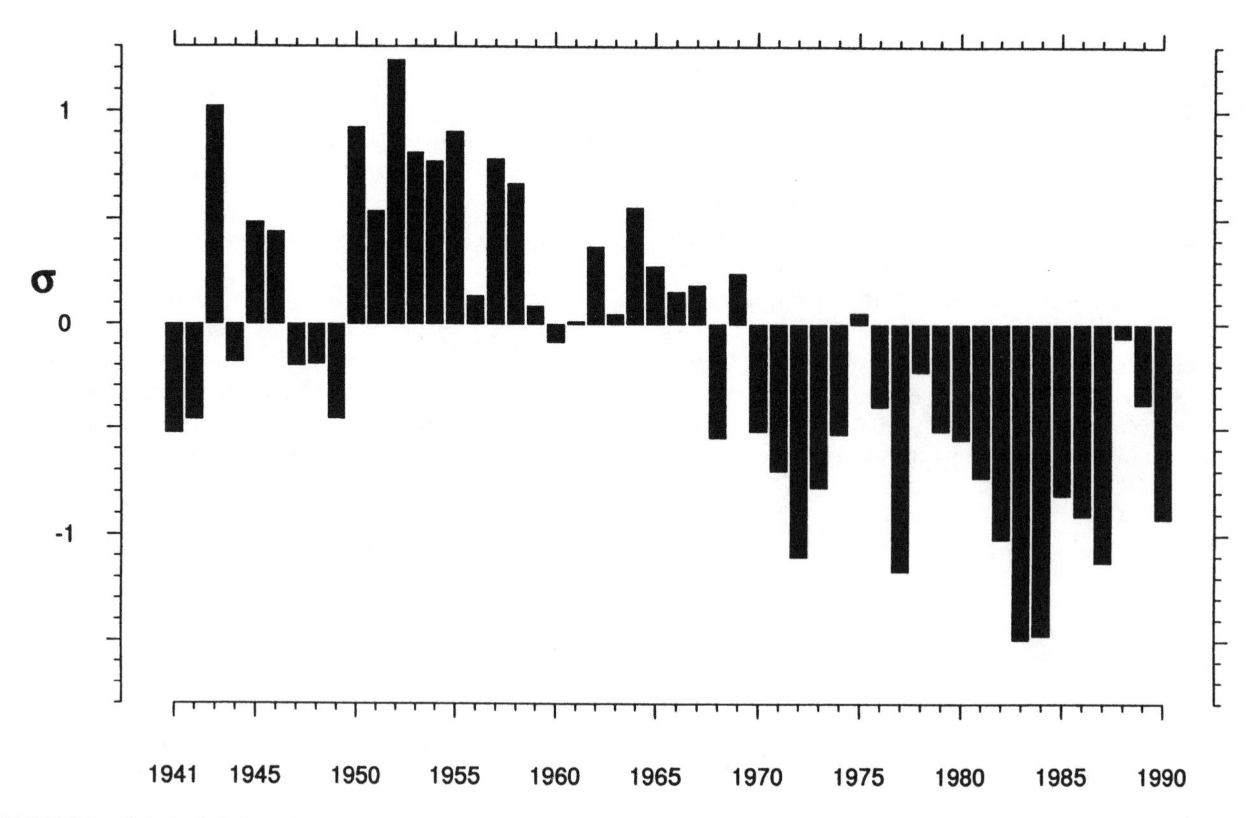

FIGURE 2-3 Sahel rainfall as a function of time. Time series of yearly average normalized April-to-October departures for 20 sub-Saharan stations located between 11° and 18° W of 10° E. (From Ropelewski et al., 1993; reprinted with permission of the American Meteorological Society.)

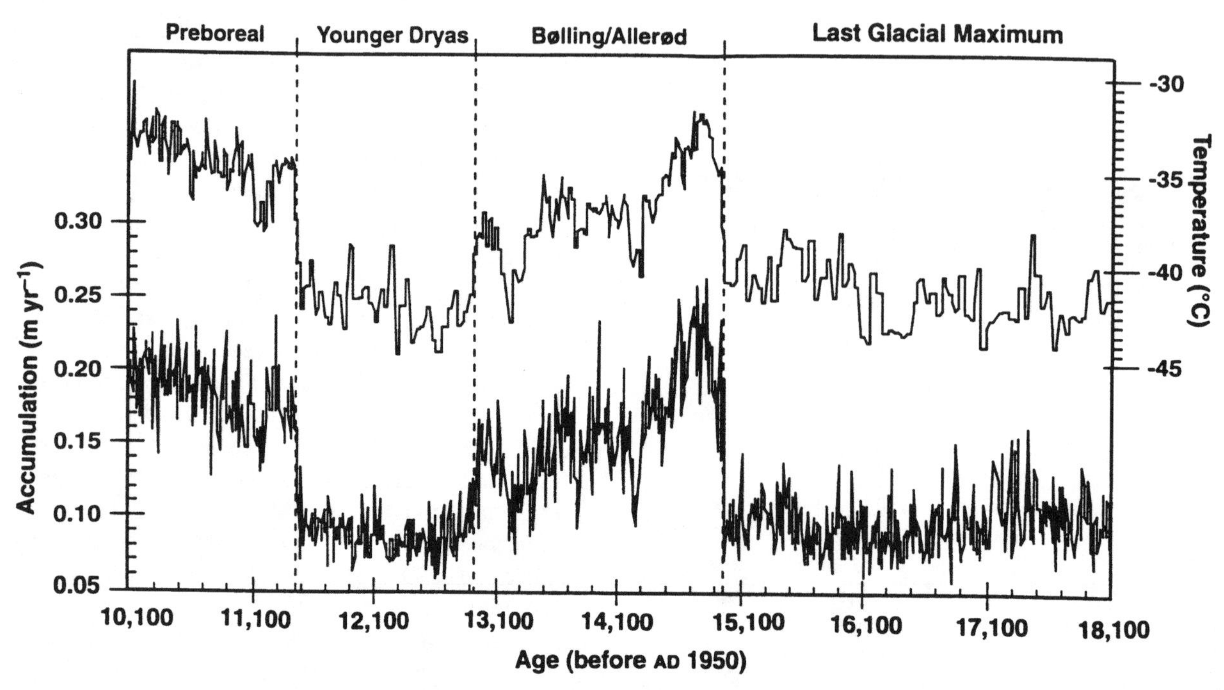

FIGURE 2-4 Snow accumulation in Greenland as a function of time from GISP 2 ice cores (bottom trace) vs. $\delta^{18}O$ converted to temperature (top trace). (From Kapsner et al., 1995; reprinted with permission of Macmillan Magazines, Ltd.)

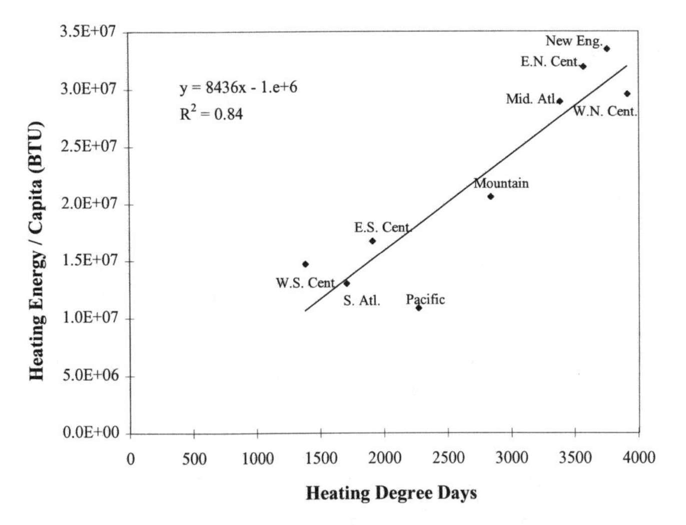

FIGURE 2-5 Relationship between temperature (heating degree days) and the energy consumption for residential heating by U.S. census regions. More energy is generally consumed for home heating in the colder regions of the United States. Pacific includes WA, OR, CA, AK, and HI; Mountain: ID, MT, WY, NV, UT, CO, AZ, and NM; W.N. Cent.: ND, MN, SD, IA, NE, MO, and KS; E.N. Cent.: WI, MI, IL, IN, and OH; W.S. Cent.: OK, AR, TX, and LA; E.S. Cent.: KY, TN, MS, and AL; New Engl.: ME, NH, VT, MA, CT, and RI; Mid. Atl.: NY, PA, and NJ; S. Atl.: WV, MD, DE, VA, NC, SC, GA, and FL. Residential heating data from 1993 EIA Residential Energy Consumption Survey (EIA, 1995). (Figure courtesy of P. Schultz, National Research Council.)

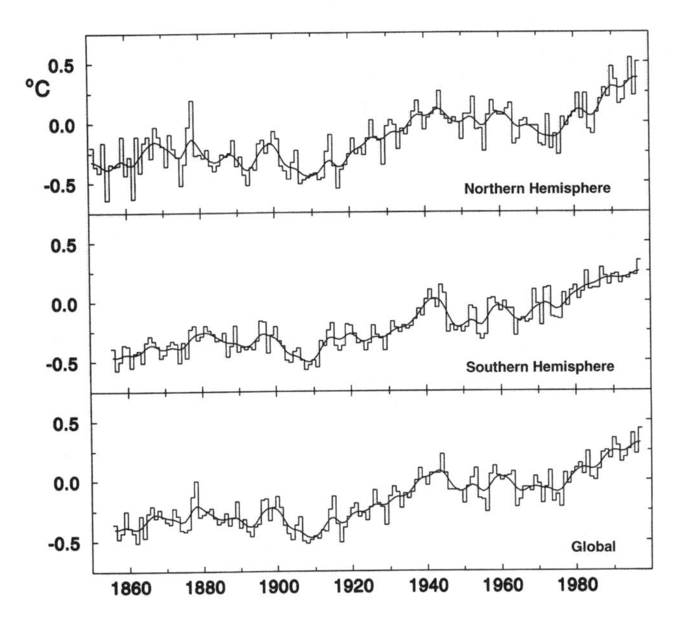

FIGURE 2-6 Combined land-surface and sea surface temperature (°C) from 1861 to 1994, expressed as anomalies from the 1961-1990 average. (From IPCC, 1996a; reprinted with permission of the Intergovernmental Panel on Climate Change.)

temperatures have risen more episodically, with the bulk of the warming concentrated in the periods about 1920-1930 and about 1975-1995. Globally, the net increase over the last century amounts to about 0.5° C in mean annual temperature (Ghil and Vautard, 1991; Jones and Briffa, 1992).

In the United States, the overall pattern of temperature change (Figure 2-7) has been similar to the global pattern, but geographically there are strong regional differences (Dettinger et al., 1995; Karl et al., 1996). Over the last 100 years there has been a strong increase in temperature over most of the northern and western sections of the country, but in the southeast and south central regions, temperatures have declined (Figure 2-8). Such variations reflect regional circulation anomalies that are otherwise obscured by large-scale averaging.

The period of warming in the twentieth century followed

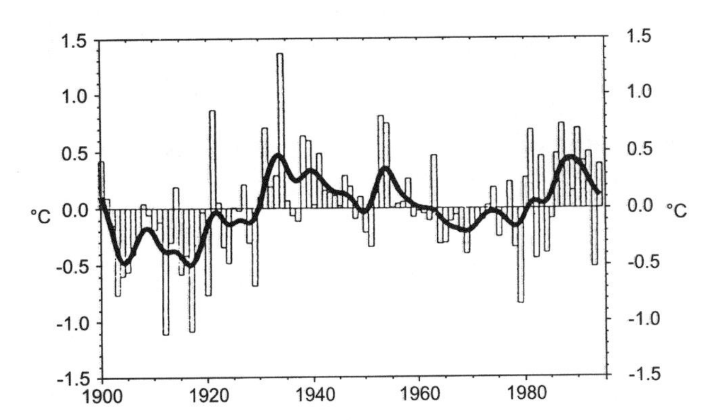

FIGURE 2-7 Departures from the long-term mean of area-averaged annual temperature over the contiguous United States (1900-1994). The dark line is a nine-point binomial filter. (From Karl et al., 1996; reprinted with permission of the American Meteorological Society.)

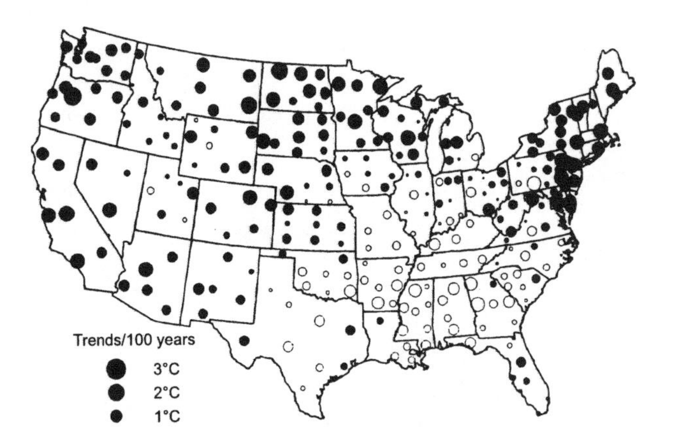

FIGURE 2-8 Trends in mean annual temperature (1900-1994) expressed as °C per hundred years. Closed circles represent warming, open circles cooling. (From Karl et al., 1996; reprinted with permission of the American Meteorological Society.)

a time of lower temperatures globally, which is often referred to as the "Little Ice Age" (Jones and Bradley, 1992). Some authors argue that this period began in the fourteenth century; others find stronger evidence for a significant mid-sixteenth-century decline in temperature (Bradley and Jones, 1992). Like modern warming, Little Ice Age cooling showed geographic variability; although the period is not well documented, in mid- and higher-latitude regions it seems to have been characterized by episodes of temperatures generally cooler than the present, lasting as long as a century, interrupted by only slightly warmer periods. The generally cooler conditions had profound environmental consequences in many parts of the world; glaciers advanced in mountain regions and in Europe, and the much colder winters froze rivers and canals, frequently bringing the transportation systems of the day to a halt. Snowfall was higher and snow stayed on the ground for longer periods in many regions (Lamb, 1972).

Historians have argued that the Little Ice Age coincided with a time when the balance between food supply and a greatly expanded European population was precarious, making European society highly vulnerable to crop failures. The more frequent presence of cold air caused unexpected freezes and heavy rains, devastating agriculture and leading to crop failures. As a result, large numbers of people died from malnutrition or starvation. Many countries are still vulnerable to anomalous climatic conditions, especially where agricultural production barely balances the needs of a growing population.

Glacial deposits in the mountainous western United States and Canada, tree-line and tree-ring studies, and anecdotal historical evidence indicate that many parts of North America also experienced Little Ice Age conditions similar to those of western Europe. Indeed, the well-known abandonment of Norse settlements in Greenland is thought to have been at least partly a consequence of colder climatic conditions and a shorter growing season than the original settlers had experienced in the preceding milder episode, often referred to as the "Medieval Warm Epoch." Whether this warmer episode was of more than regional significance remains unknown (Hughes and Diaz, 1994).

For the Northern Hemisphere as a whole, the intervals of lowest temperature in the last 600 years were in the late fifteenth, late sixteenth, and entire nineteenth centuries (Figure 2-9), although individual records show deviations from this large-scale average. Warmer conditions were more common

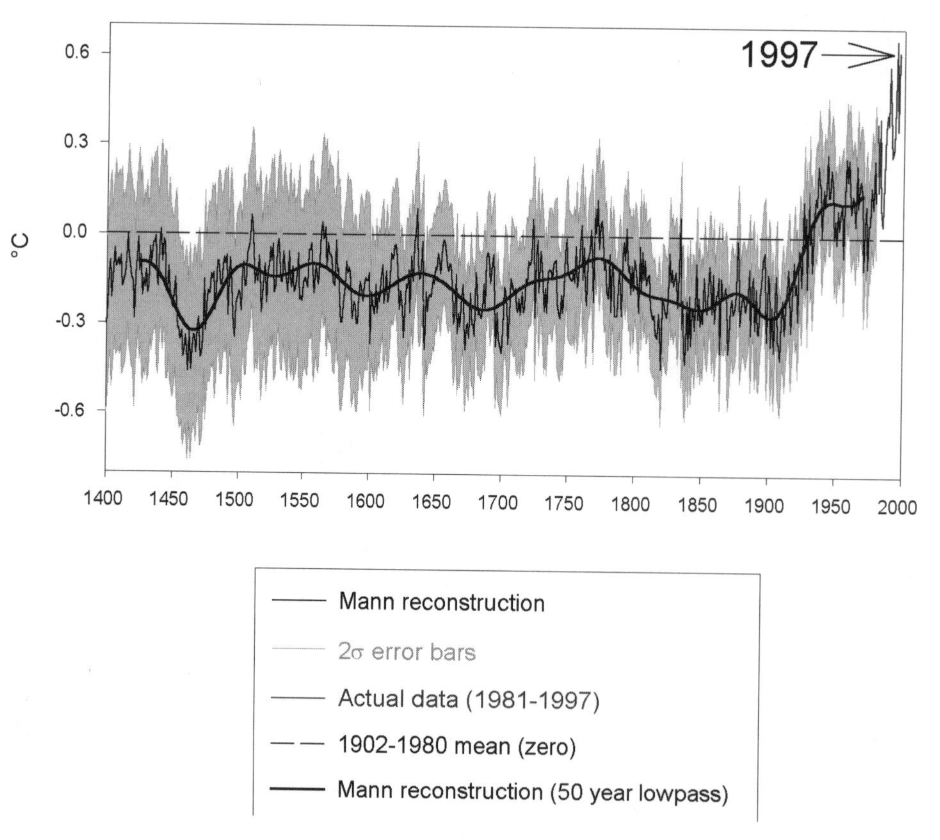

FIGURE 2-9 Reconstruction of mean annual temperature for the Northern Hemisphere (heavy lines), based on a network of paleoclimate data calibrated against instrumental data in the twentieth century. The lighter lines form an envelope describing the estimated uncertainty of individual yearly temperature estimates (at the 95 percent confidence level). The low-frequency solid curve is a 50-year low-pass-filtered version of the individual yearly values. (After Mann et al., 1998; reprinted with permission of Macmillan Magazines, Ltd.)

in the early sixteenth century and in most of the eighteenth century, although for the hemisphere as a whole, conditions comparable to the decades from 1920 onward had not been experienced for at least the previous several hundred years (Bradley and Jones 1993; Mann et al., 1998). The unusual nature of temperature conditions during the twentieth century, especially over the last 20 years or so, deserves emphasis; indeed, several years in the 1990s were warmer than any other time in at least 600 years (Mann et al., 1998). Recent low-latitude ice-core studies show in dramatic fashion that in those areas temperatures in the last decade were higher than they had been for at least 1,500 years in those areas (Thompson et al., 1993b). The loss of permanent ice caps and glaciers in the tropical Andes is of great significance for agriculture and the water-supply networks of Ecuador, Peru, Bolivia, and Chile. It has a negative impact on hydroelectric power production in those countries as well. Priceless paleoclimatic records are also melting away, often before they can be collected and analyzed.

It is important to recognize that the globally extensive instrumental recording of climate data, which forms the basis for our limited view of "global warming" began during one of the coldest periods of the last few centuries (Figure 2-9). More extensive long-term records of global temperature changes would allow a more rigorous statistical assessment of the current warming. It is clear that decadal-scale (and lower-frequency) temperature variations were also evident for many centuries prior to the nineteenth century, while greenhouse gases exhibited relatively little change. The fact that such temperature variations must have been unrelated to global-scale anthropogenic effects focuses our attention on the need to understand what drives such "natural" variability. Better records of past climate variations and of potentially important forcing factors, such as solar-irradiance changes or explosive volcanic eruptions, are needed. In particular, it would be especially valuable to examine the enigmatic Medieval Warm Epoch—a time often cited as having been warmer than the twentieth century, but so far not shown to have been warmer over an entire hemisphere, much less over the entire world (Hughes and Diaz, 1994).

SOLAR RADIATION

Sunlight is one of the bases of life on this planet. It provides the energy needed to warm the Earth and evaporate water; its geographical distribution drives the Earth's climate system. Its vertical absorption and reflection play a major role in stratifying the atmosphere, affecting atmospheric circulation, winds, clouds, and a multitude of energy-balance feedbacks. Sunlight is an essential ingredient in the creation of the protective ozone layer, and is a direct source of natural, readily harnessed energy. It provides the light animals need to see and plants need for photosynthesis. It is basic to the food chain of most life.

The link between solar variation and climate or weather has been known for millennia; in the fourth century BC, Theophrastus correlated sunspots with rainy weather. Recent measurements of solar output have shown decadal-scale variations of less than 1 percent (Lean et al., 1992; Hoyt and Schatten, 1993; Zhang et al., 1994), but we know that it has been orders of magnitude greater in past eons, and presumably could be again. Variations in solar-cycle length and Northern Hemisphere climate anomalies parallel each other over the last 100 years (Friis-Christensen and Lassen, 1991; Labitzke and van Loon, 1993), and solar activity and sea surface temperature track each other well over the past 130 years (Reid, 1991; White et al., 1997b).

Solar ultraviolet radiation affects society through its role in health. It promotes the production of vitamin D, yet it is implicated in human skin cancer, suppression of the skin's immune system, and damage to plants (see, e.g., Coohill, 1991, and Krupa and Jager, 1996). The sun provides ultraviolet radiation in the wavelength region 280-320 nm (called UV-B) that is damaging to most plants and animals. For example, there is accumulating evidence of damage to Antarctic ecosystems (Smith et al., 1992) in the aftermath of the opening of the Antarctic ozone hole. Solar UV-B radiation's impact on ecosystems in turn affects the natural cycles of water, carbon, and other nutrients.

The stratospheric ozone layer that shields the Earth by absorbing most of the UV-B radiation has changed. We do not have reliable decadal records of UV-B received at the Earth's surface and how it has varied through time. However, as Figure 2-10 shows, we do have clear observational evidence demonstrating that lower ozone columns result in higher UV-B irradiation, as predicted by models (WMO, 1995). Thus it has been possible to derive accurate calculated levels of surface UV-B irradiance, and hence of human exposure, under clear-sky conditions (Figure 2-11); note that an increase is evident throughout the middle latitudes.

The sun is also the source of photosynthetically active radiation (PAR), which is in the wavelength region 0.4-0.7 μm. The amount of PAR received at the Earth's surface is controlled by cloud cover—not only by the cloud droplet concentration and cloud thickness, but also by the spatial and temporal distribution of clouds, which varies considerably (see Figure 2-12). When the amount of cloud cover over a region changes for a long period of time, and the amount of PAR shifts, the growth of ecosystems may be altered. Such changes are thus important for agriculture and for natural or managed systems, such as forests. Since the middle of this century such long-term changes in cloud cover have in fact been observed (Figure 2-12). In addition to trends of increasing cloud cover in some regions, concomitant changes in other variables have also been detected, such as the change in cloud amount that accompanied a step-like increase in evaporation over the former Soviet Union around 1976. All such records are subject to systematic shifts as observers and instruments change; without knowledge of

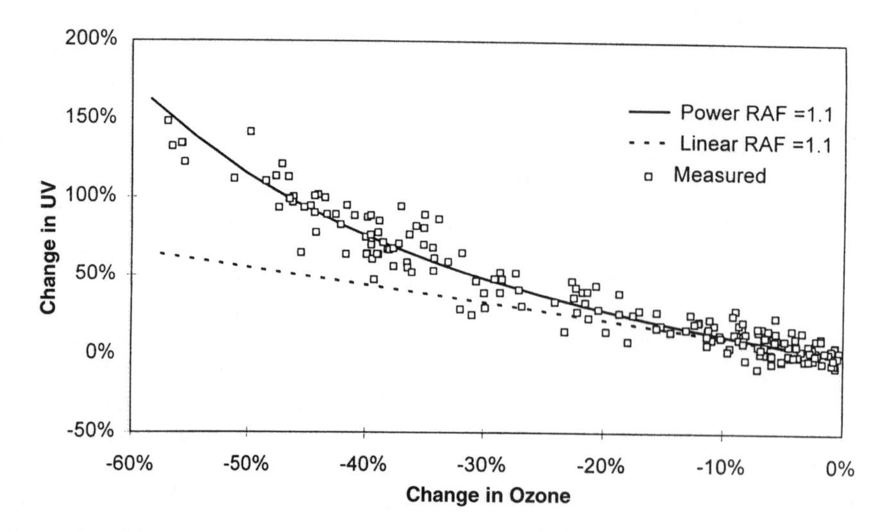

FIGURE 2-10 UV-B radiation reaching the Earth's surface as a function of ozone column, from both measurements and models. (RAF is a radiation amplification factor.) Measurements were made at the South Pole between February 1991 and December 1992. (From Booth and Madronich, 1994; reprinted with permission of the American Geophysical Union.)

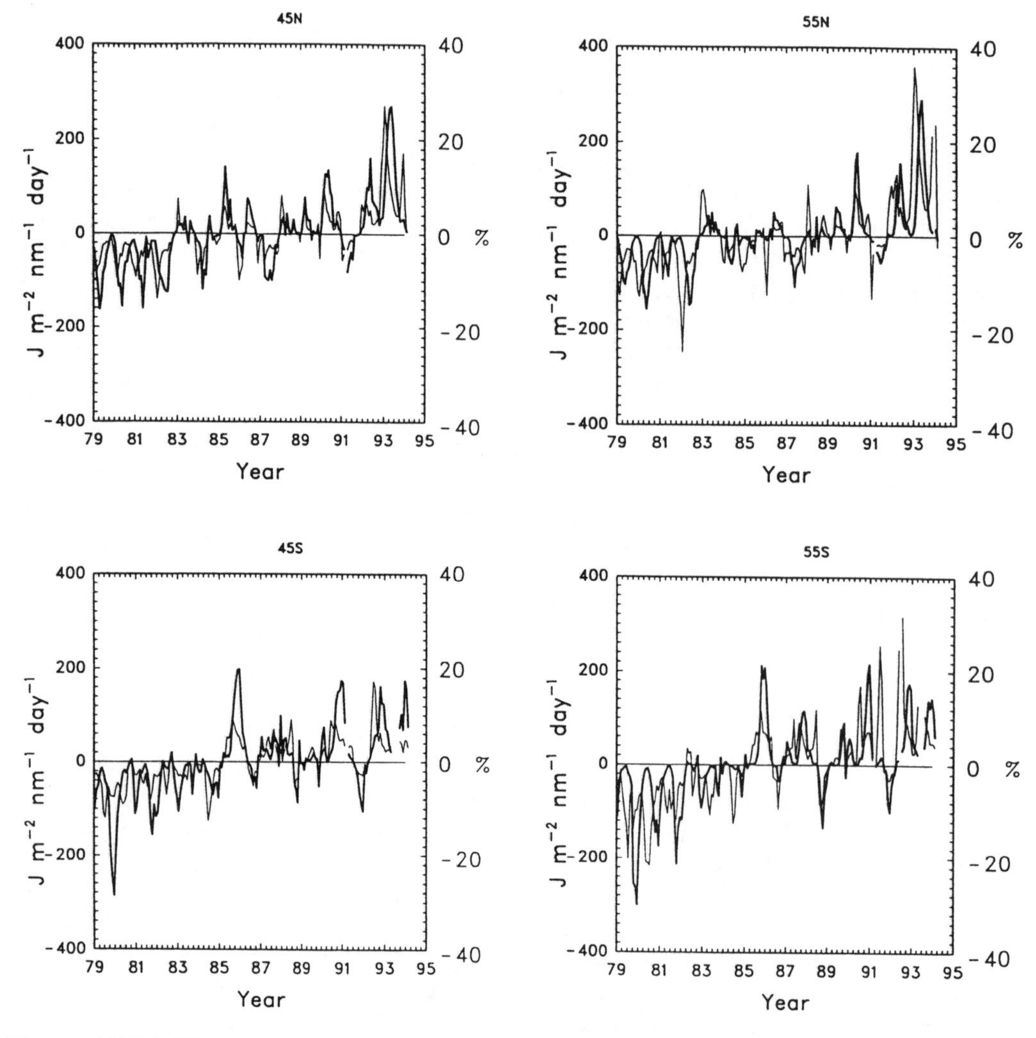

FIGURE 2-11 History of UV-B (310nm) at the Earth's surface for mid-latitudes over 15 years beginning 1979, based on the ozone/UV-B relationship in Figure 2-10 and observed ozone-column data. (From WMO, 1995; reprinted with permission of the World Meteorological Organization.)

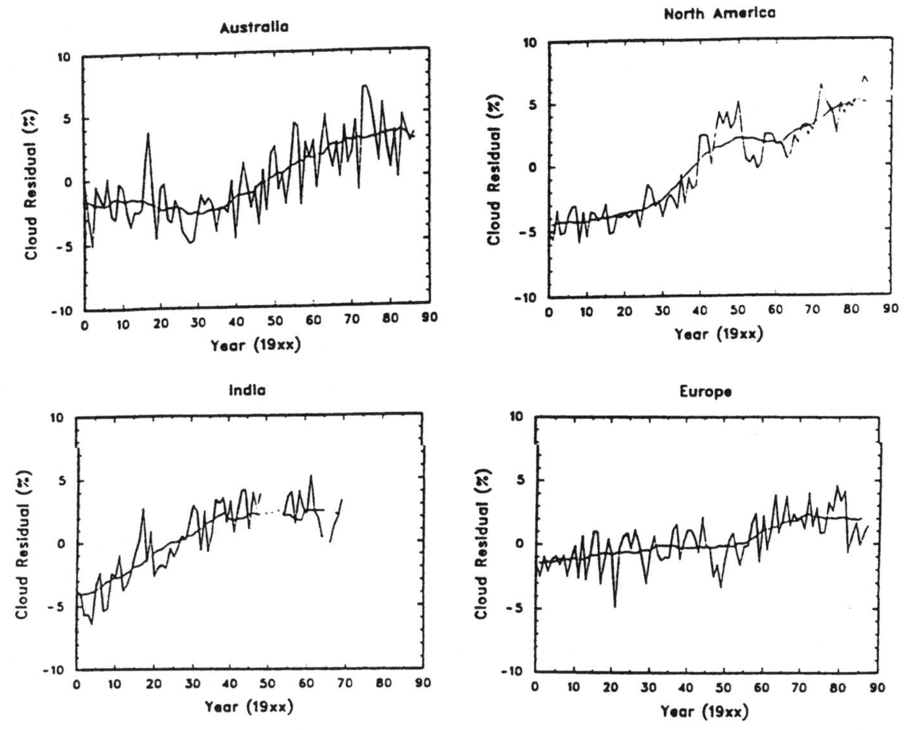

FIGURE 2-12 Cloud-cover changes over Australia, North America, India, and Europe since 1900, expressed as percent deviations from the long-term mean. (From McGuffie and Henderson-Sellers, 1988; reprinted with permission of the Canadian Meteorological and Oceanographic Society.)

these occurrences it is difficult to assemble a reliable deccen record of climate change.

The ability of solar radiation to influence the Earth's temperature is controlled by several factors that have been shown to vary over decades and centuries. Among them are variations in the total solar output, changes in the amount of sunlight reflected from clouds and aerosols in the lower atmosphere, increases in aerosols in the atmosphere because of volcanic injections, and trapping of the Earth's thermal radiation by the greenhouse gases. These factors are discussed in greater detail in Chapter 5; we present here a single example of a direct impact on society caused by changes in total sunlight.

Like volcanic aerosols, tropospheric aerosols generated by industrial pollution can play an important role in the amount of radiation reaching the Earth's surface. In certain regions, such as the Sichuan Province of China, there are clear trends of decreasing visibility. Reduced visibility is often regarded as a quality-of-life issue, affecting principally our appreciation of our natural surroundings. However, in Sichuan the industrial haze is so severe that it has led to a substantial reduction in the annual mean solar energy reaching the surface over the past four decades. Records from the eastern United States indicate interdecadal changes in light extinction associated with sulfur emissions (Figure 2-13). These changes in visibility affected the solar flux at the ground, and thus likely also surface temperature.

STORMS

Whereas the distributions of freshwater, temperature, and radiation affect us on a daily basis, the less frequent but more destructive side of nature is often realized in relatively short-lived episodes of severe weather associated with storms. Storms generally owe their destructive power to high winds, associated lightning, and high rates of precipitation in the form of rain, hail, sleet, and snow. In coastal regions, the high winds drive heavy waves and surges that cause flooding and beach erosion, often inflicting considerable, sometimes irreversible, damage on coastal communities and ecosystems. The entire landscape can be permanently altered. Storms are responsible for widespread personal-property destruction, deaths, and financial stress. The financial stress can be both personal (uninsured loss, increased premiums, higher taxes to cover government aid) and institutional (massive insurance payouts, federal and state disaster relief). Storms also influence recreation, power usage, and health. Thus, any change in their frequency of occurrence over deccen time scales can have considerable influence on economies (consider the substantial outlay of the U.S. insurance

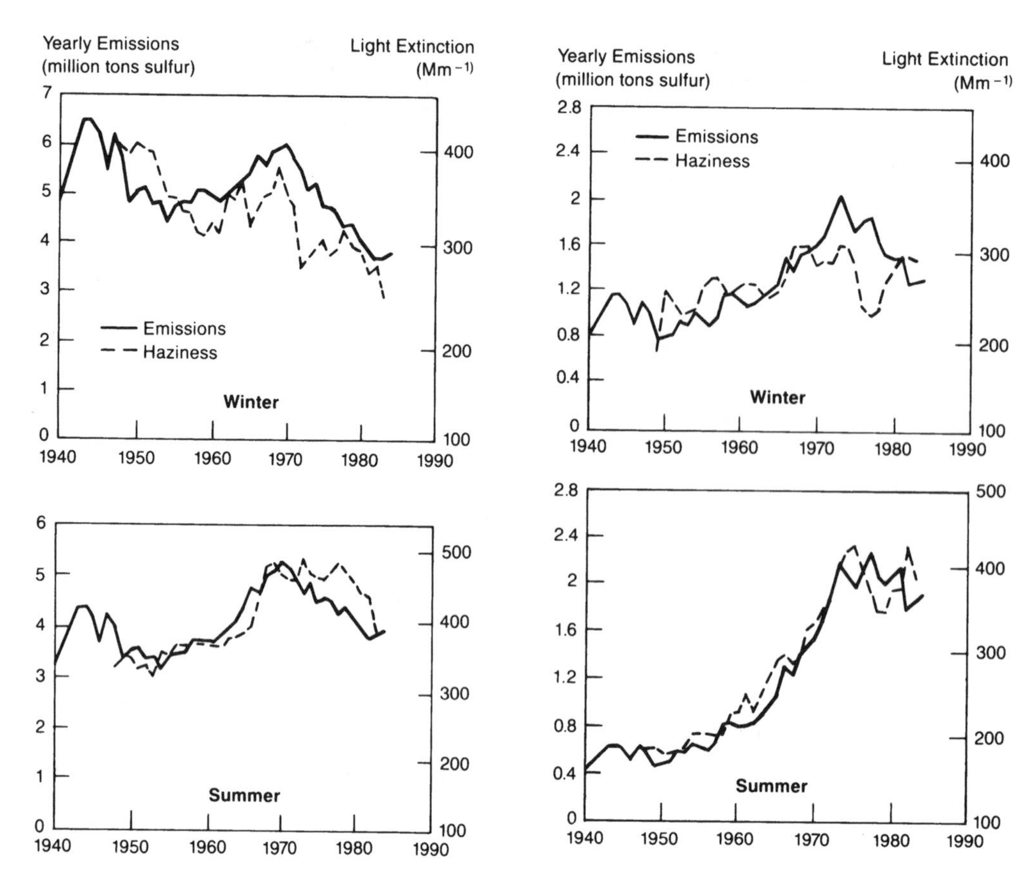

FIGURE 2-13 Comparison of historical trends for sulfur emissions and haziness for the northeastern (left) and southeastern (right) United States (From NAPAP, 1991.)

industry in recent years related to major storms) and societies.

The word "storm" refers to an array of extreme weather phenomena that display a wide range of spatial scales and owe their existence to a variety of physical mechanisms (see, e.g., Palmen and Newton, 1969). Consequently, their impact is manifested through a variety of means and geographic extents. For example, in the mid-latitudes, wintertime extra-tropical cyclonic storms extend over hundreds to thousands of kilometers, although their severe weather is generally associated with smaller-scale features such as fronts and squall lines embedded in the large-scale disturbance (see, e.g., Wallace and Hobbs, 1977). In the subtropical and mid-latitude land areas, severe storms occur in all seasons (including summer) as a result of mesoscale (length scale of tens of kilometers) organization of convection. Forming relatively localized features, such as squall lines and the so-called multi- and super-cell storms, these storms are accompanied by strong wind gusts, tornadoes, lightning, hail, and flash floods. Tropical ocean regions spawn cyclones known as hurricanes and typhoons. These storms are somewhat smaller in size than mid-latitude cyclones, but their intensity tends to exceed that of the latter. Because they have a relatively large radius of influence where high winds combine with large amounts of precipitation, tropical storms are among the most dreaded natural phenomena. The largest destructive potential of these storms is in coastal regions, where population density is often high.

Several studies, focusing predominantly on the Atlantic where the historical records are longest, have indicated the existence of decadal-to-centennial fluctuations in storm intensity and distribution in the tropics and extra-tropics. Multidecadal changes in Atlantic hurricane activity were noted by Gray (1990) and Landsea et al. (1996). These changes consist of a higher-than-normal occurrence of intense hurricanes and hurricane days between the mid-1940s and the mid-1960s, and lower-than-normal occurrence between the mid-1960s and the early 1990s (Figure 2-14a,b). Despite the general decrease in the number of Atlantic hurricanes, the damage totals from these storms have increased dramatically since the 1940s (Figure 2-14c). The increasing damages are related to the greater vulnerability arising from the growing population densities and property values in coastal regions (Pielke and Landsea, 1998). While monetary damages have increased, hurricane-related deaths have decreased over the past century (Figure 2-14c). This decline is

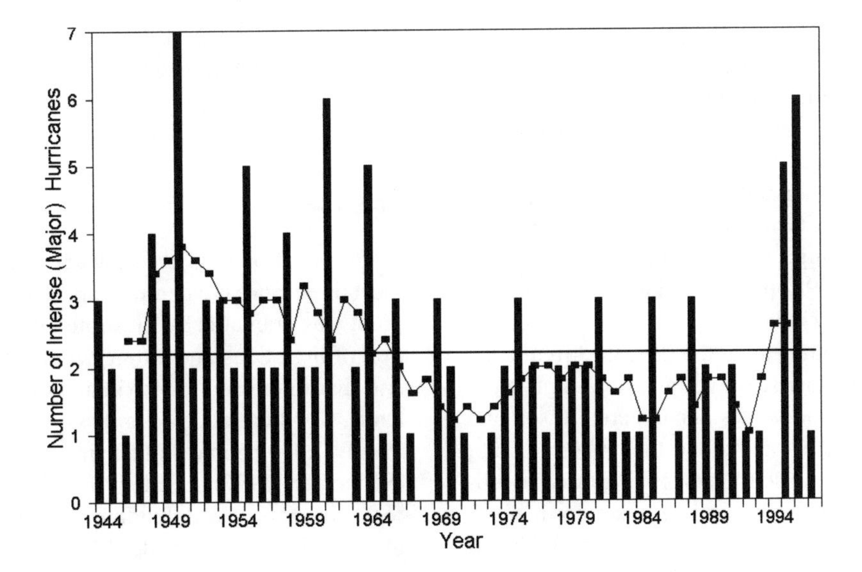

b

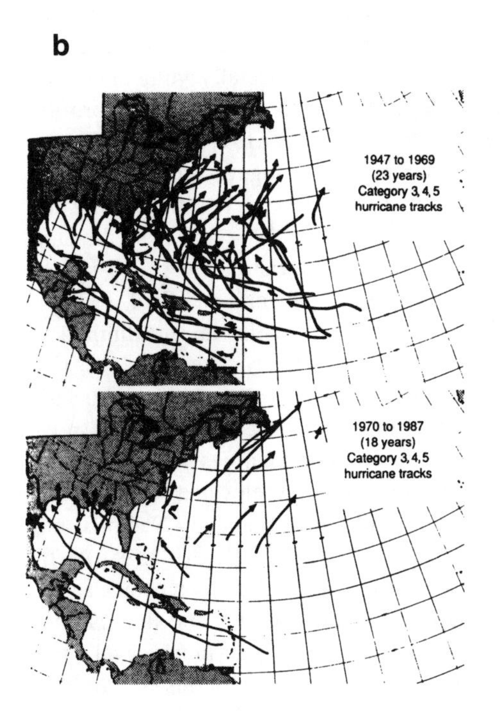

c

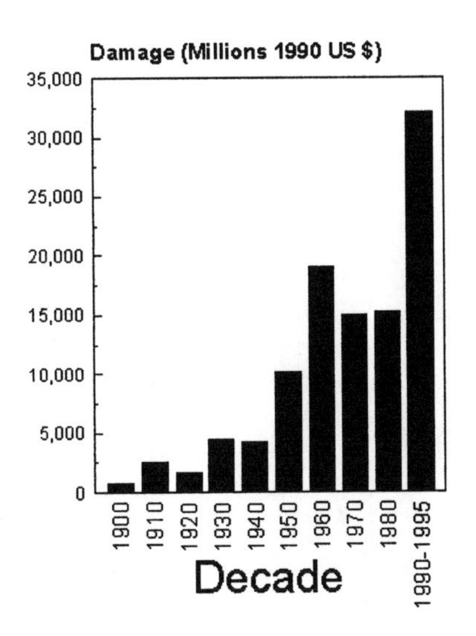

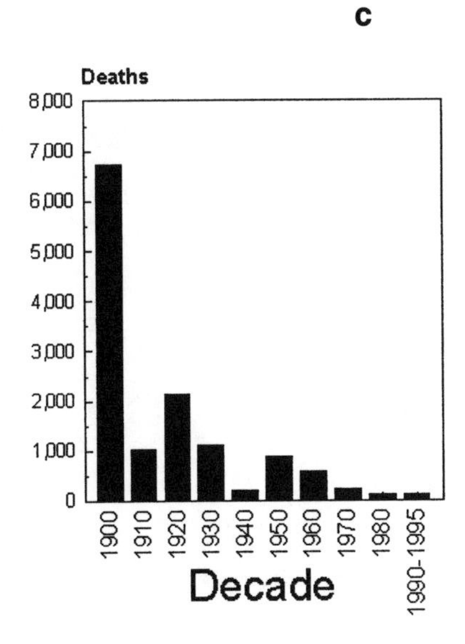

FIGURE 2-14 (a) Number of intense Atlantic hurricanes per year since 1944 . The squares show the 5-year running mean. (After Landsea and Gray, 1992; reprinted with permission of the American Meteorological Society). (b) Atlantic hurricane tracks for the periods 1947-1969 and 1970-1987. (From Gray, 1990; reprinted with permission of the American Association for the Advancement of Science.) (c) Hurricane-related damage costs and deaths in the U.S. since 1900; damages normalized to 1990 dollars. (From Hebert et al., 1996; reprinted with permission of NOAA.)

directly related to improvements in the National Weather Service's ability to forecast hurricane tracks and intensities, as well as to improvements in the dissemination speed and coverage of hurricane warnings. The declining number of intense hurricanes has also contributed to the decrease in mortality.

In the extra-tropics, historical evidence of long-term changes in storminess is found in written records and in patterns of beach erosion and destruction in western European countries bordering the Atlantic (Lamb, 1995, and references therein). At the end of the medieval times and during the Little Ice Age—that is, essentially between the early 1200s and the late 1600s—coastal flooding and reports of other coastal damage suggest abnormally strong (in both frequency and intensity) storm activity in the eastern North Atlantic. Increases in North Atlantic wave heights since the early 1960s (Günther et al., 1998) provide evidence of more intense storminess (Figure 2-15).

These observations complement the increased frequency of reports of extremely low sea-level pressure, high wind conditions, and large ocean waves encountered at sea and along western European shores since the 1960s (Lamb, 1995). The change in ocean storminess is supported by observations of increases of the wintertime mean north-south pressure gradient across the North Atlantic since 1960 or so (Flohn et al., 1990; Bacon and Carter, 1993; Hurrell, 1995; WASA Group, 1998), and the link between the large-scale pressure gradient and wave statistics (Bacon and Carter, 1993; Kushnir et al., 1997). Long-term changes in storminess have also been observed over North America (Hayden, 1981). Such changes have also been linked to large-scale climate variability (Dickson and Namias, 1976).

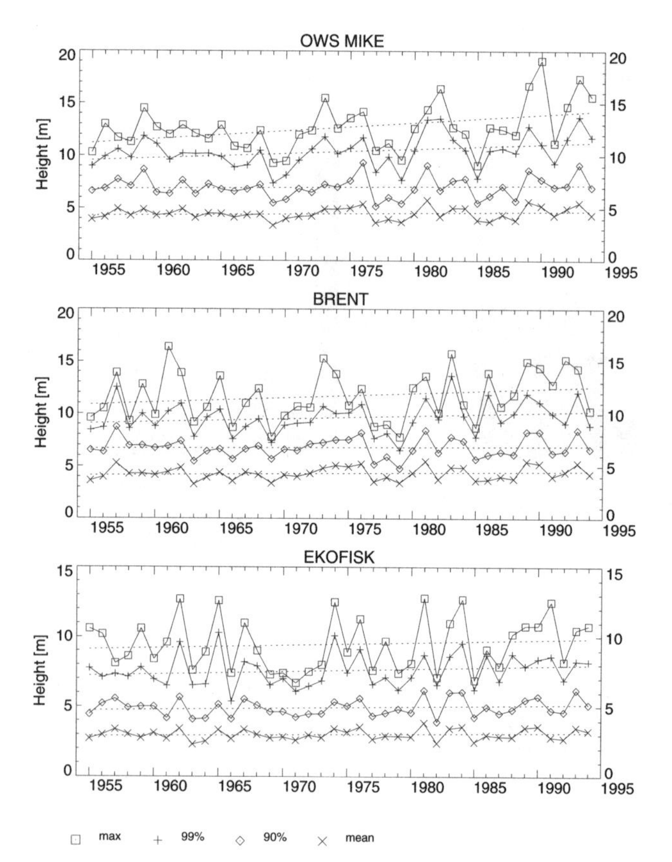

FIGURE 2-15 Annual wave-height maxima, means, and 99th and 90th percentiles at three North Atlantic sites, as derived from wave hindcasts. Note that the mean remains fairly constant, but the maxima show a marked rise and suggest a decadal pattern. (From Günther et al., 1998; reprinted with permission of the Gordon and Breach Publishing Group.)

SEA LEVEL

Coastal flooding by seawater inundation is one of the most direct and threatening impacts of raised sea level. This point was dramatically illustrated in 1953 when a combination of high tides and strong winds forced a local sea-level rise that inundated approximately one-sixth of the Netherlands, flooding land 64 km from the coast, killing approximately 2,000 people and leaving another 100,000 homeless (Schneider, 1997). Today, the Netherlands expends U.S. $30 million a year defending itself against such encroachment by the sea (de Ronde, 1993), in addition to the tremendous infrastructure already in place for this defense. South Florida, Chesapeake Bay, and several other U.S. locations are also losing land to immersion, including some settled islands. Sandy beaches can be especially vulnerable to a small sea-level rise, depending on the offshore depth profile. Bruun's rule (1962), which allows one to estimate beach erosion as a function of sea-level rise, suggests that a rise of 50 cm can erode approximately 50 m of beach, roughly the width of Virginia Beach. Between 1962 and 1985 the U.S. Army Corps of Engineers renourished over 700 km of beaches at a cost of $8 billion (Schwartz and Bird, 1990).

The physical effects of sea-level rise are not limited to simple inundation, however. In many cases sea-level rise exacerbates other changes (natural or anthropogenic) occurring in the coastal environment. Consequently, rising (relative) sea level affects coastal communities via inundation, erosion of land, saltwater intrusion, elevated water tables, and increased flooding and storm damage (Nicholls and Leatherman, 1994). In fact, the IPCC (IPCC, 1996b) estimates that worldwide approximately 46 million people a year experience flooding due to storm surges, and this number would double if there were a 50 cm rise in sea level. For example, consider the impact of erosion in this country due to changes in relative sea level. The Mississippi delta has the largest rate of land loss in the United States. Wetlands have been lost there at a rate of up to 100 km^2 per year in this century, primarily through loss of sediment and land subsidence. Sea-level rise will only aggravate this problem.

In addition to causing direct flooding, a higher mean sea level elevates the base for storm surges, which can have devastating impacts on human life and property. For example, 300,000 Bangladeshis died in a 1970 storm surge. Sea-level rise in the Ganges-Brahmaputra delta of Bangladesh has been exacerbated by local subsidence of the land, which can exceed 20 mm per year (Alam, 1996). Higher sea level also slows rainwater drainage, leading to increased risk of riverine flooding. In September 1987 river flooding affected nearly half of Bangladesh, where population and poverty compound the problem—land is so scarce that people repopulate land on which the previous tenants have died in a flood. Maldives, a small island country in the Indian Ocean lying mostly less than two meters above sea level, could be totally inundated, as could numerous other small island countries scattered throughout the tropical and subtropical world.

Furthermore, saltwater can intrude landward either through the ground or up waterways, though the latter process is much more rapid. Both processes are sensitive to sea level and to the supply of freshwater at the coast. During the droughts of the 1960s saltwater advanced 53 km up the Delaware River, forcing some industries near Philadelphia to seek water imported from the Susquehanna River Basin. Today, farmers along the edges of Chesapeake Bay are losing one to two rows of corn per year to saltwater intrusion resulting from land subsidence and sea-level rise. A concern in the Netherlands is that in drought summers the freshwater needed to combat saltwater intrusion and flush the land of salt may not be available. Likewise, in Bangladesh, heavy monsoon rainfall is needed to desalinate the topsoil so that rice can be grown in a region occupied by saltwater during the dry season.

These examples illustrate the vulnerability of societies worldwide to variations in sea level. Recent analyses suggest that global sea level has been rising at a rate of 1.2-2.0 mm per year for at least a century (Douglas, 1995; Unal and Ghil, 1995), amounting to a rise of up to 20 cm. While there has been no detectable acceleration of the rise in this century, there is evidence that the rate of rise was smaller in previous millennia (Gornitz, 1995). Considerable regional variability is superimposed on the global sea-level-rise trend; some regions have experienced falling sea level (e.g., Scandinavia) and others have seen a rise larger than the global mean (e.g., the southeast coast of the United States).

Decadal-scale regional variations during this century can also be seen. Such variations are apparent in the tide-gauge records for individual locations such as San Francisco (Figure 2-16). Records like these may reflect a number of processes, including vertical movements of land, redistribution of water in the oceans, and global sea-level changes. Decadal fluctuations tend to be synchronous along long sections of a particular coast, and can sometimes be linked to wind-stress variability (Sturges and Hong, 1995). High-frequency fluctuations are often associated with local winds, though oceanic wave propagation allows a larger spatial influence (Sturges, 1987). Low-frequency atmospheric variability influences coastal sea level on longer time scales, and planetary waves can propagate these changes over very large spatial scales, influencing regions well away from the local source of the disturbance. For example, sea-level changes

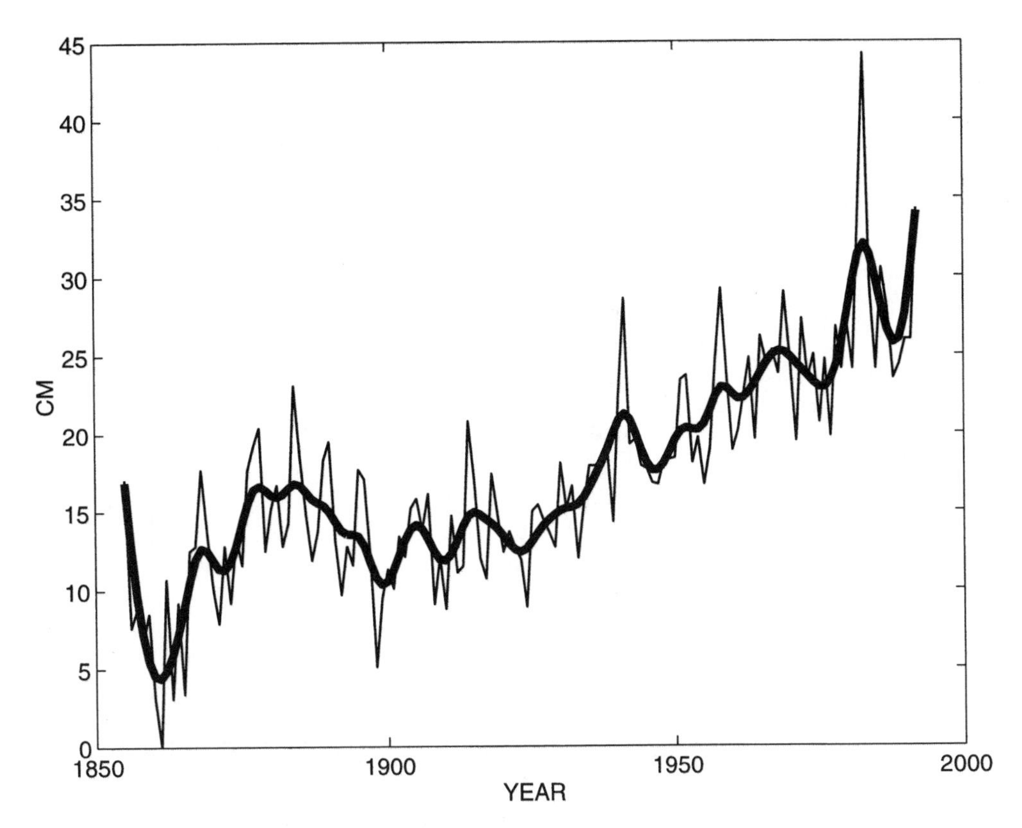

FIGURE 2-16 Variations in sea level at San Francisco since 1855, from data in Spencer and Woodworth (1993). (Figure courtesy of M. Winton, NOAA/GFDL.)

on the order of 10-50 cm are associated with the El Niño/Southern Oscillation (ENSO) phenomenon, and satellite altimetry has documented the propagation of these changes across the Pacific (Jacobs et al., 1994). At Bermuda (Figure 2-17), decadal variations on the order of 10 cm have been traced to variability of wind-stress curl over the open ocean to the east (Sturges and Hong, 1995).

ECOSYSTEMS

Ecosystems influence society directly and indirectly; our very existence depends in fundamental ways on the ecosystems of the Earth, of which we are a part. In fact, the dependence of all life forms on the physical environment, as well as their interdependence, are embodied in the concept of ecosystems, reflecting intricate communities of primary producers, grazers, and predators that are adapted to each other as well as to particular regimes of temperature, moisture, radiation, and other factors. The ecosystems in turn influence, both locally and globally, the physical and chemical environment in which they exist. For example, they moderate the flow of radiatively active gases, sequester carbon, and alter the atmospheric moisture content and the Earth's albedo; all of these functions directly influence other climate attributes. Ecosystems also influence society through other means. Our food consists of organisms taken from terrestrial and marine habitats. Ecosystems harbor the genetic diversity of life; the continuity of our food supply partially depends on the existence of wild strains that may possess resistance against emerging plant diseases and pests. Wild varieties offer the potential for higher crop yields, for new cultivars adapted to changed climatic conditions, or for entirely new crops. New medicines have often been developed from existing biological compounds.

The carriers of human illness respond to climate change when the ecosystem in which they live is affected by that change. It is believed that the outbreak of the deadly Hantavirus in 1993 in the southwestern United States was caused by a period of drought succeeded by heavy rains. The drought induced a decline in all animal populations, but deer mice, carriers of the virus, recovered far more quickly than their predators (they increased tenfold when the rains returned); this greatly increased their contact with humans (Levins et al., 1994). Coastal algal blooms, a breeding ground for the bacterium *Vibrio cholera* that sometimes finds its way into the food chain, have been linked to the cholera outbreak in South America in 1992 (Epstein, 1993). A threefold increase in malaria incidence in Rwanda in 1987 has been ascribed to record high temperatures and rainfall (Loevinsohn, 1994).

Other important contributions of ecosystems to our well-being are the creation of soils over thousands of years, and the establishment of an environment that is varied and enjoyable. The latter often translates directly into economic value via money spent on recreation. Soils are, of course, indispensable for agriculture, but they also function as reservoirs of water, carbon, and nutrients. However, ecosystems did not come into existence solely for the benefit of humans; they also harbor pathogens and toxic compounds, and climate-related changes in the soils may lead to changes in these threats.

Ecosystems change over most time scales. On decade-to-century time scales, there is considerable natural variability as well as anthropogenic change. For example, in southern Ontario the analysis of pollen has shown that dominant beech

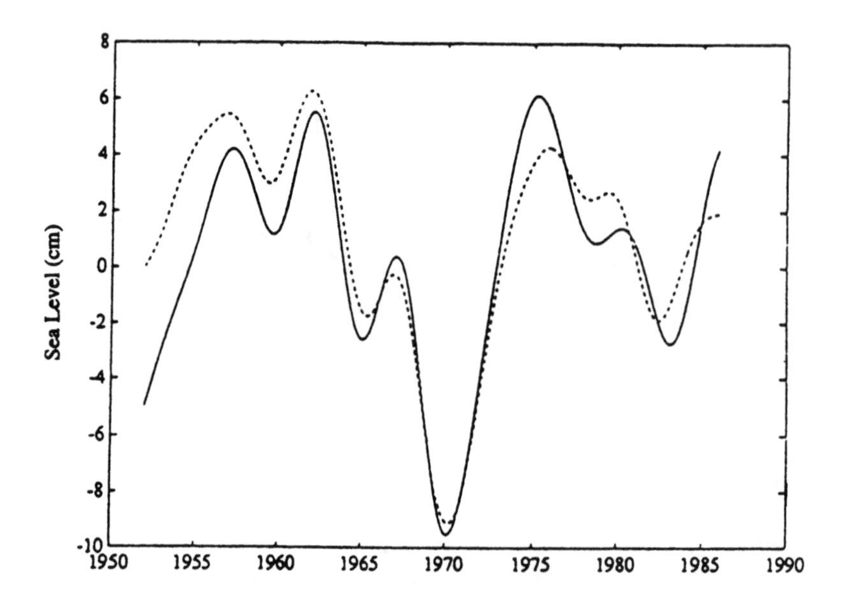

FIGURE 2-17 Sea-level fluctuations for Bermuda. The solid line is the observed record, adjusted to constant atmospheric pressure. The dashed line is a hindcast made by a simple model forced with wind-stress observations. (From Sturges and Hong, 1995; reprinted with permission of the American Meteorological Society.)

trees were replaced first by oak, then by pine trees, during the Little Ice Age, only a few hundred years ago (Campbell and McAndrews, 1993). Also, the monitoring of fluctuations in Northwestern salmon catch since the early part of this century (see Figure 2-18) has revealed a remarkable coincidence with decadal changes in the North Pacific sea surface temperature (Mantua et al., 1997). Decadal-scale changes in the extent of the Sahel have also been documented

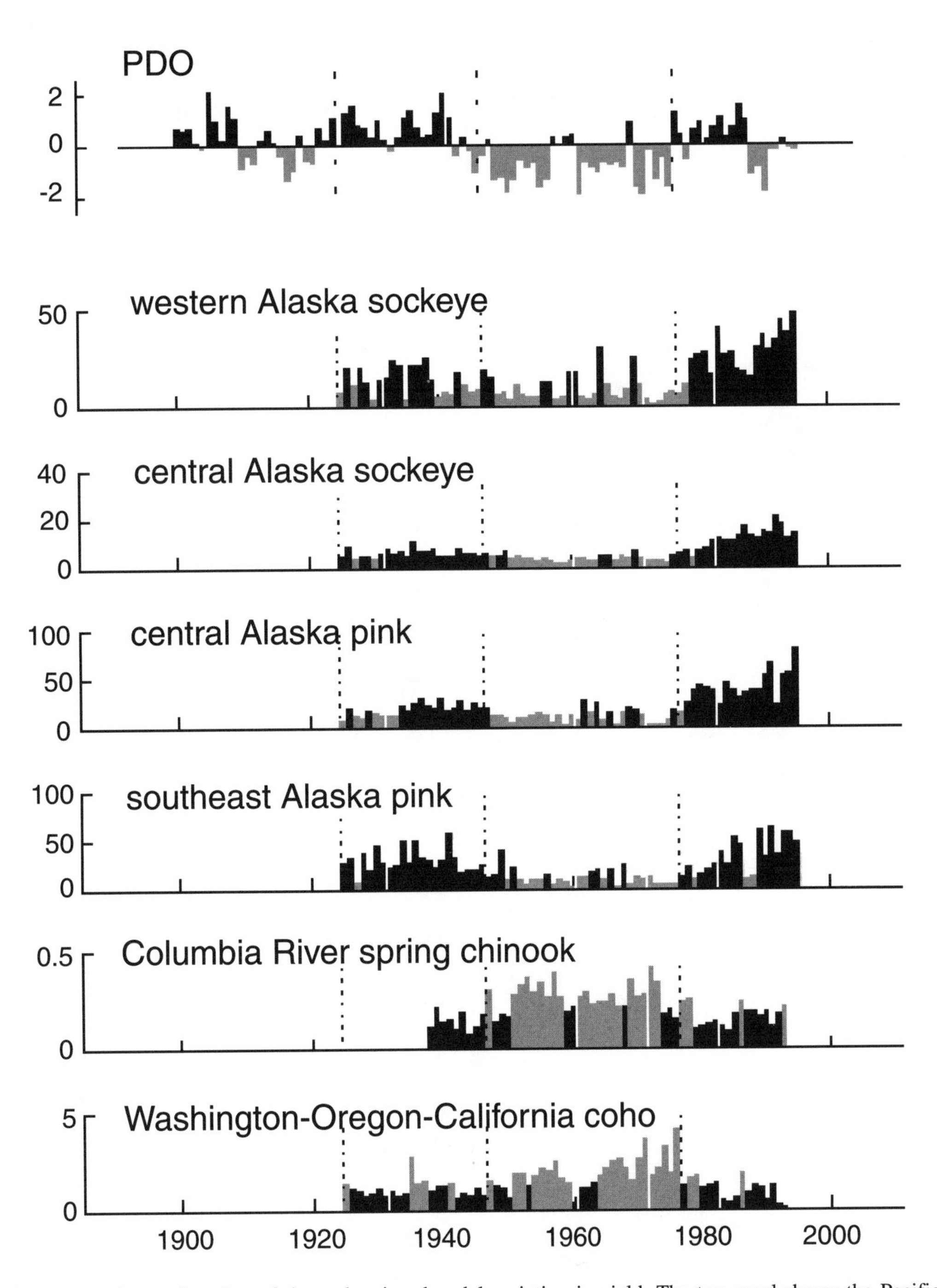

FIGURE 2-18 Salmon catch as a function of time, showing decadal variation in yield. The top panel shows the Pacific (inter)Decadal Oscillation (PDO) index. The panels below show catch records for various types of Pacific salmon. For Alaska catches, the black (grey) bars denote values that are greater (less) than the long-term mean. The shading convention is reversed for the bottom two panels. Light, dotted vertical lines indicate the PDO reversal times at 1925, 1947, and 1977. (From Mantua et al., 1997; reprinted with permission of the American Meteorological Society.)

(see, e.g., Tucker et al., 1991). Barry et al. (1995) document a shift in the community structure of invertebrate fauna in a California rocky intertidal zone consistent with observations of warming sea surface temperature between 1932 and 1993. In addition to these regional changes, the observed increase in the intra-annual amplitude of atmospheric CO_2 concentration may indicate that the cycle of net primary production and respiration is increasing in vigor on a hemispheric scale (Keeling et al., 1996a).

Natural and anthropogenic habitat destruction and fragmentation are the greatest contributors to the extinction of species. Consider the entirely new "ecosystem" that has recently grown to occupy several percent of the land area of the United States. It is composed of a patchwork of asphalt, concrete, and structures, with some accompanying reforestation; all of these influence and are influenced by other ecosystems, society, and climate. Fire and other disasters such as floods and storms are part of a naturally varying environment to which ecosystems will adapt, but this new "ecosystem" will be more challenging.

3

Modes of Climate Variability

Climate can be loosely defined as the ensemble of weather. As such, it is inherent to the atmosphere, but is affected by interactions between the atmosphere and the ocean, the biosphere, the land surface, and the cryosphere. Because climate varies on all time scales (see, e.g., Mitchell, 1976; NRC, 1995), an appropriate mean can serve as a reference state for the study of variability on shorter time scales, while itself changing on longer time scales. In practice, whatever the definition used for the mean, an anomaly is the difference between the instantaneous state of the climate system and that mean. Climate variability and change are characterized in terms of these anomalies. As the study of climate progresses, it is becoming increasingly apparent that the variations are not randomly distributed in space and time, but often appear to be organized into relatively coherent spatial structures that tend to preserve their shape—or assume a limited number of related shapes—while their amplitude, phase, and sometimes geographic position change through time. Though the precise nature and shape of these structures, or patterns, vary to some extent according to the statistical methodology employed in the analysis, consistent regional characteristics that identify the patterns still emerge. Therefore, in studying climate variability and change, the examination of spatio-temporal patterns is a natural development. Such study is also consistent with the IPCC Second Assessment (IPCC, 1996a), where it was noted that much of our attention in recent years has shifted from the analysis of changes in mean global temperature to that of changes in the spatial distribution of temperature and other climate variables, reflecting the anticipation that climate may vary in both space and time.

We do not yet have an exhaustive inventory of global and regional patterns, nor do we understand their mechanisms, relationships, temporal characteristics—such as persistence or periodicity—or full implications for climate prediction. Still, study of the most thoroughly investigated spatio-temporal pattern—that which dominates the tropical Pacific and is associated with the El Niño/Southern Oscillation (ENSO) phenomenon—led to the first successful numerical climate predictions, while yielding considerable insight regarding the climate system, the nature of its air-sea couplings (Cane et al., 1986) and its scales of teleconnectivity (NRC, 1996). Many of the other patterns, while less well documented or studied, appear to affect regional climate, as well as agricultural yields and regional fish inventories, and appear to be related to the frequency of hurricanes, variations in the ocean's thermohaline circulation, and other things. These patterns vary over a broad range of space and time scales, and their relative phasing can dominate global and regional temperature variations. They often show regional and global teleconnections, involve a number of distinct climatological variables, and apparently focus different forcings and processes into single coherent responses. Because of these attributes and co-varying relationships, it is hoped that their further study may ultimately yield benefits for dec-cen climate predictability similar to those obtained for seasonal-to-interannual predictions through the study of ENSO. Spatio-temporal patterns thus provide one obvious avenue by which the search for a predictable climate signal—that is, the extraction from the complex climate system of a finite set of regular components—should be pursued.

The literature is replete with descriptions of patterns, covering a broad range of climatological variables and spatial and temporal scales. Several of these patterns have received considerable attention in recent years, and their names are now firmly established in the climatological lexicon. The purpose of this chapter is to provide a brief description of the more widely discussed patterns that have been observed to vary on decadal or longer time scales. This chapter thus serves as a glossary, albeit incomplete, for the remainder of the text, while describing a representative selection of patterns with their characteristics, couplings, and relationships. A number of issues related to improving our understanding of the role of spatio-temporal patterns in climate change and variability over dec-cen time scales are presented as well.

CLIMATE PATTERNS IN THE ATMOSPHERE

North Atlantic Oscillation

The North Atlantic Oscillation (NAO) is usually defined through the regional sea-level pressure (SLP) field, although it is readily apparent in mid-tropospheric height fields. Its influence extends across much of the North Atlantic and well into Europe (Figure 3-1a). Like other patterns to be discussed here, it has a basically fixed spatial structure. The NAO's amplitude and phase vary over a range of time scales from intraseasonal (van Loon and Rogers, 1978) to interdecadal (Wallace et al., 1992); the largest amplitudes typically occur in winter. Figure 3-1b shows more than 100 years of NAO variability.

The NAO is often indexed by the difference in SLP between Iceland, representing the strength of the Icelandic (or Newfoundland) climatological low, and the Azores or Lisbon, near the central ridge of the Azores high. Correlation of the NAO index with surface air temperature and sea surface temperature (SST) further reveals the extent of the atmospheric connection between the North Atlantic and the northern portion of Europe, and part of northern Asia (Hurrell and van Loon, 1996; Hurrell, 1995). Typically,

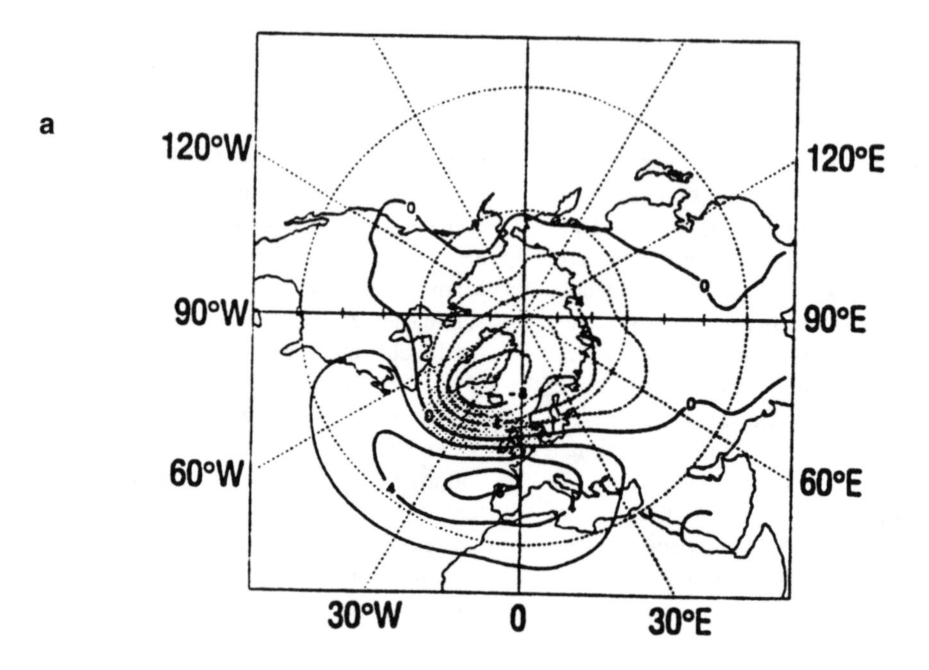

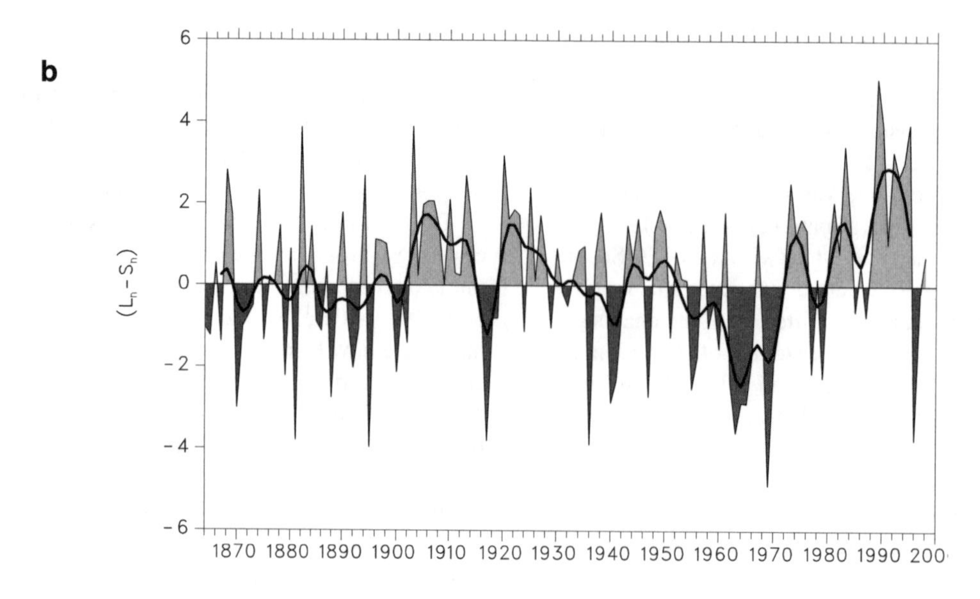

FIGURE 3-1 (a) Differences between sea-level pressures in high and low NAO-index years, showing the region of NAO influence. (b) Variation in the NAO (December-March) index since 1864; the heavy line represents a filtered version of the data. (Both figures from Hurrell, 1995; reprinted with permission of the American Association for the Advancement of Science.)

when the index is high the Icelandic low is strong, which increases the influence of cold Arctic air masses on the northeastern seaboard of North America and enhances the westerlies carrying warmer, moister air masses into western Europe in winter (Hurrell, 1995). Thus, NAO anomalies are related to downstream wintertime temperature and precipitation across Europe, Russia, and Siberia (Hurrell and van Loon, 1996; Hurrell, 1995). They have also been linked (see, e.g., Dickson et al., 1996) to changes in the thermohaline circulation in the North Atlantic (Lazier, 1988), the cod stock in the northwest Atlantic Ocean (Dickson et al., 1988), the mass balance of European glaciers (Pohjola and Rogers, 1997), the Indian monsoon (Dugam et al., 1997), and the atmospheric export of North African dust (Moulin et al., 1997).

Pacific-North American Pattern

The Pacific-North American (PNA) pattern represents a large-scale atmospheric teleconnection between the North Pacific Ocean and North America. It appears as four distinct cells in the 500 hPa (hPa are equivalent to mb) geopotential height field near Hawaii, over the North Pacific, over Alberta in Canada, and over the Gulf Coast of the United States. Wallace and Gutzler (1981) defined an index for the phase of this teleconnection pattern through a weighted average of 500 hPa normalized-height anomaly differences between the centers of the four cells. (Figure 3-2a shows the region and extremes of influence of the PNA pattern, and Figure 3-2b shows 15 years of variation in the PNA index.) The PNA is reflected in SLP (Rogers, 1990) as well, however, and can

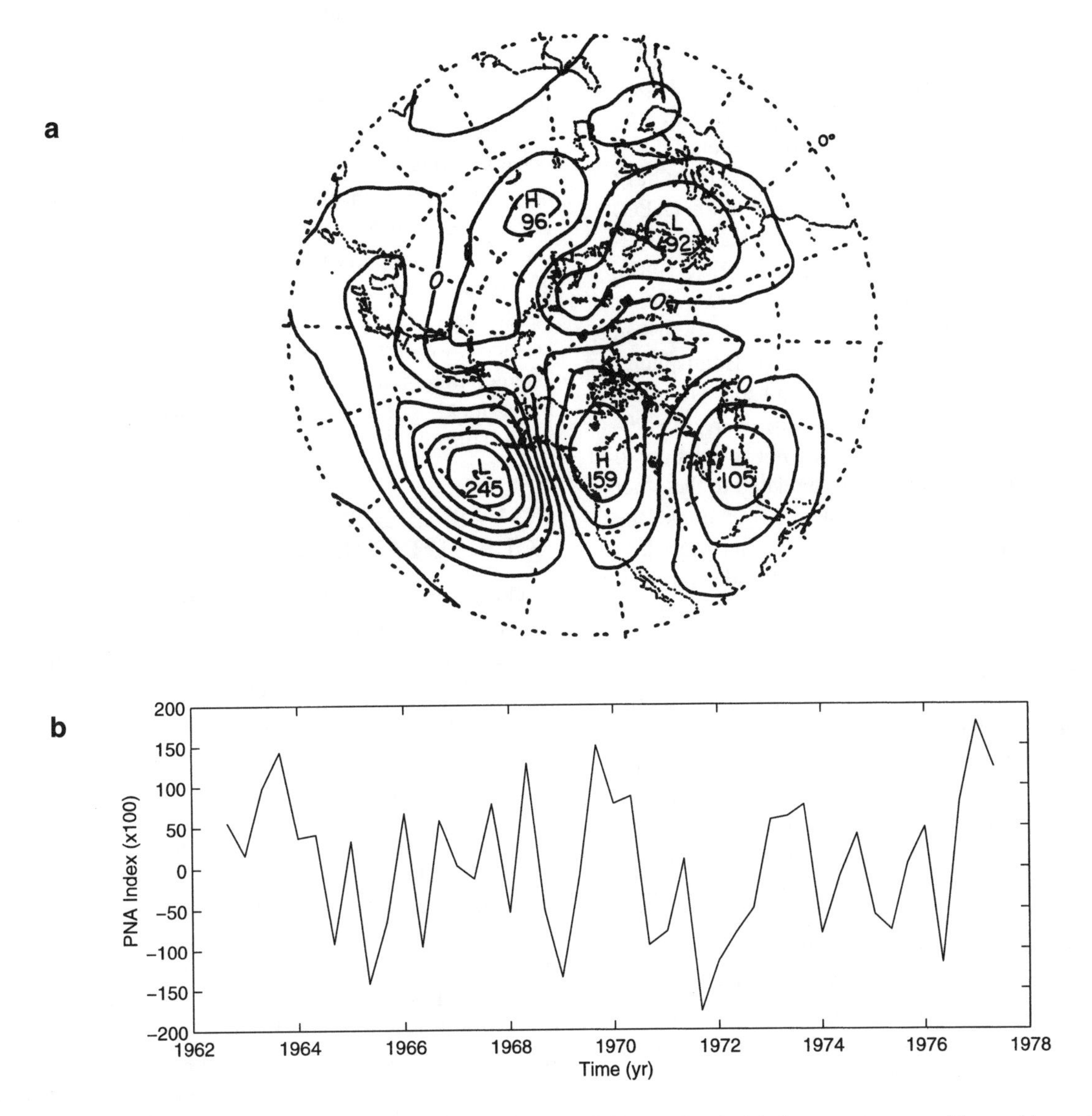

FIGURE 3-2 (a) Composite of the difference in the 500 hPa height field associated with the ten extreme positive and ten extreme negative values of the PNA index for 1962-1977. (b) Time series of the monthly mean value of the PNA index. Only Dec.-Jan.-Feb. values are shown, with the year tick marks at the Jan. values. (Both figures from Wallace and Gutzler, 1981; reprinted with permission of the American Meteorological Society.)

be depicted by the North Pacific Index (NPI) (Trenberth, 1990). The NPI is defined as the averaged SLP over a large area of the North Pacific Ocean near the center of the Aleutian low.

It has been speculated that the decadal variability of the PNA and NPI is associated with decadal changes in the tropical Pacific, as discussed below (Nitta and Yamada, 1989; Graham, 1994; Lau and Nath, 1996). Interdecadal variability of ENSO (discussed below in the section on the role of dec-cen variability in global warming) and the PNA is also thought to be responsible for a significant amount of the variance in the salmon inventory along the northwest coast of North America (Mantua et al., 1997).

OTHER PATTERNS OF INTEREST

Regional Patterns

The NAO and PNA patterns introduced above, while predominant in the literature and displaying variability on decadal time scales, represent but two of those identified. A number of other regional atmospheric or SST patterns have been analyzed, such as the North Pacific Oscillation (NPO) (Walker and Bliss, 1932; Rogers, 1981), West Pacific Oscillation, West Atlantic Pattern (Wallace and Gutzler, 1981), and Pacific Decadal Oscillation (Mantua et al., 1997). Other atmospheric patterns, such as the Pacific South American pattern, have been identified in the Southern Hemisphere; the data are frequently too sparse in time and space to allow more detailed analyses of these patterns, however. In addition, other structures exist that may or may not be considered climate patterns, although they are often related to the other patterns or presented in a similar manner. For example, the Asian monsoon, while predominantly a seasonal signal, is strongly correlated with ENSO and shows decadal variability as indexed by precipitation and wind speeds over India. Some investigators treat it as another distinct form of decadally varying pattern.

The number of regional patterns identified in the Northern Hemisphere is on the order of 10 (Wallace and Gutzler, 1981; Esbensen, 1984; Barnston and Livezey, 1987), which raises a question as to whether they are all unique. If atmospheric variability amounted to a continuum (i.e., no phase preference existed, a situation comparable to "white noise" in the frequency domain), then statistical analyses, such as those used in teleconnection studies, would produce a finite number of teleconnections (see, e.g., Wallace, 1996). Thus it has been suggested that the multiplicity of patterns is partially the result of a regional continuum (Kushnir and Wallace, 1989; Wallace, 1996), and that not all teleconnections are indeed unique phenomena.

Cold Ocean-Warm Land

Finally, one other "pattern" warrants introduction here. This is the "cold ocean-warm land" or COWL pattern (Wallace et al., 1995). The COWL pattern is *not* a fundamental mode of climate variability as defined through the decomposition of climatological variable fields, nor is it a particular climate phenomenon; rather, it simply represents a distinct geographic distribution of near-surface temperature anomalies predominantly reflecting the contrast in thermal inertia between land and ocean (Wallace et al., 1995, Broccoli et al., in press). In particular, the COWL pattern is a Northern Hemisphere winter phenomenon that is a manifestation of the dominant effect of the continental land masses on the mean surface air temperature during the cold season. It shows considerable high-frequency variability (e.g., monthly), because the air over the ocean responds slowly to change because the ocean's large heat capacity makes it slow to change, whereas the response of air over land is considerably faster because the land's small heat capacity permits more rapid response. These differences lead to rapid change over land in concert with large-scale shifts in the atmospheric circulation cells, while change over the oceans is much slower and more attenuated. Despite the apparent short-term memory of the COWL pattern, it displays long-term variability as well, which is of particular importance to the global warming experienced over the last 20 years. The lower-frequency variability of the COWL pattern seems to be related to long periods of simultaneous surface warmth in northwestern North America and the Eurasian continent, and can be identified with the simultaneous phase-locking of the PNA and NAO patterns (Wallace et al., 1995; Hurrell, 1996).

CO-VARIABILITY IN THE CLIMATE SYSTEM: COUPLED PATTERNS

Coupled patterns have expressions in at least two climate-system components, but they are not presumed to be causally related. The term therefore includes coupled modes, but also refers to patterns in each component that are simply coherent.

Tropical Atlantic Variability

Numerous studies have found a robust relationship between SST anomalies in the tropical Atlantic and changes in soil moisture, albedo, and surface-roughness over North Africa (see Nicholson, 1989, for a brief overview). SST anomalies appear to be responsible for a large part of the variability in tropical rainfall over Africa and Brazil. Observational studies by Hastenrath and Heller (1977) and Lamb (1978) were the first studies that linked the variability in tropical Atlantic SST to variability in the rainfall over the Sahel and Nordeste Brazil, respectively (see also Markham and McLain, 1977; Lamb et al., 1986; Hastenrath and Greischar, 1993; Rao et al., 1993). (Figure 3-3 illustrates the correlation between tropical Atlantic convection and Nordeste Bra-

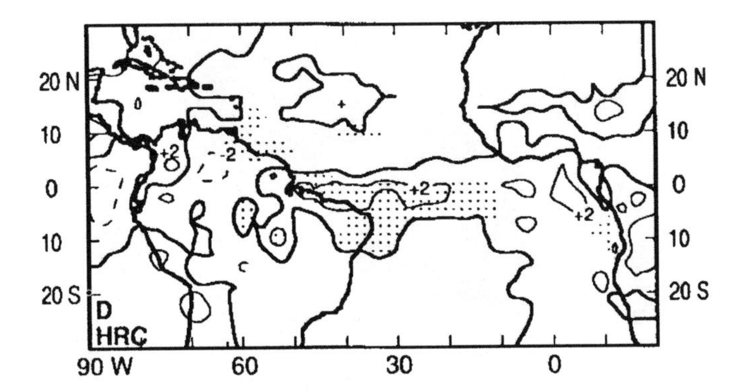

FIGURE 3-3 March-April field of convective activity in the tropical Atlantic sector. Shown are the differences in the number of days per month with highly reflective clouds for Nordeste Brazil wet (1984-1986, 1989) minus dry (1980, 1982, 1983, 1990) years. Significance of differences at the 5 percent level, as determined from t-test, is indicated by dot raster. (From Hastenrath and Greischar, 1993; reprinted with permission of the American Geophysical Union.)

zil precipitation.) Together these observational studies indicate that rainfall variability is associated with changes in the position and structure of the Intertropical Convergence Zone (ITCZ), a band of surface wind convergence characterized by strong and frequent convective activity, which in turn is extremely sensitive to variations in the meridional SST gradient. (See also Lamb, 1978; Shinoda, 1990; Rowell et al., 1992; and Nobre and Shukla, 1996, for discussion of this topic.) Similarly, atmospheric general-circulation model (GCM) studies using prescribed SST anomalies also show the importance of tropical SST anomalies in generating and controlling rainfall anomalies in Brazil and Africa (Moura and Shukla, 1981; Mechoso et al., 1990; Hastenrath and Druyan, 1993; Folland et al., 1986; Owen and Folland, 1988; Palmer et al., 1992; Rowell et al., 1992; Semazzi et al., 1993).

The tropical Atlantic Ocean shows a coherent structure in SST variability. The dominant pattern of SST, as defined by empirical orthogonal function (EOF) analysis, often shows a warm pool in the tropical North Atlantic and a complementary cool pool in the tropical South Atlantic, or vice versa. These centers of action seem to vary coherently over decadal time scales but independently on shorter time scales (Houghton and Tourre, 1992; Mehta and Delworth, 1995; Chang et al., 1997). Consequently, this phenomenon is sometimes referred to as the Atlantic Tropical Dipole, although the lack of a clear consensus on the actual dipole nature of the pattern leaves many simply referring to it as the decadal tropical Atlantic SST variability. This low-frequency SST phenomenon shows concurrent anomalies in the rainfall over Brazil and northern Africa (Figure 3-4a). Periods of greater-than-normal rainfall were experienced over northeast Brazil in the 1960s, and periods of lower-than-normal rainfall in the mid-1970s to early 1980s (Figure 3-4b). Gray (1990) suggests that the decadal changes in the SST in the subtropical North Atlantic may also be responsible for the changes in the distribution and intensity of hurricanes in this region.

While the physics of the dipole climate oscillations in the

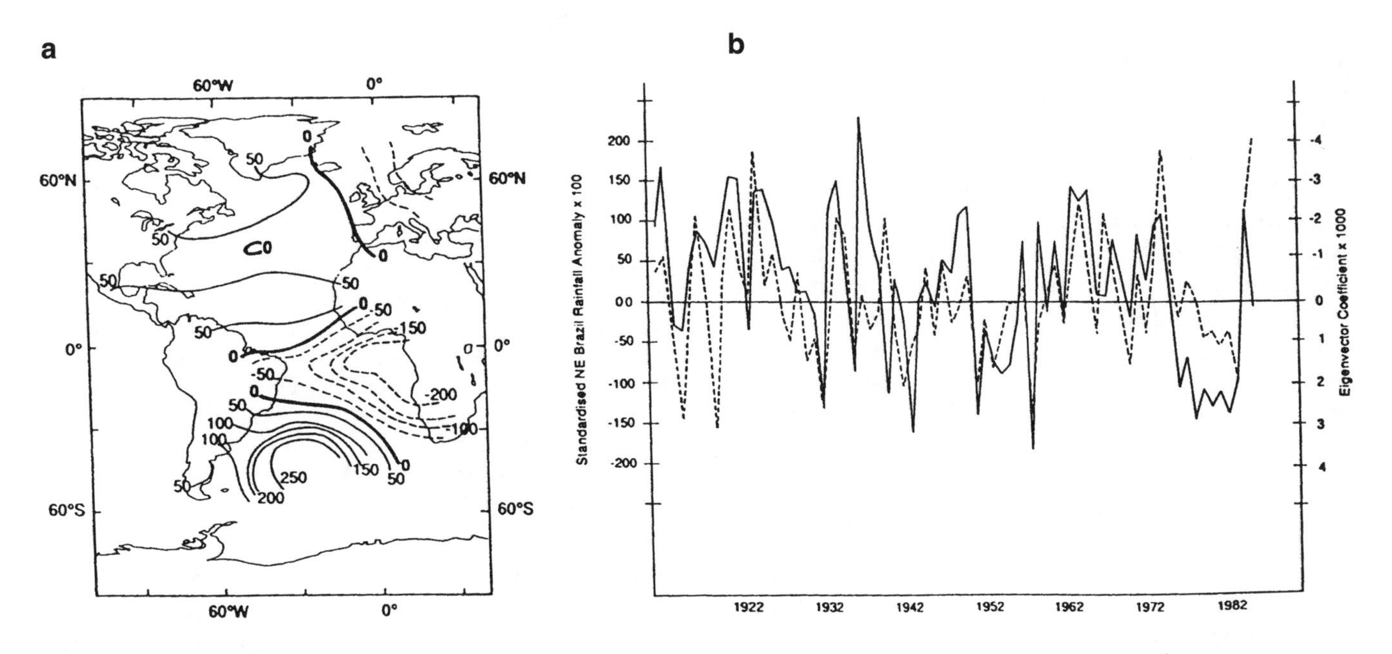

FIGURE 3-4 (a) The spatial pattern of the EOF of Atlantic SST that is strongly related to rainfall in Nordeste Brazil and western Africa. (b) Time series (solid line) of the March-to-May values of the EOF shown in (a) and the north Nordeste Brazil rainfall anomalies (dashed line). (From Ward and Folland, 1991; reprinted with permission of John Wiley and Sons, Ltd.)

tropical Atlantic are not yet understood, recent results from Chang et al. (1997) demonstrate that an intermediate-level coupled atmosphere-ocean model of the tropical Atlantic does support decadal oscillations that have a structure similar to the observed dipole phenomenon. Finally, the climate in and around the Atlantic is also affected by variability in the tropical Pacific on both the interannual (see, e.g., Folland et al., 1986; Hastenrath et al., 1987; Enfield and Mayer, 1997) and decadal (Zhang et al., 1997) time scales.

As in the tropical Pacific, within 5° of the equator (the so-called tropical waveguide), ocean dynamics seem to play a significant role in the generation of SST anomalies on the interannual time scale and in the equatorial waveguide (Zebiak, 1993). The SST anomalies in the waveguide are linked to interannual variations in rainfall along the Guinea coast (Wagner and da Silva, 1994). However, little is known about the cause of the low-frequency variability in the SST of the tropical and subtropical Atlantic that is associated with the large-scale rainfall anomalies and with the variations in hurricane activity.

North Atlantic Variability

Kushnir (1994) examined the multidecadal variability in the observational record of SLP, SST, and surface wind velocity in the North Atlantic basin, and found two examples of warm and cold epochs in the twentieth century; each of these epochs lasted more than a decade. Warm periods are characterized by positive SST anomalies around southern Greenland and negative anomalies along the northeastern U.S. seaboard (upper panel of Figure 3-5). The concurrent SLP and wind anomalies indicate a southern displacement of the Icelandic low and a relaxation of the winds in the subtropics (lower panel of Figure 3-5) coinciding with a decrease in NAO. Kushnir (1994) concluded that the variability demonstrated on these time scales in the observational record was governed by a basin-scale interaction between the large-scale oceanic circulation and the atmosphere.

Deser and Blackmon (1993) found that SST in the North Atlantic subpolar basin varied concurrently with the atmospheric surface wind anomalies. These atmosphere and ocean anomalies span the twentieth-century record and display a roughly 10-year period. Deser and Blackmon also note that the spatial relationship between SST and wind anomalies in these quasi-decadal cycles is consistent with what could be expected theoretically if the phenomenon were inherently due to coupling between the atmosphere and the ocean. They point out, however, that there is a high negative correlation between the SST anomalies and the anomalies in sea-ice extent in the Baffin Bay/Labrador Sea region. Furthermore, the sea-ice anomalies lead the SST and wind anomalies by a few years.

In addition to the above phenomena, an event that began in the late 1960s has drawn unusual attention from the ocean community. A significant surface freshwater anomaly ap-

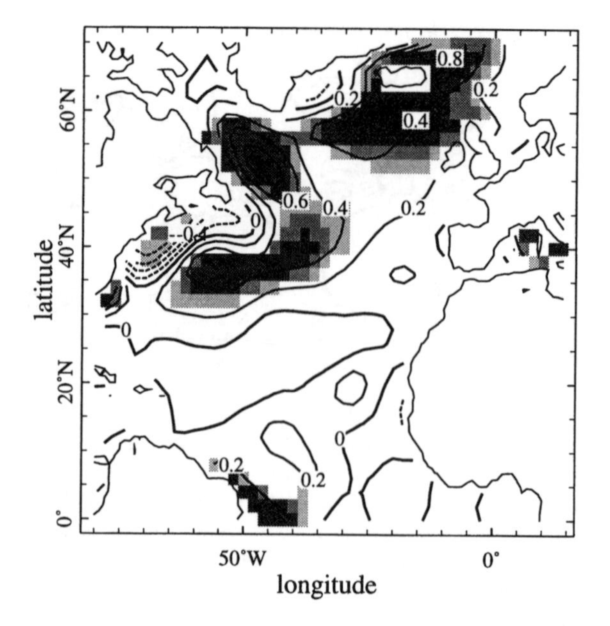

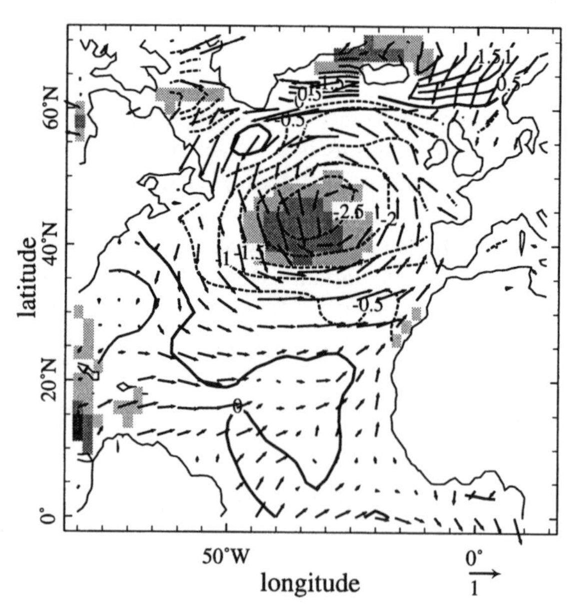

FIGURE 3-5 Upper panel: The difference between the annual winter Atlantic SST averaged from 1950-1964 (warm years) minus the winter average from 1970-1984 (cold years). Contour interval is 0.2°C. Lower panel: as above, but for SLP and winds. Contour interval is 0.5 mb. The arrow at the bottom of the panel is 1 m s^{-1}. Distribution of the t-variable corresponding to SST and SLP differences is denoted in three levels of gray: light for 2.0-2.5, medium for 2.5-3.0, and dark for 3.0-3.5. (From Kushnir, 1994; reprinted with permission of the American Meteorological Society.)

peared in about 1969 in the Labrador Sea. Now known as the "Great Salinity Anomaly" (GSA), this feature can be traced moving eastward across the subpolar gyre, into the Norwegian Sea, and ending up near Fram Strait more than 10 years later (Dickson et al., 1988). Aagaard and Carmack (1989) hypothesize that the GSA was born from an increase

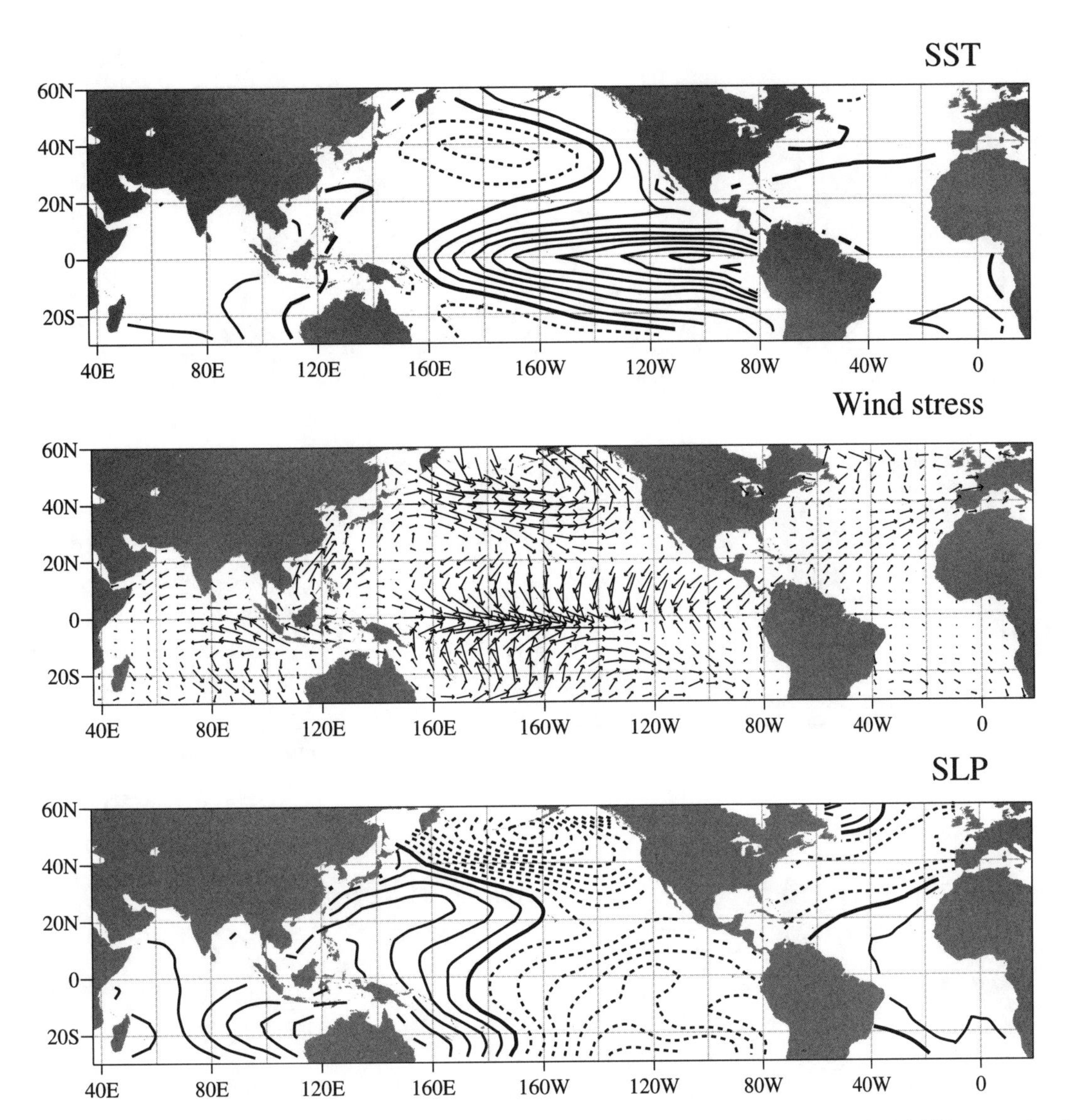

FIGURE 3-6 Global SST pattern, wind stress, and sea-level pressure that is related to the interannual variability associated with ENSO, based on linear regression between a high-pass-filtered "cold-tongue index" (CT) and global SST. (From Zhang et al., 1997; reprinted with permission of the American Meteorological Society.)

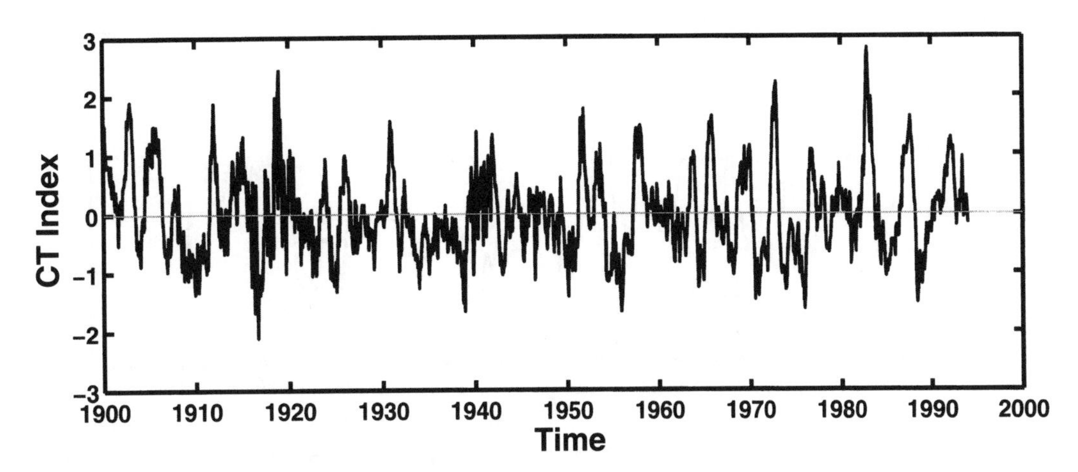

FIGURE 3-7 Time series of a cold-tongue index, corresponding to the SST pattern displayed in Figure 3-6. (From Zhang et al., 1997; reprinted with permission of the American Meteorological Society.)

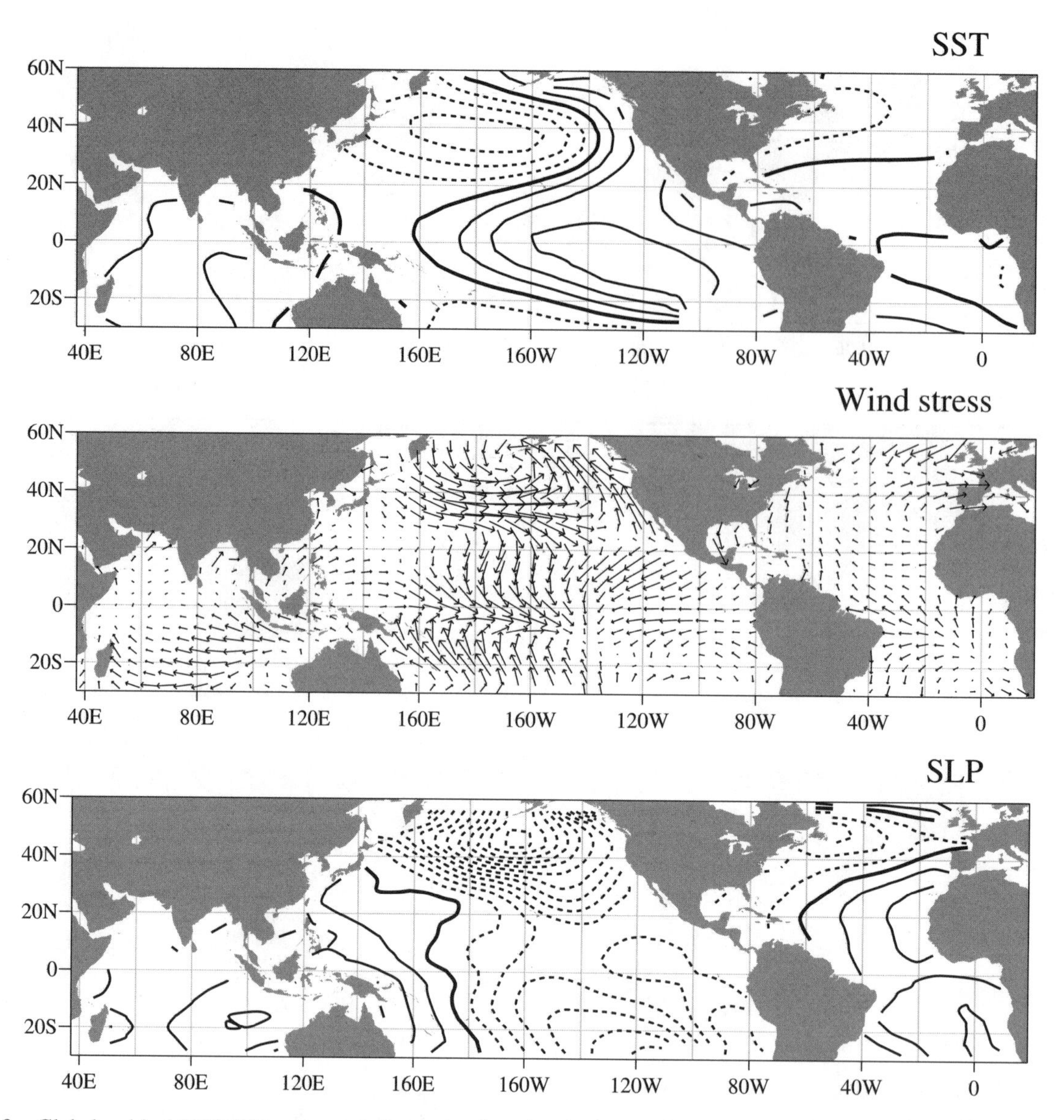

FIGURE 3-8 Global residual (GR) SST pattern, wind stress, and sea-level pressure, from which the linearly related ENSO variability has been removed. (From Zhang et al., 1997; reprinted with permission of the American Meteorological Society.)

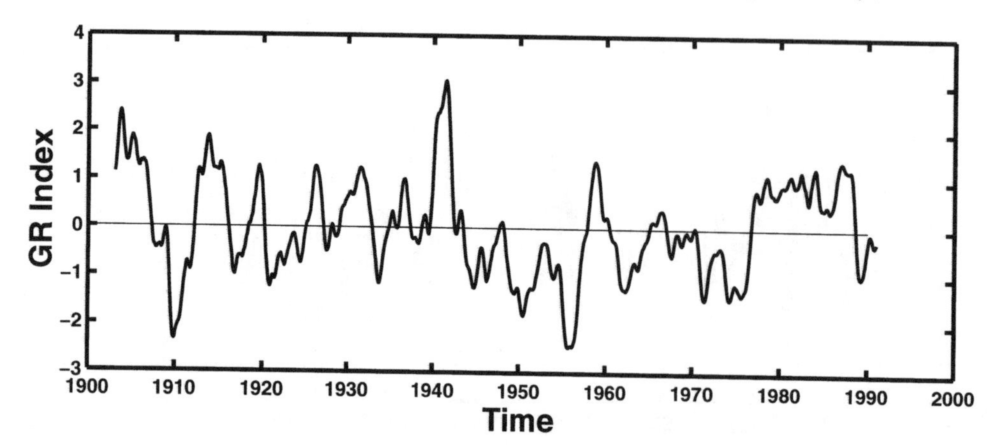

FIGURE 3-9 Time series of the global-residual index, corresponding to the SST pattern displayed in Figure 3-8, which indicates that the ENSO-like pattern in that figure is associated primarily with decade-to-century-scale variability. (From Zhang et al., 1997; reprinted with permission of the American Meteorological Society.)

in sea-ice export from the Arctic (see also Hakkinen, 1993, and a related paper by Mysak et al., 1990). There is evidence that the GSA's stabilization of the upper water column interrupted deep-water production in the North Atlantic (Lazier, 1988). Thus, the GSA is thought to be an example of phenomena that may lead to changes in the thermohaline circulation in the ocean, which may then lead to feedbacks to the atmosphere on much longer time scales.

Pacific Decadal ENSO-like Pattern

Tanimoto et al. (1993) and Zhang et al. (1997) demonstrate that the time variability of the leading EOF of the global SST field is separated into two components: one identified with ENSO variability (Figure 3-6) showing predominantly interannual variability (Figure 3-7), and the other a linearly independent "residual" (Figure 3-8) dominated by dec-cen-scale variability (Figure 3-9). The two components exhibit remarkably similar spatial signatures in global SST, SLP, and wind-stress fields, with the SST field in the residual pattern being less equatorially confined in the eastern Pacific than the interannual pattern, and having a larger extratropical signature in the North Pacific. In fact, the residual pattern is very similar to the leading EOF of North Pacific SST that Mantua et al. (1997) have called the Pacific (inter)Decadal Oscillation (PDO). The residual pattern's SLP signature is also stronger in the extratropical North Pacific, and its wintertime 500 hPa height anomaly (Figure 3-10) more closely resembles the PNA pattern than the ENSO pattern shown in Figure 3-6. The amplitude time series of the ENSO-like pattern in the residual variability reflects much of the low-frequency variance in the data as well as some of the interannual variability, including a notable regime shift in 1976-1977 (Quinn and Neal, 1984) and an equally remarkable shift of polarity in the 1940s (see Figure 3-9).

This ENSO-like pattern in SST appears to be teleconnected to anomalies in the mid-latitude atmosphere and ocean of the North Pacific (Figure 3-6). The decadal ENSO-like anomalies of Figure 3-8 are also teleconnected throughout the tropics, with large concurrent changes in tropical Atlantic and Indian Ocean SST (Zhang et al., 1997), as well as the North Pacific (see, e.g., Kumar et al., 1994). Unlike the previous patterns, which are clearly defined through EOF analyses of atmospheric or ocean property fields, this pattern appears only when the data (atmosphere or ocean fields) are time-filtered prior to the analysis. However, because of its large spatial influence and its apparent relationship to the shorter-time-scale ENSO phenomenon, it has received considerable attention in recent years.

The last few decades have represented a warm phase of this climate anomaly, which has preceded a significant reduction in the alpine glaciers throughout the tropics (Thompson et al., 1993b; Diaz and Graham, 1996). In addition, the streamflow and snowpack in the northwest and southwest of North America (Cayan and Peterson, 1989; Cayan, 1996) are well correlated with this time series of the decadal ENSO-like climate phenomenon.

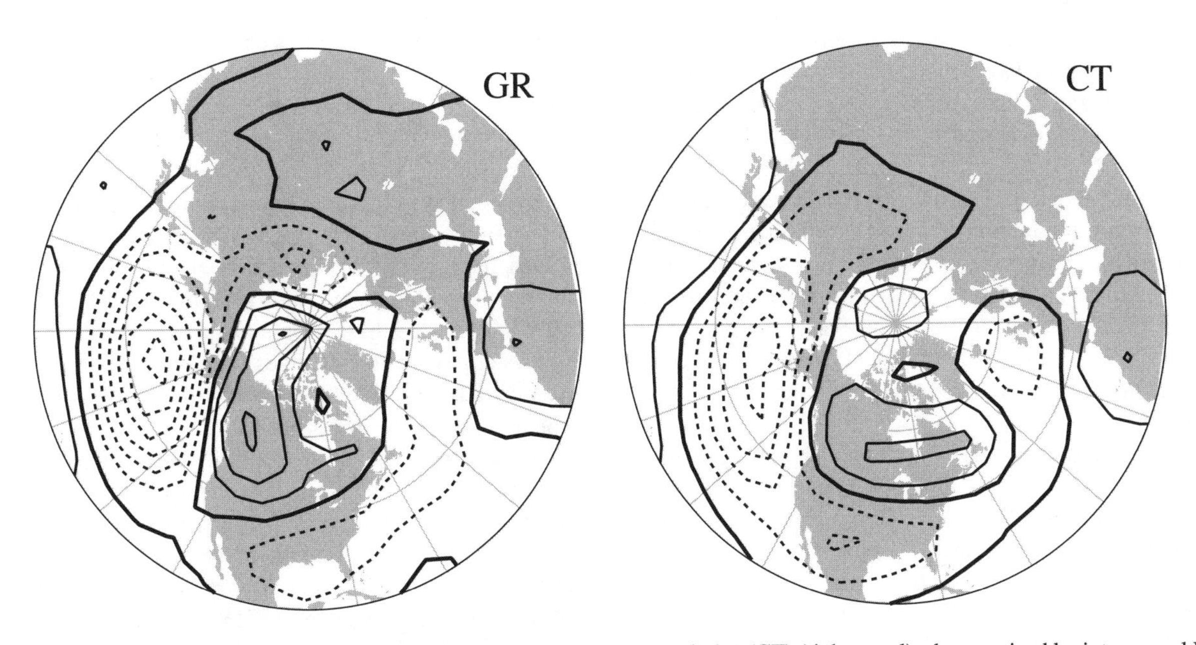

FIGURE 3-10 The 500 hPa height anomalies associated with the cold-tongue index (CT) (right panel), characterized by interannual ENSO variability, and the global-residual index (GR) (left panel), characterized by ENSO-like decadal variability. (From Zhang et al., 1997; reprinted with permission of the American Meteorological Society.)

Pacific Summer Variability

Norris and Leovy (1994) point out that there are long-term trends in summer SST in the North Pacific over the period 1952-1981, and that the changes in SST are significantly anti-correlated with the long-term trends in the maritime stratiform cloudiness. They note that the strongest (anti)correlations between SST and maritime stratiform cloudiness are found co-located with the largest climatological SST gradients. Norris and Leovy suggest that these trends may result in part from the persistence of SST anomalies from winter to summer.

Arctic Variability

When Walsh et al. (1996) examined the long-term record of SLP in the Arctic basin, they found that a large decrease in the annual mean SLP occurred in the mid-to-late 1980s. The anomalously low SLP persisted through 1994, the end of the record they examined. The SLP anomalies are largest in the central Arctic, decreasing toward the adjacent coastal regions around the Arctic Ocean; they are present year-round but are greatest in the winter season. The annual mean pressure changes are larger there than anywhere else in the Northern Hemisphere.

Walsh et al. argue that it is unlikely that the observed change in the Arctic atmospheric circulation could be associated with low-frequency variability in the extrapolar regions. They note, however, that the observed changes in the atmospheric circulation can be expected to lead to changes in the transport and compactness of the sea ice in the Arctic basin. Specifically, the mean anticyclonic motion of the ice pack should decrease, and the sea ice should now be more divergent than it was during the 1970s and early 1980s. Indeed, there have been two remarkable large-scale anomalies in Arctic sea ice in the past decade: the extraordinarily thin sea ice that was experienced by the SHEBA (Surface Heat Budget of the Arctic project; see Moritz and Perovich, 1996, for a description) in the winter of 1997 (McPhee et al., 1998), and the offshore contraction of the sea ice off Siberia in 1990. The latter, which is unprecedented in the record, has recently been linked to changes in the pan-arctic atmospheric circulation (specifically, the wind stress) by Serreze et al. (1995) and Bitz (1997). Cavilieri et al. (1997) note that the areal extent of sea ice has decreased by about 6 percent between 1978 and 1996.

The large trends in the Arctic atmospheric circulation and in the NAO (see, e.g., Figure 3-1) during the past 30 years appear to be related (see Thompson and Wallace, 1998, and references therein). Thompson and Wallace argue that these large-scale decadal trends are best described in terms of a planetary-scale mode of variability, which they referred to as the Arctic Oscillation (AO), whose regional extension into the North Atlantic accounts for the phenomena attributed to the NAO. The AO involves fluctuations in the strength of the polar vortex that extend from the surface upward into the lower stratosphere, and occur on time scales ranging from weeks to decades. The "high index" (strong westerlies) polarity of the AO is characterized by mild wintertime temperatures over most of Eurasia poleward of 40°N. Stratospheric involvement in the AO is most clearly apparent during late winter and early spring, when wave-mean-flow interactions at these levels are strongest. Thus, the trend in the AO, and hence in the NAO and Arctic circulation, may be viewed as a systematic bias in one of the atmosphere's most prominent modes of internal dynamic variability. Whether it is occurring in response to anthropogenic forcing has yet to be determined.

Thompson and Wallace (1998) found that the AO, as represented by the first empirical orthogonal function (EOF) of SLP, changed rapidly in amplitude during the mid- to late 1980s. This change is consistent in timing and sense with the changes noted by Walsh et al. (1996). Furthermore, this atmospheric change coincides with a number of additional changes noted in the sea ice and upper ocean. (The precise timing of the changes is unknown for most of the upper-ocean variables, given the relatively sparse data available for the Arctic region.) For example, data collected through the early 1990s by Morison et al. (1998), Carmack et al. (1995, in press), McLaughlin et al. (1996), and Steele and Boyd (in press) all show that the position of the central Arctic upper-ocean front, which separates the Atlantic and Pacific waters, has shifted relative to the 1950-1989 climatological position (as established using the Environmental Working Group atlas built from U.S. and Russian hydrographic observations released through the Gore-Chernomyrdin joint commission); the region dominated by Atlantic water has expanded by nearly 20 percent. These same data, as well as those of Anderson et al. (1994), Rudels et al. (1994), and Quadfasel (1991) suggest a warming of the Atlantic layer occurring at this same time. Other studies are finding changes in the surface winds, sea ice, and other upper-ocean characteristics, such as pycnocline properties. Four summers with the most extreme minimum Arctic sea-ice coverage (Maslanik et al., 1996) have occurred since this change took place, and anomalously thin ice has also been reported (McPhee et al., 1998). Together, these ocean, ice, and atmosphere observations suggest that the changes in the late 1980s in the Arctic may have involved the entire vertical column from the upper ocean to the stratosphere.

There are, however, theoretical reasons to expect large variability in Arctic systems because of the coupling among the polar atmosphere, ocean, and sea ice. First, the Arctic sea-ice thickness is thought to be extremely sensitive to changes in vertical heat transport in the ocean (see, e.g., Maykut and Untersteiner, 1971); only modest circulation changes in the ocean would be required to induce variability in the ocean heat transport that would have a significant impact on the thickness and spatial extent of the Arctic sea ice, and hence on the albedo of the Arctic. In addition, it has recently been argued that the observed high-frequency (subseasonal) variability in atmospheric energy transport into the Arctic may lead to large variability in Arctic sea-ice thickness in the Arctic that occurs primarily on decadal and multidecadal time scales (Bitz et al., 1996).

Variability of the sea ice in the Arctic is a defining aspect of the Arctic climate system. In addition, the ramifications of changes in Arctic circulation and sea ice are clearly important for understanding the circulation of the subpolar North Atlantic Ocean mentioned earlier. Thus, while the connections between the AO and NAO, and the polar and extra-polar regions, are still unclear, their co-variability suggests that these coupled Arctic variations are intimately tied to extra-polar regions.

Antarctic Variability

A completely different kind of pattern involving sea ice, surface winds, SST, and SLP has been found in the Southern Ocean. Specifically, the Antarctic Circumpolar Wave (ACW) is characterized by co-varying deviations in monthly climatological averages of these variables along the Antarctic polar front, near the winter marginal ice zone (White and Peterson, 1996). It is also highly coherent with temporal variations in ENSO (White and Peterson, 1996) and the Indian Ocean monsoons (Yuan et al., 1996), although the underlying physics are not yet understood. It is predominantly an interannual phenomenon, but, as with ENSO, it shows longer-period variability. It is not clear how the ACW is related to, or interacts with, the dominant mode of variability of the zonal mean flow in the Southern Hemisphere (Hartmann and Lo, 1998), or the standing-wave patterns of van Loon and Jenne (1972) and van Loon et al. (1973), though the superposition of these various patterns in space suggests that their interaction is feasible.

THE ROLE OF DEC-CEN VARIABILITY IN GLOBAL WARMING

It is clear that the global warming experienced over the past 20 years is distinguished by an enhanced warming in winter that was not evident in previous decades, dominated by a strong warming over Northern Hemisphere land, and compensated for to some degree by a lesser cooling over parts of the Northern Hemisphere oceans. Despite its decadal persistence, this pattern is consistent with the basic COWL pattern described above. Its geographic aspect is readily apparent in the global surface-temperature data when the last 20 years are compared to the previous 20 years (Color plate 1). A pattern similar to the COWL pattern (though not identified as the COWL pattern) is produced by numerous modeling studies simulating anthropogenic inputs, and thus is considered by some to be one component (of several) of the so-called "greenhouse fingerprints" (Wigley and Barnett, 1990; Santer et al., 1995). Its presence in the actual observational data has therefore been accepted as additional evidence of anthropogenic warming (IPCC, 1996a).

Figure 3-11 shows that when the monthly-average Northern Hemisphere surface-temperature time series for this century is adjusted to eliminate the influence of the COWL pattern, two things become apparent (Wallace et al., 1995): (1) a large fraction of the month-to-month variability, which is particularly apparent in the colder half of the year, is removed; and (2) a significant fraction of the accelerated warming in the cold-season months that has been apparent since the mid-1970s (Figure 3-11, middle panel) is also removed, and the summer and winter trends become comparable (bottom set of curves).

Further investigation suggests that much, though not all, of the accelerated warming that has occurred since the mid-1970s is attributable to the COWL pattern. Much of the COWL pattern itself can be explained by the synchronous polarity of the NAO and PNA patterns during this accelerated period of warming (Hurrell, 1996). That is, over the

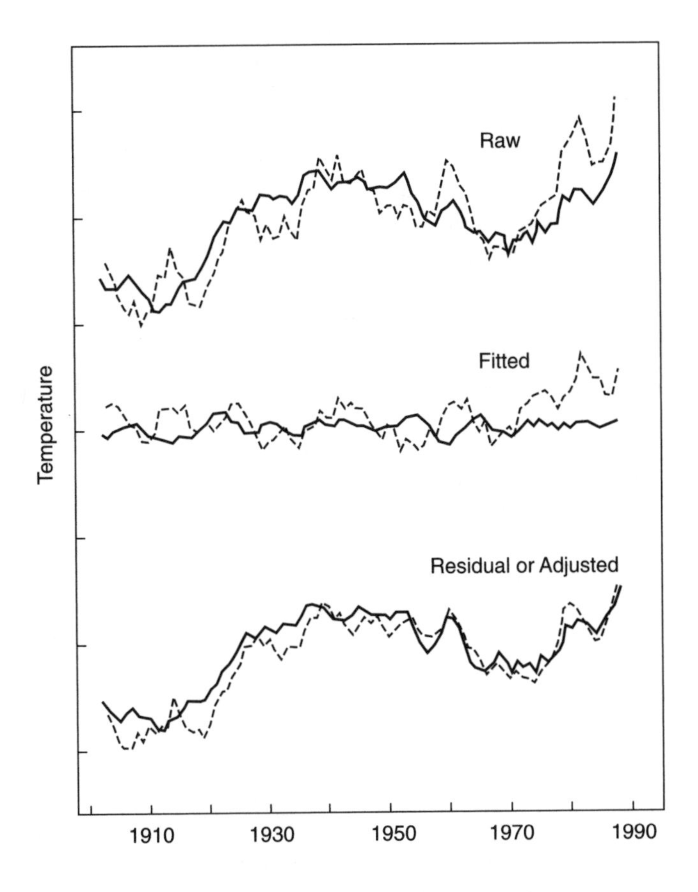

FIGURE 3-11 Top, smoothed monthly-average surface-temperature time series for the Northern Hemisphere. The solid line is the warming-months average, the dashed line the cooling-months average. Center, the portion of the series attributed to the COWL pattern. Bottom, the time series adjusted by eliminating the COWL contribution. (From Wallace et al., 1995; reprinted with permission of the American Association for the Advancement of Science.)

last 20 years, the NAO and PNA patterns (the latter represented by the North Pacific Index) both seem to show a seemingly unusual persistent tendency, on average, to occupy states that favor a warming of Europe and Northern Asia by the NAO, and warming of North America by the PNA. When this warming is removed, the global warming trend of the last 25 years is similar to, though slightly larger than, the warming that occurred over several decades in the early part of this century (from 1910 to 1940, say). Broccoli et al. (in press) find that their coupled model reproduces this warming attributed to the COWL pattern. They also find that such an accelerated warming is highly anomalous (exceeding the 99th percentile) in simulations that do not contain greenhouse warming. The extent to which the COWL pattern is anthropogenically caused or influenced is an outstanding issue, and one crucial to all aspects of the greenhouse-warming puzzle.

Another critical question is whether anthropogenic increases in radiative forcing play a role in amplifying the intensity, frequency, and/or duration of ENSO events. For example, several works have shown that the coupled ocean-atmosphere system, particularly in the tropical Pacific Ocean, plays an important role in regulating the mean climatological state (see, e.g., Neelin and Dijkstra, 1995; Sun and Liu, 1996; Clement et al., 1996; Seager and Murtugudde, 1997; and Cane et al., 1997). Sun and Trenberth (1998) extend this concept by suggesting, on the basis of an observational study, that the El Niño events are an effective means for removing heat from the tropical Pacific and may even arise as a necessity for removing excess heat associated with increased solar heating and the greenhouse effect. This suggestion is also consistent with a number of coupled-modeling studies of increased radiative forcing or increased CO_2 that have shown more energetic and more extreme ENSO events (Sun, 1997; Timmermann et al., 1998; and an ENSO-like pattern of response (see, e.g., Meehl and Washington, 1996, and Knutson and Manabe, 1995, in press). Much of our current uncertainty regarding the role of the coupled ocean-atmosphere system in greenhouse warming is the result of uncertainties involving the role of the cloud-radiative feedback, which underscores the need to improve model representations of this critical process. Substantial uncertainties regarding model response of the tropical Pacific to greenhouse warming also arise in the modeling of the equatorial thermocline.

Such a response has been suggested by Trenberth and Hoar (1995, 1997). Through analyses of more than 100 years of climate data they concluded that the record of the Southern Oscillation Index (SOI) in the period since 1976 is significantly different from the earlier portion of the record. Though this research preceded the powerful 1997-1998 El Niño event, they found a recent tendency for more frequent El Niño events and fewer La Niña events (contrary to the model results of Timmermann et al. (1998)). Although Trenberth and Hoar posit that this change may be caused by the observed increase in greenhouse gases, it is not currently possible to say definitively what underlies this change in character of El Niño events (NRC, 1996). Trenberth and Hoar's conclusions are predicated on a statistically stationary, linear-time-series model of the SOI time series. Rajagopalan et al. (1997) offer an alternative viewpoint using a time-series model whose parameters are allowed to vary slowly over time. This analysis highlights significant dec-cen variability both in the probability of El Niño and La Niña states, and in the probability of transitions between these states. Their results suggest that the recent tendency to more frequent and persistent El Niño events may not be nearly so unusual as Trenberth and Hoar indicate. These differences underscore the need to discriminate better between natural and anthropogenic causes for dec-cen variability, and to think further about what statistical methodology is appropriate for such analyses. By improving our understanding of how anthropogenic climate change interacts with natural processes such as El Niño, we will improve our ability to detect global warming signals, and perhaps gain predictive insight into the dec-cen-scale modulation of phenomena that have traditionally been viewed in a seasonal-to-interannual context.

Finally, it is interesting to note that the analysis of Huang et al. (1998) of 135-year-long indices has revealed a relationship between the NAO, ENSO, and PNA, albeit a complex one. They find that an enhanced positive phase of the PNA is likely to occur with the positive phase of the NAO. Furthermore, at frequencies of 2 to 4 years, the NAO is coherent with ENSO from 1960 to 1990; this coherence is most apparent during strong to moderate El Niño events. (This result is consistent with the findings of Rogers (1984).) As noted earlier, the COWL spatial pattern also correlates with positive NAO and PNA indices. Greenhouse warming thus carries the potential for altering all four patterns. Our understanding of the climate system's likely response to anthropogenic forcing might benefit from considerable attention to the relationships between greenhouse warming and these (and other) natural modes of variability.

FUNDAMENTAL ISSUES AND QUESTIONS

The patterns and coupled modes occupy large spatial areas, describe significant climate variance, and bridge high-, mid-, and low-latitude zones. Despite the uncertainties about their roles in anthropogenic global warming and natural climate change, they represent an obvious avenue through which coherent climate variations and change may be propagated globally. The IPCC Second Assessment (IPCC, 1996a) noted that much of our attention has recently shifted from the analysis of mean global temperature to that of its spatial distributions, anticipating that climate change might manifest itself irregularly in space and time—yet patterns have appeared. The identification of coherent patterns with coupled modes that explain significant fractions of the spatial and temporal variability offers hope that there may be a signal in what appears to be just "noise." Furthermore, the apparent persistence of these patterns in time, even allowing for a slow evolution, provides additional hope that these signals may be exploited to help us understand and predict future climate variability and change. Also, the apparent relationship between specific climate-pattern dispositions and regional climate characteristics lends support to the notion that understanding long-term trends in the patterns may enable us to make short-term predictions for some regions.

To realize these potentials, considerable effort must also be invested in improving our general understanding of the patterns and coupled modes: their mechanisms (dynamic and thermodynamic, natural and anthropogenic), couplings, feedbacks, and sensitivities. These are truly cross-disciplinary issues, requiring a strong interdisciplinary approach. Specifically, we must answer the following questions:

- *What is the longevity of the patterns and their spatial/temporal variance?* The patterns offer tantalizing evidence that some fraction of the Earth's climate shows spatially and temporally coherent structure with some degree of (predictable) persistence in a time-averaged sense. However, the fundamental patterns themselves may be transitory phenomena reflecting the current configuration of a slowly changing climate. Consequently, we need to understand the dynamics between climate patterns and the general state of climate. For example, do the patterns and climate evolve in a systematic manner? At some level of change do the patterns become simple artifacts of the change, as opposed to predictors of the change?
- *What is the best way of characterizing the known patterns, and are there additional patterns of interest?* Specifically, how can we most effectively define their salient features, their co-varying components and coupled modes (including regional influences and correlation with or control of the climate attributes discussed in Chapter 2), their sensitivities to analysis technique, and their spatial distribution and broader teleconnections? Likewise, robust and optimal indices of these coherent atmospheric-circulation patterns need to be established, because some of the indices employed to represent the patterns, while convenient, do not capture much of their spatial and temporal complexity. For example, the Bermudan-Azorean high remains relatively stable in its spatial orientation, but the Icelandic low often migrates southward to Newfoundland, and the North Atlantic SST pattern tends to show a rotation around the North Atlantic basin (see, e.g., Hansen and Bezdek, 1996) that a simple dipole index, such as the NAO, relating two fixed points, cannot capture. The new findings of Thompson and Wallace (1998), showing that the first EOF of Arctic SLP may be an even better indicator of Eurasian climate change than the NAO, further underscores the need to examine the optimal means of classifying the relevant modes of climate. Therefore, while indices have proven useful in their ability to simplify the temporal history of complex patterns and demonstrate their broad spatial coherence and importance, additional research is required to better characterize the patterns and isolate their significant characteristics. That is, more robust indices that efficiently describe the fundamentally important characteristics of the patterns must be developed. What patterns and coupled modes exist in data-poor regions, and what are their spatial and temporal characteristics?
- *Which patterns represent true dynamic modes, and which ones are simply statistically consistent structures, or geographically forced distributions?* That is, are the patterns fundamental modes of climate variability reflecting coupled, internal, and external dynamics and thermodynamics? Or are they simply the reflection of structures that are intrinsic to the atmosphere (i.e., determined by the land-sea distribution and internal atmospheric dynamics)? In the latter case, they stochastically force the ocean (see, e.g., Hasselmann, 1976), integrate the response, and provide a feedback to the atmosphere by "reddening" the spectrum of the patterns in the atmosphere without affecting atmospheric dynamics (Barsugli and Battisti, 1998). Or are they the consequence of statistics or chaos, representing attractors of random but spatially consistent distributions? Understanding the mechanisms underlying these patterns will be fundamental in our assessment of how they can ultimately be used in long-term forecasting and prediction of climate variability and change.
- *What are the mechanisms responsible for generating, maintaining, and modifying the patterns? What role do these mechanisms play in the spatial propagation of regionally initiated variability or change, and what are their critical dependencies?* To understand how a change in the state of a pattern in one location may dictate the regional climate in some more remote location, it is necessary to understand the mechanisms that control the spatial and temporal evolution of the patterns and their broader influences or teleconnections. This knowledge will also provide an indication of how a local disturbance may influence the dominant regional pattern, leading to broader propagation of the anomaly, and thus influencing the controlling components of the climate system.
- *What is the relationship between decadal-to-centennial patterns and global warming?* What part of the COWL warming pattern is due to natural variability versus anthropogenic modulation of these naturally occurring patterns? Will there be extended periods of time in which they display similar, relatively persistent polarity quite by chance, or is COWL a manifestation of anthropogenic warming through the polarity of the natural climate modes? (See discussions by Wallace et al. (1995) and Hurrell (1996).) Is the residual warming—that is, warming apart from the COWL contribution—natural variability or anthropogenic warming? What is the relationship between the COWL pattern, greenhouse fingerprint, and natural climate patterns—that is, how do the natural modes of the climate system respond to different changes in forcing, natural or anthropogenic? Are there unique characteristics or

response modes? What controls the degree and nature of the spatial distributions and interactions of the modes, and how do they co-vary?

It is clear that a more comprehensive understanding of the variability of these patterns on decade-to-century time scales is absolutely essential if we are truly to distinguish between anthropogenic change and natural climate variability, or understand their interaction. If an enhanced greenhouse effect strengthens or phase locks existing natural modes of variability, the study of dec-cen variability and the study of anthropogenic change are inextricably combined. It will be impossible to understand one without the other.

4

Mechanisms and Predictability

We have seen that climatic attributes directly affect human activities and viability, and that these attributes are directly connected to the major processes acting in the Earth's climate system. To exploit beneficial climate variations and mitigate the effects of the harmful variations, we must first understand the operation of the system and then project what could happen in the future.

By now it is well known that some features of our climate can be forecast beyond the two-week limit of atmospheric deterministic predictability. This chapter details the nature of predictability on dec-cen (and shorter) time scales, and endeavors to distinguish between the predictable aspects of natural variability and the predictable consequences of possible anthropogenic changes. It examines how predictability is related to the mechanisms of climate variability and assesses the possibility of predicting climate on dec-cen time scales, especially the attributes discussed in Chapter 2. Last, it speculates on the applications of such predictions should they be attainable.

THE NATURE OF CLIMATE PREDICTION

Weather Prediction

The state of the atmosphere and ocean is governed by what is generally agreed to be a set of deterministic equations: If the initial state is exactly known, the future evolution of the system is determined for all time. On the other hand, the atmosphere and ocean are rife with instabilities and nonlinearities that imply that the climate system is chaotic: Two sets of initial conditions that appear to be very close to one another (but not identical) will evolve along trajectories that inevitably diverge. Their divergence cannot continue forever, though, given the bounds imposed by the system's finite energy. As a result, trajectories continuously approach, as well as diverge from, each other, generating the system's chaotic, albeit deterministic, behavior.

A numerical weather-forecast model also has deterministic solutions, but any small error in the initial state guarantees that the resulting forecast trajectory will diverge from a trajectory that began with the "correct" initial state. The inevitable growth of the initially specified error, and the subsequent mixing of ever-changing trajectories, limits predictability. The pioneering work of Lorenz (1963, 1969), and the research it spurred over the years, established that the doubling rate of errors for large-scale atmospheric flows is on the order of 2-3 days (Lorenz, 1982), so that the global atmosphere is predictable only on scales of two weeks or so, given the currently achievable accuracy of initial-state determination. Similar studies for the detailed predictability of the oceans are in their infancy, but the slower growth and saturation times of oceanic instabilities suggest potentially longer predictability times, on the order of months rather than weeks. Still, all these times are much shorter than the dec-cen time scale of interest here.

Climate Prediction

The obvious question is, if the ultimate limit of detailed prediction for atmosphere and ocean weather is on the order of weeks to months at best, how could we possibly expect to predict climate on time scales of years or decades? The answer follows directly from the definition of climate: Climate is the statistics of the atmosphere (and other components of the climate system). We have known for a long time that atmospheric statistics are determined entirely by the boundary conditions of the atmosphere; every atmospheric modeler uses this paradigm. The boundary conditions for an atmospheric model (in particular SST and land and sea ice) are specified, and the atmospheric circulation and hydrologic cycle are allowed to come into equilibrium with these specifications. Other boundary quantities are then determined by internally coupling the atmospheric model to a land model, in particular land-surface moisture and vegetation, and land snow and ice cover. Whether or not climate is uniquely determined by the boundary conditions is still undetermined, but for the purposes of argument, we will assume here that it is.

If, therefore, we can predict the boundary conditions of the atmosphere at a specific time, particularly SST and sea ice, we will have some information about the statistics of the atmosphere at that time. We will not be able to predict the precise state of the atmosphere, because it can vary in equilibrium with the predicted boundary conditions, but we will know something about its average conditions. We may be able to predict the average monthly or seasonal precipitation over a region even if we cannot say on what specific day the precipitation will fall. Knowing only a mean value at a given time can still be helpful if the associated variability is small. Predictions of tropical boundary conditions at a certain time are likely to be useful because tropical climate variability is low, while predictions of mid-latitude boundary conditions would be less useful because mid-latitude variability, especially during winter, is high.

Two more important points need to be made. First is the distinction between initialized and uninitialized prediction. To make a prediction about a specific time in the future, say the summer of 2009, there must be some connection to the actual conditions now. We call this estimation of the actual beginning state of the system "initialization" (while recognizing that this term is sometimes used elsewhere to mean the act of bringing a model system to a state of equilibrium without estimating its current conditions). If we do not make this initial estimation, we will not be able to forecast the time at which the climate will assume a given state, though we may still draw conclusions about its statistics (that is, changes of its mean and the nature of its variability). The difference between initialized and uninitialized prediction becomes important in discussing greenhouse-warming predictions versus ENSO predictions. The second important point is the potential for making empirically or statistically based (analog) predictions. Sufficient information is available from past climate records to allow predictions to be made (with specified uncertainty) whenever specific climate states exist that have in the past been accompanied or followed by particular regional or local climate conditions. "Climate state" is defined here, as in NRC (1975), as the average of the complete set of atmospheric, hydrospheric, and cryospheric variables over a specified period of time in a specified domain of the earth-atmosphere system.

For climate prediction on all time scales, whether initialized or not, the tool for predicting the boundary conditions of SST and sea ice is the coupled climate model—a model that consistently links the atmosphere, ocean, and ice together in responding to a specified external forcing.

SHORT-, MEDIUM-, AND LONG-RANGE CLIMATE PREDICTION

There is no accepted terminology describing the various time scales for prediction. This report will use "short-range climate prediction" to denote prediction on time scales up to interannual, "medium-range climate prediction" to mean prediction at decadal time scales, and "long-range climate prediction" (sometimes called "greenhouse prediction") for prediction on centennial time scales—the scale of a human lifetime.

Short-range climate prediction is an established enterprise: Skill has been demonstrated for predicting the SST changes in the tropical Pacific that are characteristic of the ENSO phenomenon on lead times of 6 to 12 months. Atmospheric properties elsewhere may then be inferred from these forecasts. These predictive skills, which vary as a function of several factors (including season, model type, and decade), have been well documented (Battisti and Sarachik, 1995; Glantz, 1996; Latif et al., 1998). ENSO prediction is initialized prediction (in the sense defined above), so a real-time observing system in the tropical Pacific was put in place by the TOGA research program. It has been kept in place even though TOGA has ended, which should permit us to develop our skill further.

Long-range climate prediction has so far been limited to predicting forced climate change in response to the anthropogenic addition of radiatively active gases and aerosols to the atmosphere. Because this type of prediction is essentially uninitialized, it cannot predict the actual state of the boundary conditions at some specific future time. It can, however, be used to derive the statistics of the boundary conditions (and therefore the statistics of the atmosphere in equilibrium with the statistics of the boundary conditions) at some future time. Thus, initialized short-range climate prediction can predict the SST in the tropical Pacific for January of 1999, say, while greenhouse predictions can only say that annually averaged SST will be warmer in the year 2050 by some specified amount, or within a certain range. Such greenhouse predictions are still valuable if the forcing changes the mean boundary conditions enough for a difference beyond natural variability to be apparent; again, small shifts of the mean may be noticeable in the tropics where the variability is low, while larger shifts may be masked in mid-latitudes where variability is high. Long-range forecasts permit the assessment of shifts in average precipitation, or length of the growing season, or changes in patterns of runoff; as indicated by Karl et al. (1996), even subtle shifts in the mean state can have considerable implications for the frequency and magnitude of extreme climate events.

Medium-range climate prediction, prediction on time scales of a decade or so, is the most problematic type of prediction. Its value as uninitialized prediction is limited: The year-to-year variability of climate, together with the relatively slow approach of the climate system to equilibrium with anthropogenically added radiatively active atmospheric constituents, limit the value of prediction of the statistics of boundary conditions a decade in advance. Even this type of prediction may be useful under certain circumstances, however. When regional changes are fast and crossing the threshold of a new climate state can be predicted, preparations can be made for change even if the exact time of occurrence is not known.

The possibility of initialized medium-range climate prediction is a real and intriguing one. As the fully coupled system is allowed to evolve freely over the course of a decade from its initial state, water parcels from deeper parts of the ocean reach the surface and imprint themselves on the SST field. The question, of course, is whether or not the evolving imprint of the initial ocean state on the SST can survive both the mixing in the ocean (as parcels wend their way to the surface) and the inevitable noise from high-frequency atmospheric forcing.

PREDICTION AND MECHANISMS

The existence and nature of climatic predictability depends on the nature of the mechanisms responsible for the variability. We can distinguish two basic types of mechanisms for decadal variability: variability forced by processes external to the climate system, and variability generated by processes internal to the climate system.

Externally Forced Atmospheric Variability

In the external-forcing category are forcings by varying solar output; by addition of aerosols due to major volcanic eruptions, biomass burning, and industrial sources; and by the addition of radiatively active gases to the atmosphere. Their effects are discussed in detail in Chapter 5, "Atmospheric Composition and Radiative Forcing."

It is generally true that variability generated by external forcing is unpredictable when the forcing itself is unpredictable; the decadal variations of solar radiative output and the geodynamics of future eruptions of volcanoes are both poorly understood. When volcanic aerosols are added to the stratosphere, there is a period of a year or two in which the aerosols stay in the stratosphere, so their radiative and chemical effects can be predicted over the following year (see, e.g., Fiocco et al., 1996). The addition of radiatively active gases to the atmosphere produces mean warming of the Earth, regional changes of mean temperature and precipitation, and possible changes to natural climate cycles, such as ENSO and the Pacific North American pattern, but prediction of these mean changes relies on our ability to know the past and future emissions of these gases. Those radiatively active gases that also contain chlorine or oxides of nitrogen affect the stratospheric burden of ozone, so to the extent that the concentration of these gases is known and the chemistry of ozone is understood, the concentration of ozone may be predicted.

Internally Forced Atmospheric Variability

We may classify the internal decadal-variability mechanisms into three distinct categories: those arising from high-frequency forcing of the slow components of the climate system by the more rapidly varying atmosphere; those arising from slow internal variations in the ocean, atmosphere, cryosphere, or biosphere; and those arising from the coupling of components of the climate system that individually would not have such an effect (Sarachik et al., 1996). Specific mechanisms associated with each of the components of the climate system will be presented in Chapter 5.

Chapter 3 described a variety of patterns of variability, and some of their co-varying aspects. In this section we discuss the extent to which these patterns, or other broad-scale features of the atmosphere, appear to be coupled to other parts of the climate system, predominantly the ocean. These links are important to predictability; when a fast and a slow component are coupled, and the latter has mechanistic control of the pattern, the longer time scales of the slower component can be capitalized on to make more distant forecasts or predictions of the faster component (in this case the atmosphere) than would otherwise be possible.

The Hasselmann (1976) theory of climate variability is a convenient starting point for understanding the relevance of different climatic time scales to climate prediction. Hasselmann's theory asserts that the atmosphere produces, through instabilities of various types, high-frequency variability that presents itself as weather. At these frequencies, the variability may be considered random. When a slower-reacting reservoir, such as the ocean, is forced by such high-frequency variability, the high-frequency variability is damped in the slower component. This basic climate mechanism of Hasselmann seems to account for a great deal of observed variability (see, e.g., Frankignoul and Hasselmann, 1977) and of modeled natural variability in long, coupled climate simulations (Manabe and Stouffer, 1996).

If the Hasselmann mechanism were the only operative mechanism—say, the atmosphere acting on the ocean—the temporal extent of the predictability would be the autocorrelation time of the sea surface temperature. In this case, the best forecast of SST would be a forecast of persistence (i.e., no change) that fades to the norm with a time constant consistent with the autocorrelation function. Over many parts of the world ocean, persistence is on the order of months. If, however, the SST generated by the Hasselmann mechanism feeds back to the atmosphere in a coherent way, coupled modes may result, increasing predictability notably. Similar couplings may exist with other parts of the climate system, such as the perennial ice or snow fields, though this possibility has not been explored to any great extent.

Extension of these ideas to the stochastically forced coupled atmosphere-ocean system suggests that such coupling will act to significantly enhance the very-low-frequency variance in the atmosphere (see, e.g., Barsugli and Battisti, 1998). This is the simplest theory accounting for the presence of very-low-frequency variability in the climate system, and is usually assumed to be correct in the absence of other information. The Hasselmann mechanism would also determine the maximum level of predictability of climate anomalies, if the coupling between the atmosphere and the ocean (or biosphere or cryosphere) did not support patterns of its own. However, there are several climate phenomena on the decadal time scale that are most likely the result of processes other than those inherent in the Hasselmann mechanism. The best documented of these coupled phenomena are discussed below, and the possible mechanisms are summarized.

The essential reason for studying the mechanisms of decadal variability (aside from their intrinsic scientific interest) is that determining which mechanisms are operative will determine the extent to which climate can be predicted. Some mechanisms (e.g., external forcing by volcanic eruptions) we will assume to have no predictability, and thus to offer no improvement in predicting climate. Some mechanisms have moderate predictability (e.g., random forcing of SST by the atmosphere), and some have significant predictability (e.g., SST variations in tropical Pacific caused by coupled atmosphere-ocean modes). These are well worth exploring.

Coupled Modes

Chapter 3 described a large-scale decadal mode of variability in the Pacific that has ENSO-like SST characteristics in the tropical regions and strong out-of-phase covariation of SST in the North Pacific. The mechanisms responsible for this variability are not yet known. Various investigators have postulated various reasons for it, including:

- inherent nonlinearities in the physics of ENSO (e.g., Munnich et al., 1991);
- interaction between ENSO and the seasonal cycle (e.g., Jin et al., 1994, 1996);
- interaction between ENSO and other unstable coupled atmosphere-ocean modes (e.g., Mantua and Battisti, 1994);
- stochastic forcing of a linearly stable coupled system (e.g., Penland and Sardeshmukh, 1995); and
- low-frequency changes in the shallow equatorial thermohaline circulation that may lead to changes in the amplitude and frequency of the interannual ENSO variability (e.g., Pedlosky, 1987; McCreary and Lu, 1994; Gu and Philander, 1997).

Whatever the cause, a significant portion of the low-frequency variability in the global climate system is ENSO-like in structure, as can be seen in Figure 3-7. Decadal-scale changes in the state of the tropical Pacific atmosphere-ocean system might also affect the predictability of the higher-frequency ENSO variability. For both of these reasons, the Pacific is an important focus for understanding dec-cen variability in the climate system.

In addition to the ENSO-like coupled phenomenon, there is ample evidence of variability in the mid-latitude North Pacific atmosphere-ocean system on interannual time scales. Specifically, wintertime variability is largely forced locally by the atmosphere; about half of the variance in interannual SST anomalies is forced by ENSO, and communicated to the North Pacific Ocean via the atmospheric teleconnections. The variability in the North Pacific climate on decadal and longer time scales is not yet fully documented. It is known that, over the last half-century, a substantial portion of the variability in the atmosphere-ocean system on these time scales is associated with the global ENSO-like structure displayed in Figure 3-7.

Recently, Latif and Barnett (1996) have suggested a mechanism by which coupling between the atmosphere and upper ocean that takes place in the mid-latitude North Pacific basin may give rise to climate variability in the North Pacific and North America on the multidecadal time scale. The spatial structure of the model anomalies in the Latif and Barnett study is remarkably similar to that of the mid-latitude anomalies in the global ENSO-like structure of the observations displayed in Figure 3-7, which is also primarily a low-frequency pattern of variability. However, the observed anomalies clearly involve the tropical atmosphere-ocean system. Hence, if the mechanism suggested by Latif and Barnett does indeed operate in nature, it will be necessary to sort out how much of the variance in the mid-latitudes comes from forcing in the tropical Pacific that is teleconnected to the mid-latitudes, and how much from atmosphere-ocean interactions that are specific to the North Pacific.

Delworth et al. (1993) found multidecadal variability in the North Atlantic of the coupled atmosphere-ocean GCM of Manabe and Stouffer (1988). The pattern of the SST anomalies associated with this variability is somewhat similar to that found in the observations by Kushnir (1994) and in simpler coupled models (Chen and Ghil, 1996). However, the mechanism associated with the variability in both types of atmosphere-ocean models seems to be internal to the ocean thermohaline circulation, and not inherently a coupled atmosphere-ocean phenomenon.

Sarachik et al. (1996) review the current theories of mechanisms for producing dec-cen climate variability from an ocean and ocean-atmosphere modeling perspective. These theories include: stochastic forcing of the ocean by white-noise, synoptic-scale variability of the atmosphere (e.g., Hasselmann, 1976; Mikolajewicz and Maier-Reimer, 1990), internal ocean variability (e.g., Delworth et al., 1993; Chen and Ghil, 1995), coupled ocean-atmosphere modes (Hirst, 1986; Latif and Barnett, 1994), and ENSO variability (Trenberth and Hurrell, 1994; Wallace et al., 1995). Yin and Sarachik (1994) proposed an oceanic advective and convective mechanism. A completely different type of interannual (in the North Atlantic) to decadal (in the North Pacific) variability has been modeled by Jiang et al. (1995) via the nonlinear dynamics of the double-gyre circulation's strength, heave, and wobble (see also Cessi and Ierley, 1995, and Speich et al., 1995) and by Spall (1996) via the nonlinear dynamics of the Gulf Stream and the deep western boundary current.

Prospects for Climate Prediction

As stated, the nature of the mechanisms causing dec-cen variability will determine whether or not it is possible to make deterministic climate predictions on these time scales. It is clear that the mechanisms for some of the most notable decadal phenomena are unknown at present. If the natural decadal phenomena are forced externally, they cannot be predicted a decade in advance because those forcings are assumed to be unpredictable. If the Hasselmann mechanism is the operative one, then the scale of predictability is limited to the order of the autocorrelation time of the slower component of the system, usually the ocean. If the ocean is the driver of long-term climate-system variability, however, the fact that its circulation is sluggish offers some hope for initialized decadal-scale prediction: Initial experiments with low-resolution coupled models (Griffies and Bryan, 1997) have indicated that errors imposed on the initial (internal) state of the North Atlantic grow slowly enough that SST and sea level may be predictable as much as a decade in advance. Skill in forecasting equatorial Pacific SST, which has been demonstrated out to two years (Chen et al., 1995), may be extended to longer forecast periods, perhaps approaching a decade.

The best hope for initialized prediction is offered by the existence of true atmosphere-ocean modes. Depending on the degree of coupling and the internal ocean phenomena that set the time scale for the oscillation, the initialization of those ocean phenomena captures the wherewithal for the coupled evolution in part of its cycle, and guarantees that parts of its evolution will continue into the future. ENSO on shorter time scales is an example of this; the initial state is the thermocline's configuration, which, when specified at the initial time, determines the future evolution of the upper tropical ocean and hence its temperature structure. This kind of prediction also seems to work for the modal structures involved with the Atlantic subtropical dipole (Chang et al., 1998).

The possibility of long-range (uninitialized) greenhouse prediction has been demonstrated by many coupled predictions over the years (see IPCC, 1996a, for a complete review). As yet, however, only the grossest measures of warming have been used. Also, it is clear that other climate system components besides temperature must change in response to a significant change in greenhouse gases—for example, the warming may be kept to a minimum by changes in the atmospheric moisture distribution. These changes, such as those in the hydrologic cycle or cloudiness, may actually be more important to society than the temperature effect. Also, the regional usefulness of current predictions is limited both by the uncertainty of the prediction of the mean itself and by the regional variability that masks modest changes in the mean. The general usefulness of uninitialized greenhouse prediction is defined by the magnitude of the forced response relative to the existing variability. These factors, combined with the impossibility of demonstrating the correctness of such a prediction until the extremely long prediction time has passed, make the use of such predictions particularly challenging.

THE USES OF CLIMATE PREDICTION

This section considers the uses of decadal and longer prediction, both initialized and uninitialized, should the skill of such prediction turn out to be significant. In order to determine whether the climate attributes that affect humankind (discussed in Chapter 2) are likely to be predictable, we must first demonstrate that some related physical quantity (e.g., SST) is predictable. While we do not know enough yet to do this properly, we can indicate what is known and what still needs to be known in order to make predictions of each of the climate attributes.

Uses of Medium-Range Climate Prediction (Initialized)

The discussion of decadal-scale prediction here deals with initialized predictions only. There are two ways they can apply: directly (for the region from which the prediction is derived) or remotely (when teleconnections exist between that region and another).

Precipitation and Freshwater Availability

At this time, there are a few predictable phenomena that clearly affect rainfall over vulnerable populations: The Atlantic subtropical dipole that influences rainfall in Brazil and Africa (Hastenrath and Heller, 1977; Lamb, 1978; Hastenrath, 1990) is one, and the extreme ENSO states influencing western U.S. rainfall events (Cayan and Peterson, 1989) are another. Other phenomena affect rainfall and are probably predictable (e.g., the large-scale variability over the northern Pacific), but so far they have not been shown to be predictable on decadal time scales.

The variation of rainfall in the Brazilian Nordeste clearly has both interannual and decadal components. It has been related to the location of the ITCZ over the tropical Atlantic, which in turn has been related to SST variations in both the tropical and subtropical Atlantic and Pacific (most recently in Uvo et al., 1997). The possibility therefore exists that prediction of SST will lead to better estimates of the location of the Atlantic ITCZ and of consequences of future rainfall in Brazil and in the Sahel. SST has come to be predicted better on short (seasonal-to-interannual) time scales in the tropical oceans; the population dislocations in the Nordeste caused by years with low precipitation during the growing season have gradually been reduced by actions taken on the basis of predictions made a few months to a year in advance (Moura, 1994). Short-term forecasts for Morocco have proven quite successful as well. These are clear examples of how useful prediction of future climate can be to society.

The indication by Chang et al. (1998) that the decadal variation of tropical to subtropical Atlantic SST anomalies may be predicted years in advance offers the possibility of longer-range planning for the allocation of water resources to agriculture, drinking water, and hydroelectric power for northeastern Brazil and northwestern Africa. A prediction of dry conditions for the next five years might shift resource planning from power generation to irrigation, while a prediction of wet conditions might do the opposite. To the extent that precipitation in the Sahel and Morocco could be predicted, resources could be directed to agricultural infrastructure on a forecast of good rainfall, or to food-relief infrastructure on a prediction of poor rainfall.

Temperature

Temperature is the variable most relevant to the intensity and timing of cryospheric melting. To the extent that rivers are fed by glaciers and/or snowpack, their streamflow is partly determined by temperature. A prediction of increased average temperature would imply earlier melt, and therefore an early peak in streamflow. For those regions requiring summer water, the earlier peak may leave the summer unreasonably dry. Believable decadal (half-decadal) forecasts of temperature and precipitation—and, more important, streamflow, particularly in arid and semi-arid regions such as the U.S. West—will be critical to water managers in developing new operating rules that are better adapted to the changing conditions.

High temperatures over land also affect vegetation and soil moisture, especially during summer, both directly, by increasing evaporation, and indirectly, by affecting precipitation. A decrease in soil moisture has a feedback effect that increases surface temperature, as does reduced evapotranspiration by vegetation. Believable early warning of a prolonged period of increased temperature and decreased precipitation, especially in the prime agricultural lands of the Northern Hemisphere, would provide opportunities to avert global food insecurity and attendant disruptions: Irrigation infrastructure could be built, markets could be stabilized by futures hedging, drought-resistant seed could be stocked, and so on.

Changes in atmospheric temperature over the oceans eventually work their way into the interior of the ocean, where they affect ocean volume and may lead to noticeable changes in sea level. Although multiple decades generally pass before the impact of temperature change is felt, approximately 50 percent of the sea-level rise in this century may be attributed to the global warming over this time period.

Storms

Advanced numerical models can simulate and predict the short-term evolution of storms with a remarkable degree of success. However, the science of describing and predicting the statistical properties of storms on time scales of a month or longer is still in its infancy. Of major interest to climatologists is the relationship between extreme weather events (which are associated both with tropical and extra-tropical storms and with climatic factors such as regional SST) and the strength and position of the semi-permanent features of the atmospheric circulation. It has been demonstrated in several observational studies that a close relationship exists between the intensity and distribution of storms and climate variability. Changes in tropical cyclone activity have been observed to be linked to variations in regional SST and SLP (Shapiro, 1982; Emanuel, 1987), as well as to the entire state of the tropical atmosphere associated with ENSO (Gray et al., 1992; Nicholls, 1985). Extratropical cyclonic storms are linked to the state of the "teleconnections" in the monthly mean circulation (Lau, 1988; Rogers, 1990). Similarly, the high-frequency storm events exert an influence on low-frequency climatic fluctuations (Gray, 1979; Lau and Nath, 1991).

Existing observations are not adequate to elucidate the climatology of severe storms and their link to climate variability. The limited extent of the instrumental record and inconsistencies in reports of storm activity are major obstacles to a satisfactory resolution of these issues. Attempts have been made to extract useful information on severe weather from GCMs, but the results are at best rudimentary (Broccoli and Manabe, 1990; Haarsma et al., 1993). Once the evolution of large-scale climatic factors can be predicted, it is plausible to envision the emergence of an ability to predict the degree of storminess. In fact, schemes to predict up to a year in advance the characteristics of an upcoming tropical-cyclone season, given atmospheric and ocean variables, have already been demonstrated with some success (Nicholls, 1985; Gray et al., 1992).

Ecosystems

Temperature changes will alter the stability of the oceans. Changes to the stability-sensitive upwelling of nutrient-rich waters, as well as direct thermal effects (among other things), may modify the patterns of biological productivity in the world's oceans (IPCC, 1996b). Only recently has it been shown that the salmon fisheries of the Pacific Northwest undergo decadal variability in phase with an index of the North Pacific SST (Mantua et al., 1997; see Figure 2-18). Currently Alaska is producing a prodigious amount of salmon, while the Columbia River basin is producing very little. Just knowing this variability exists is useful; much time, effort, and money are spent in reviving salmon fisheries, when in fact their decline might be due to natural factors rather than fishing practices. A prediction of this pattern several years in advance would allow resources to be applied more efficiently.

Uses of Long-Range Climate Prediction (Uninitialized)

This section deals with uninitialized prediction on interdecadal time scales, and in particular the response of climate to the anthropogenic addition of radiatively active constituents (gases and aerosols) to the atmosphere. This topic has been treated in considerable detail by the IPCC which was established to assess available scientific information on climate change, assess the environmental and socio-economic impacts of climate change, and formulate response strategies. Indeed, the three volumes of the second IPCC (1996a,b,c) assessment devote some two thousand pages to doing just that, so only a few salient points connected with the climate attributes of Chapter 2 are noted here.

Precipitation and Freshwater Availability

Since the earliest days of coupled modeling and its use to determine the climatic effects of increases in radiatively active gases, an unambiguous and robust result has been that the hydrologic cycle will run faster in a warmer world (Manabe and Wetherald, 1975). What has been in some dispute is where and how the additional rainfall would occur, what the impact of the increased evaporation potential on land and vegetation would be, and how the frequency and intensity of extreme precipitation events would be altered. In most models, the increase in surface irradiance leads to increases in summer evaporation and reduced soil moisture (with subsequent vegetation stress). The greater evaporation also implies higher surface temperatures in continental interiors, which will reduce precipitation in those regions. The excess precipitation occurs over oceans and in high latitudes (see, e.g., IPCC, 1996a) where it is not beneficial to human activity. There are also suggestions that extreme precipitation events would increase (see, e.g., Trenberth, in press).

The patterns identified earlier (NAO, PNA, the Atlantic dipole, and ENSO-like decadal variability) all influence rainfall, by controlling the location of the storm tracks, by controlling the location of the ITCZ, or, in the case of ENSO, by controlling rainfall directly through SST expression or remotely through teleconnections. Thus it becomes important to know how these patterns of variability change as radiatively active constituents are added to the atmosphere. Unfortunately, current climate models cannot answer this question unambiguously.

Since much of the distribution of precipitation and its subsequent return to the ocean is topographically determined, it becomes necessary to more fully resolve model orography and streamflow. This cannot be done with the current generation of models that predict the response to radiative gases; for practical reasons, they cannot have high resolution if they are to be run for a hundred years. There is increasing evidence (see, e.g., Giorgi and Marinucci, 1996) that embedding high-resolution models within global-prediction models gives better distributions of precipitation, and offers the hope that future predictions of climate response to radiatively active gases will be specific enough for more accurate planning. In addition, statistically-based climate-downscaling methodologies are showing promise (Hewitson and Crane, 1996; Zorita et al., 1995).

The implications of having accurate medium-range forecasts of precipitation are immense. They are particularly important for water-resource management and planning, though they are relevant to a great many other aspects of society as well. More regionally accurate precipitation forecasts will be important in planning for the world's food security in the face of rising population.

Temperature

Until now, the basic question has been how the surface temperature of the globe, and of various regions, would be altered by the addition of radiatively active constituents to the atmosphere. As we have seen, this question must be broadened to include possible changes to the naturally occurring climate patterns, since such changes affect both the mean and variations of climate. Until we know whether natural cycles will change under the addition of radiatively active constituents, we will not be able to predict the regional and global temperature responses. It should also be noted that statements about detection of global warming are usually statements about detecting a shift of the mean in the presence of a naturally varying background. When the mean is partly the result of the phase-locking of various natural cycles, and these cycles themselves may be affected by the mean they help to produce, the question of surface temperature alteration must be deepened and reformulated.

Storms

Projections of changes in storminess due to anticipated changes in greenhouse-gas concentration remain debatable (IPCC, 1990; Hall et al., 1994; Lighthill et al., 1994). In the words of the IPCC (1996a): "Overall, there is no evidence that extreme weather events, or climate variability, has increased, in a global sense, through the 20th century, although data and analyses are poor and not comprehensive. On regional scales there is clear evidence of changes in some extremes and climate variability indicators. Some of these changes have been toward greater variability; some have been toward lower variability." Further discussion of changes in storms, as they relate to changes in flood frequency, is included in the "Hydrologic Cycle" section of Chapter 5.

Sea Level

Long-term prediction is particularly relevant for changes in sea level. The changes in relative sea and land levels measured at the coast by tide gauges can be usefully divided four ways, into the local and the global changes of the land and the sea level. These distinctions are important for understanding the time scales and causes of ongoing changes, and predicting their future evolution. Land levels are significantly influenced by global-scale tectonic effects—the adjustment of the Earth's mantle to the removal of the glacial-era icesheets, for example. Local changes in land level can result from altered sedimentation rates, or from subsidence due to extraction of groundwater or oil. Sea level is also subject to local changes forced by local winds, river runoff, and the passage of oceanic waves of various frequencies. The global sea level is determined primarily by the mass of water in the ocean and its temperature structure.

Particular effort has gone into understanding the global sea-level component of the tide-gauge measurement, because it is expected to change with climate. Tide-gauge records longer than 50 years are needed to eliminate spurious trends due to low-frequency variability. Once records have been corrected for post-glacial rebound, a trend over the past 100 years of about 1.8 mm per year emerges (Douglas, 1991). There is no firm evidence of an increase in the rise, nor would it be expected from the change in climate that has occurred over that period. Archeological and geological studies indicate that the variation of sea level over the previous two millennia was no more than a few tens of centimeters. The time of onset of the current rise is not known.

While uncertainty about the measured rise remains, because of the lack of global coverage and the possible influence of coastal subsidence, uncertainties about the components of the rise are far larger. Two factors contribute significantly to the change of global sea level with climate: thermal expansion of the ocean, and redistribution of water between land and sea. Surface thermal anomalies penetrate down into the ocean's interior via the wind and thermohaline-driven overturning. These circulations have a range of time scales from decadal to millennial. Existing direct observations of ocean temperature are insufficient to reveal the past global warming of the ocean, although significant local changes have been observed. Models of ocean circulation have therefore been used to calculate the thermal-expansion part of the observed sea-level rise. These models yield estimates ranging from 0.2 to 0.7 mm per year (IPCC, 1996a). This calculation is inherently uncertain, however, since the boundary conditions—wind stress, surface temperature, and salinity—are not well known. The calculation becomes even more tenuous when made for future climate scenarios with additional greenhouse gases.

Ninety-nine percent of the world's land ice is contained in the Greenland and Antarctic ice sheets. The response of these ice sheets to climate change is difficult to predict (see, e.g., Oppenheimer, 1998). Since the mass balance of these ice sheets reflects long time scales, they are likely still adjusting to past climate changes. In general, the increased supply of moisture in a warmer climate is expected to dominate the increased melting for the Antarctic ice sheet, while the reverse is expected for the Greenland ice sheet. Current observations are insufficient to detect a mass imbalance in either. Here, a climate prediction might attempt at least to determine the relative change in the mass balance of the ice sheets, when models can determine temperature and precipitation in the high latitudes.

To interpret observations of sea-level and ice-volume changes, they must be placed in the context of the past and compared with projections of the future. It is clearly of interest to know when the current rise began and whether there were past rises of comparable magnitude and duration. Paleo-studies and data "archeology" (recovery of unpublished records) can help address these issues. Most projections of sea-level response to anthropogenic forcing have been based on simple models (e.g., one-dimensional upwelling-diffusion ocean models). Sea level is fully embedded in the climate system, however, and a coupled ocean-atmosphere-ice model must be used to maintain consistency in all the elements. Furthermore, the dynamic response of the ocean to climate change gives rise to regional changes in sea level that may be of a magnitude comparable to that of the global mean change. Continued improvement of these sophisticated models will be necessary if useful projections are to be made. Such projections will prove invaluable, though, because sea-level rise can have such a large and devastating impact on the vastly developed and densely occupied coastal regions of the world.

Ecosystems

Parameterizations of climate-induced ecosystem changes are rapidly improving. To predict ecosystem changes under scenarios of elevated greenhouse gases, earlier models simply mapped the recently observed biomes to the GCM-predicted locations with similar climatic conditions. Some of the latest models include vegetation interactions with nutrients, CO_2 fertilization, and fire (VEMAP, 1995; IPCC, 1998). Recent integrated-assessment models of climate change even include climate-vegetation and carbon cycle-vegetation feedbacks, as well as the effects of changing land use (see, e.g., CIESIN, 1995). While the climate scenarios that have been explored with these models are often derived from transient coupled ocean-atmosphere GCMs, the ecosystem models themselves tend to be designed to simulate an equilibrium land-surface biosphere, rather than the transient ecosystem compositions that will precede the equilibrium state.

The veracity of potential (i.e., omitting land-use changes) vegetation-distribution predictions made from uninitialized climate forecasts is as yet unknown. Because between one-third and one-half of terrestrial biological production is used or dominated by human action (Vitousek et al., 1997), and human-induced land-use changes are difficult to anticipate, the prediction of future vegetation distributions is a difficult undertaking. However, even in the absence of local-scale ecosystem forecasting skill, there is value in assessing the large-scale response of modeled ecosystems to uninitialized climate-model forecasts. Such models can be used to assess the future, large-scale response of terrestrial carbon sinks to altered climate or to anthropogenic inputs, for instance, or the effects of large-scale, vegetation-related albedo or surface roughness changes on climate. As longer time series of the relevant vegetation data become available for testing ecosystem-climate models, and these models are more rigorously validated and improved, the value of their forecasts will increase, particularly to the societies and institutions that depend most directly on ecosystems.

5

Climate-System Components

The climate attributes that influence society, as noted earlier, are themselves influenced by a broad range of physical and biogeochemical processes, or components (including forcings) of our climate system. Therefore, to improve our understanding of how changes in these attributes manifest themselves over decade-to-century time scales, we must address the issues involving those components that will most efficiently advance this understanding.

While the existence of climate patterns offers hope that some fraction of the variability in the climate attributes may be related to the state of these patterns, ultimately we must understand the physics that control both the evolution of the climate system and the patterns themselves. A relationship between climate patterns and climate attributes may afford us some statistical forecasting capabilities, but only of configurations or types of changes already documented. Forecasting future variations demands that we understand the physical and biogeochemical interactions controlling climate response and feedbacks, and identify the slow components of the system in which predictability resides.

This chapter briefly describes our current understanding of how physics and biogeochemistry influence climate, particularly the six climate attributes outlined in Chapter 2, and presents the primary issues that must be resolved to advance most expeditiously and cost-effectively our understanding of climate change and variability on dec-cen time scales. The six sections of this chapter present the components and forcings of the climate system in discipline-based discussions. This division is somewhat arbitrary, since dec-cen-scale change and variability in the atmosphere involve considerable couplings with and feedbacks from the oceans, land, and cryosphere. Consequently, the study of dec-cen change and variability entails multi- and interdisciplinary issues, and highly coupled systems. Past study of climate and its components has generally proceeded along disciplinary boundaries, however, and the funding sources for such study have been similarly partitioned. Much as we would have liked to have organized this chapter into the new cross-disciplinary structures that will ultimately be needed for future advances in dec-cen climate research, it proved quite difficult to determine an ideal, or even acceptable, cross-disciplinary structure that would conveniently present the multitude of issues, both disciplinary and cross-disciplinary, in a logical progression. We have chosen instead to indicate by cross-referencing the relationships that may guide future cross-disciplinary organizational structures.

This chapter begins with an overview of the atmospheric composition and radiative forcing, which is fundamental to externally forced (natural and anthropogenic) variability and change. External forcing of the climate system, while not properly a component of climate, is included here. Because this document articulates a plan for addressing the science of dec-cen climate change and variability, external forcing must be included for completeness, and to provide the necessary foundation for subsequent discussion in the report. Given the thoroughness of the topic's coverage in the IPCC assessment process, and the accessibility of the IPCC reports, we do not attempt to replicate that review. Rather, we draw from it and build on it in order to provide an overview of the atmospheric composition and radiative forcing most relevant to dec-cen climate issues.

The remaining sections of this chapter focus on five distinct components of the climate system. The first two, which are closely related, involve two aspects of the atmosphere: atmospheric circulation and the hydrologic cycle. (Of course, the latter section's scope involves more than just the atmosphere, since it discusses the storage of water and its movement through the atmosphere and boundaries.) These two sections are followed by the three atmospheric boundary components from which most internal dec-cen variability originates: the oceans, the cryosphere, and land and vegetation. Interdisciplinary aspects of the components' interactions are presented throughout the sections when appropriate, and several of the broader crosscutting issues that defy traditional disciplinary categorization are presented in Chapter 6.

In each of the six sections of this chapter, the discussion is partitioned into subsections dealing with the influence of the particular climate-system component on climate attributes, the evidence of variability and change of that component on dec-cen time scales, and the mechanisms through which that component operates within the climate system. At the end of each section there is a discussion of the principal outstanding issues associated with that climate-system component, as well as an overview of some of the key observational and modeling priorities that will help resolve the outstanding issues. The discussion of the requisite observational and modeling strategies is not intended to be comprehensive; rather, it provides a broad perspective on the types of research initiatives that are most likely to be productive.

Finally, we wish to emphasize that this chapter deals with all the components of the climate system that influence dec-cen variability, whether that variability be natural, anthropogenically induced, or anthropogenically modified natural.

ATMOSPHERIC COMPOSITION AND RADIATIVE FORCING

Changes in solar output—either in terms of total radiative flux (the solar constant), or in terms of the spectral distribution of this radiation—will directly influence the radiative environment and energy budget at the Earth's surface, the response of the climate system, and the response of many life forms. Moreover, changes in the atmospheric concentration of a number of trace constituents directly influence the transfer of radiative energy throughout the atmospheric column, and therefore the energy balance in the atmosphere, including the temperature at the Earth's surface. Such direct climate influences are modified by myriad feedbacks that indirectly affect surface temperatures and radiative fluxes, the hydrologic cycle, storm frequency and intensity, sea level, and ecosystem structure and functioning. Increasing the skill with which such feedbacks can be quantified is a principal challenge for earth system science over the next decade.

The primary reason for the current widespread concern about global climate change is that human activities are increasing the greenhouse effect of the atmosphere and the tropospheric aerosol burden, and weakening the stratospheric ozone shield against ultraviolet radiation. Greenhouse gases (e.g., H_2O, CO_2, CH_4, N_2O, chlorofluorocarbons, and O_3 in the troposphere) warm the Earth's surface by trapping a portion of the outgoing longwave-radiation flux. Atmospheric aerosols tend to cause surface cooling by scattering solar radiation back into space (although they can produce the opposite effect if they consist of very dark material or if they are over a bright surface such as snow or ice), and they exert indirect effects by providing nucleation sites for the formation of cloud droplets. The net influence of the myriad feedbacks responding to changes in atmospheric gas and aerosol content has yet to be determined. Better understanding of these climatic influences will be fundamental to our ability to predict the nature and magnitude of the climate's response to anthropogenic change in any of the forcing factors.

Radiative forcing is affected not only by anthropogenic changes, but also by natural variations in the sun's output and by the input and distribution of volcanic aerosols. Largely unpredictable, these elements exert measurable influence over the Earth's radiative budget and atmospheric chemical interactions, and account for some of the natural dec-cen variability in the Earth's climate. Solar output, volcanic aerosol contributions, and atmospheric gases and aerosols thus represent the main forcings, natural and anthropogenic, of the climate system. In this respect, they are distinct from the components of the climate system discussed in the other sections of this chapter, and changes in them will drive responses in those other components. Ultimately we need to be able to differentiate climate variations driven by changes in the forcings (internal or external) from variations that are the expression of internal or coupled modes of variability, which will occur even when forcing is steady. Our efforts to understand the behavior of climate variations may be furthered by the fact that the forcings and responses may vary with latitude or regional characteristics, possibly relating specific forcings to specific responses or climatic fingerprints. For example, the stratospheric warming by volcanic aerosols in the Northern Hemisphere winter is greater in low latitudes than in high latitudes (Labitzke and Naujokat, 1983; Labitzke and McCormick, 1992). The differential heating produces a larger pole-to-equator temperature gradient, which in turn increases the zonal winds and enhances the stratospheric polar vortex. The stronger polar vortex may affect the vertically propagating tropospheric planetary waves, and so modify the tropospheric circulation and alter surface air temperature (Mao and Robock, 1998). Thus, radiative influences associated with aerosols may differ from those driven by other types of radiative forcing in the high latitudes.

Influence on Attributes

The solar radiation striking the Earth, however it may be modified by the atmosphere's components, fundamentally mediates the Earth's energy budget and climate through a complex array of feedbacks. In the process, it influences all of the climate attributes discussed in Chapter 2. These feedbacks include changing the atmospheric concentration of water vapor, itself the major greenhouse gas; changing cloudiness; changing the surface albedo due to changes in snow, ice, and vegetative cover; changing source and sink rates for carbon dioxide, methane, and nitrous oxide; changing the formation rates for tropospheric ozone and aerosols; and changing the transport and storage of heat in the oceans. Each of these feedbacks further influences the surface temperature and radiative fields, which in turn alter the evaporation of water from, and precipitation onto, land and water surfaces, as well as the water balance of glaciers, ice caps, and snow fields. Soil moisture and runoff are affected, influencing the water quality and quantity of surface waters and the salinity of surface layers of the ocean. Sea level responds to the heat content of the oceans and the distribution of heat in the oceans, as well as reflecting the proportion of the Earth's total water mass that resides in the oceans. Changes in the radiation budget also affect ocean transport and storage of heat and carbon, further modifying surface temperatures and the hydrologic cycle. Changes in energy and water fluxes may also alter the pattern or strength of pressure systems in the atmosphere, thereby modifying the tracks, intensity, and frequency of storms.

Ecosystems are influenced by changes in radiation through a variety of related processes and reactions. For example, ozone is important to ecosystems and society, in part, because it filters UV-B radiation, as mentioned in Chapter 2. Ozone depletion increases the surface flux of UV-B, which increases health and ecosystem risks. Increased UV-B has been explicitly linked to damage to marine phytoplankton (Smith, 1995), which form the base of the marine trophic system and organic carbon cycle. Industrial and natural aerosols in the lower troposphere reduce the quality of the air we breathe, increasing risks to human health and to ecosystems. Changes in atmospheric carbon dioxide directly influence vegetation through both fertilization and changes in response to water stress. A number of chemical feedbacks associated with changes in aerosol and ozone levels can affect ecosystems. For example, tropospheric ozone controls the oxidizing capacity of the troposphere and its ability to remove other pollutants. It has also been implicated in reduced crop growth (see, e.g., Reich and Amundson, 1985).

Evidence of Decade-to-Century-Scale Variability and Change

External forcings of the climate system, in a number of cases, vary on dec-cen time scales. Some of these forcings are the result of human activities (e.g., emissions of chlorofluorocarbons), some are natural in origin (e.g., solar variability), and some are both anthropogenic and natural (e.g., aerosols). The primary types of radiative forcing exhibiting dec-cen variability are outlined below; each of these three classes is represented.

Greenhouse Gases

Carbon dioxide is the most important of the greenhouse gases emitted as a result of our activities. Not only is it responsible for a little over half of the current direct anthropogenic greenhouse forcing (IPCC, 1995) but its long atmospheric residence time assures that any enhancement of atmospheric concentration will persist for many centuries. Methane, which is the second greatest contributor to direct anthropogenic greenhouse forcing, is characterized by shorter residence times, but more rapid growth in atmospheric concentration than CO_2 (IPCC, 1995). The increases in atmospheric carbon dioxide and methane over the last thousand years, as measured from ice cores and directly in the atmosphere, are depicted in Figure 5-1. The relative constancy of both gases until the turn of the twentieth century indicates that their natural variability in the atmosphere has been relatively small over the last millennium. During the last glacial maximum (about 18,000 BP) CO_2 and CH_4 were respectively about 70 percent and 45 percent of the more recent pre-industrial levels (Barnola et al., 1987; Jouzel et al., 1993; Nakazawa et al., 1993; Chappellaz et al., 1993a). Extensive analyses of sources and sinks for both of these gases (e.g., Wigley and Schimel, 1994; IPCC, 1995), leave no doubt that their steep rises during the latter part of this century, coinciding with the human population explosion, is the result of human activities. The rate of CO_2 emissions from fossil-fuel burning has increased approximately 250 percent in the past 30 years (Figure 5-2, upper curve). Although the net global CO_2 uptake rate exhibits substantial interannual variability in response to climatic variations (Figure 5-2, lower curve), it has generally increased as the concentration of atmospheric CO_2 has risen (Figure 5-1, solid curve). Atmospheric methane's rate of growth varies substantially from year to year, but that rate has generally been decreasing over the last two decades (Figure 5-3), for reasons that are not entirely clear.

Evidence from ice cores indicates a strong coupling between global surface temperature and the concentration of atmospheric methane since at least 40,000 BP (Chappellaz et al., 1993a; Severinghaus et al., 1998). It is believed that

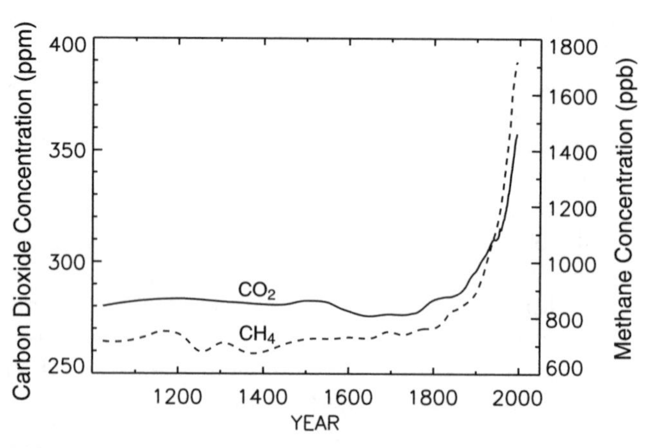

FIGURE 5-1 Atmospheric carbon dioxide and methane during the last 1,000 years. CO_2 (solid curve) refers to the vertical scale on the left; CH_4 (dashed curve) refers to the scale on the right. The CO_2 curve is based on long-term CO_2 data from Etheridge et al. (1996) and modern CO_2 data from Conway et al. (1994). The CH_4 curve is based on long-term CH_4 data from Blunier et al. (1993) and Nakazawa et al. (1993), and more recent CH_4 data from Dlugokencky et al. (1994) and Etheridge et al. (1992). (Figure courtesy of P. Tans, NOAA/CMDL.)

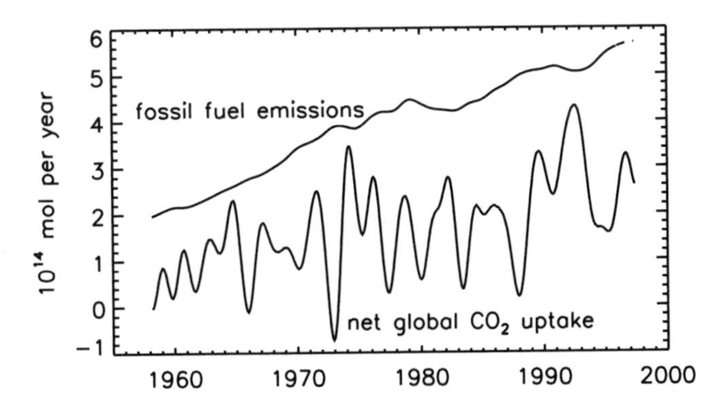

FIGURE 5-2 The upper curve represents the rate of CO_2 emissions from fossil-fuel burning (Marland et al., 1994). The lower curve represents the net global uptake rate of CO_2 by the oceans and the terrestrial biosphere. This uptake rate was derived using the assumption that the Mauna Loa CO_2 record is representative of the atmosphere as a whole. The difference between the lower and upper curves is the rate of atmospheric CO_2 increase (corrected for the seasonal cycle). (Figure courtesy of P. Tans, NOAA/CMDL.)

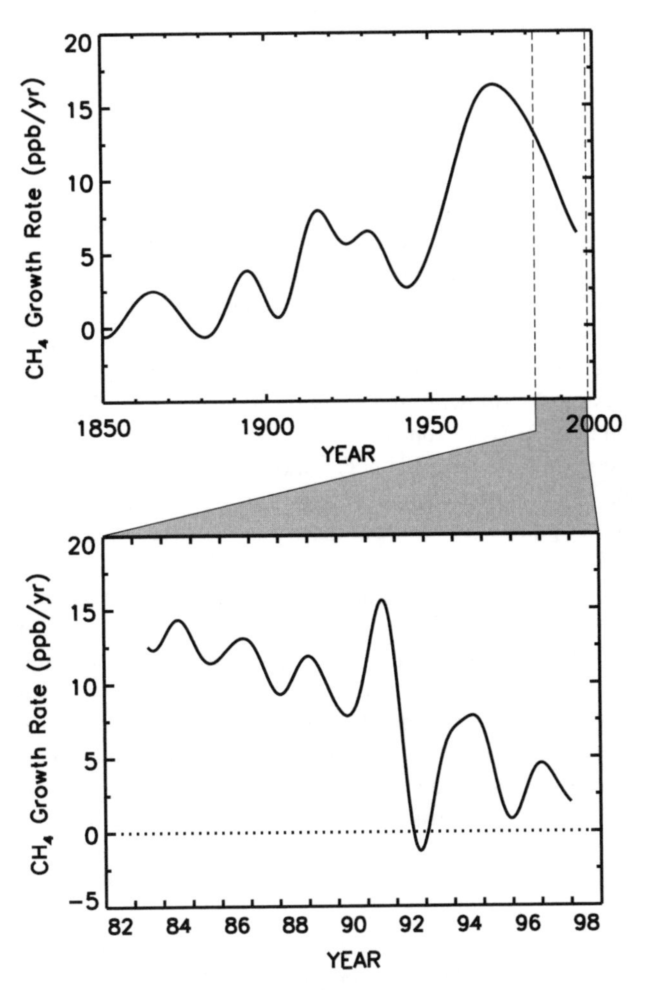

FIGURE 5-3 The rate of increase of atmospheric methane over the last 150 years. Based on data from Etheridge et al. (1996) and Dlugokencky et al. (1994). (Upper panel courtesy of P. Tans, NOAA/CMDL; lower courtesy of E. Dlugokensky, NOAA/CMDL.)

the changes in methane may have been responding, at least in part, to changes in soil moisture and wetland extent (which partially control methane emissions), driven by re-organizations of the climate system. Although the precise nature of the mechanisms that have caused temperature and methane to co-vary in the past are somewhat uncertain, these paleo-records indicate the possibility that temperature and methane may also co-vary in response to future climate changes.

Changes in tropospheric ozone, a third greenhouse gas, are not well documented. We have a limited number of discontinuous surface records that indicate tropospheric ozone may have doubled since the 1950s or at least since the nineteenth century (Figure 5-4). The data on free tropospheric ozone that are available from selected sites since 1970 show no consistent trends, however.

Stratospheric Ozone

One of the best-known changes in atmospheric composition observed over the last several decades is the dramatic

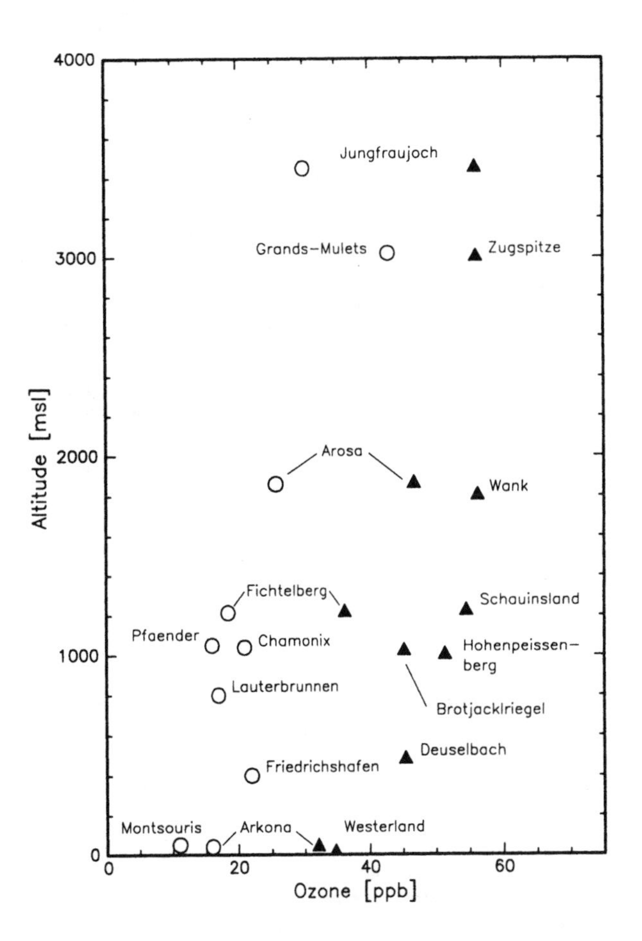

FIGURE 5-4 Measurements of surface ozone from different locations in Europe showing increasing concentrations from before the end of the 1950s (circles) to 1990-1991 (triangles) during August and September, as a function of altitude. (From Staehelin et al., 1994; reprinted with permission of Elsevier Science.)

decrease in stratospheric ozone over Antarctica. The comparison of the annual cycle in column ozone between Arctic (Resolute) and Antarctic (Halley Bay) locations is shown in Figure 5-5. Measurements in Antarctica between 1956 and 1965 showed a difference of 200 Dobson units (one DU equals 10^{-2} mm·atmosphere of column ozone) between Arctic and Antarctic springtime values (Dobson, 1966). This difference is due to the differing meteorologies of the two regions, particularly the isolation of the Antarctic vortex much later in the spring. More recent measurements, made at Halley Bay by the British Antarctic Survey, demonstrate an additional 200 DU deficit, commonly called the Antarctic "ozone hole." This dramatic decrease in Antarctic stratospheric ozone has been a regular feature since 1989; it represents a major decadal change in our planet. In the past few boreal springs, significant decreases in Arctic ozone have been noted as well (NOAA, 1995, 1996, 1997). Although these Arctic levels have generally not been much lower than typical tropical values (~250 DU), they do constitute a significant anomaly for that region.

Evidence of stratospheric ozone depletion over dec-cen time scales is also indicated in other records. Figure 5-6 shows annually-averaged deviations in ozone over Arosa, Switzerland, since the 1930s; a decline over the last two decades is apparent. Losses of total ozone (i.e., the mass of ozone vertically integrated through the entire atmosphere) have been greatest in the higher latitudes, with very little change in the tropics (WMO, 1995).

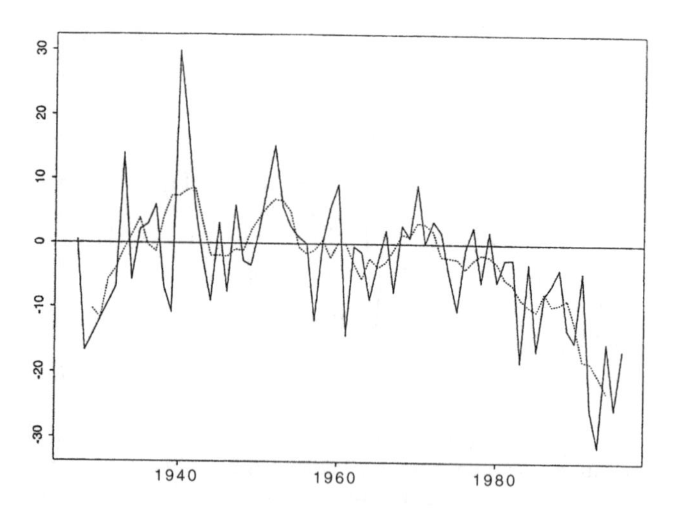

FIGURE 5-6 Anomalies from 1926-1996 of total ozone measured over Arosa, Switzerland, relative the 1926-1969 mean of 339 DU. The dotted line shows the 5-year moving average and the solid line shows the annual mean. The downward trend since 1978 averaged 1.12% per decade. (From Staehelin et al., 1998; reprinted with permission of the American Geophysical Union.)

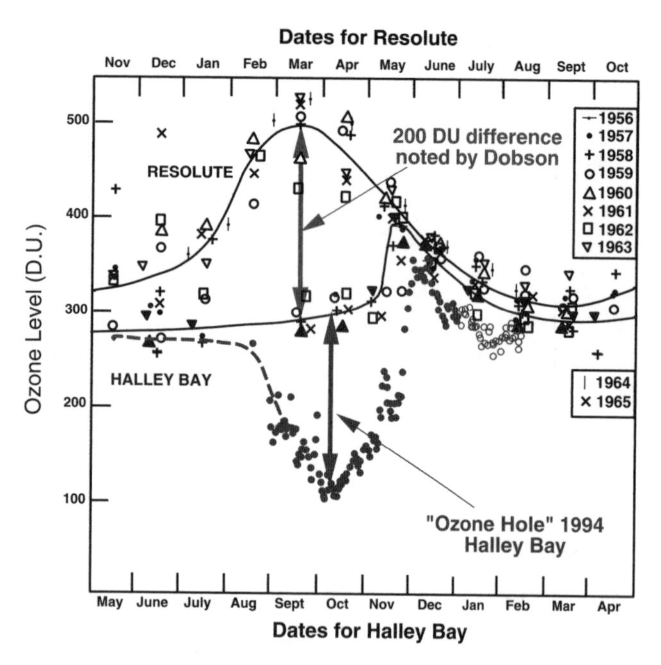

FIGURE 5-5 Annual cycle of column ozone from the Arctic (Resolute) and Antarctic (Halley Bay) for 1956-1965 and 1994. The top and middle curves are smoothed representations of the Arctic and Antarctic data, respectively. The points towards the bottom of the figure illustrate the magnitude of the ozone "hole" at Halley Bay in 1994. Units are Dobson units (DU). The Southern Hemisphere time scale (bottom axis) has been shifted by 6 months to line up with that of the Northern Hemisphere (top axis). (Figure courtesy of R. Stolarski. Halley Bay data from J.D. Shanklin of the British Antarctic Survey. After Dobson, 1966; reprinted with permission of the Royal Meteorological Society.)

The eruption of Mt. Pinatubo in June of 1991 provided a nearly hundred-fold increase in the surface area available for heterogeneous chemical processing in the stratosphere. Observations following the eruption indicated significant reductions in NO_2 (Johnston et al., 1992; Koike et al., 1994) along with increased concentrations of HNO_3 (Rinsland et al., 1994). These changes suggest that reactive nitrogen species (e.g., NO_2) were repartitioned into less reactive forms, which in turn helped to temporarily enhance the levels of active, ozone-depleting chlorine radicals (e.g., ClO) relative to those of the more inert chlorine reservoirs (e.g., HCl). Although the predicted massive ozone loss in the volcanic cloud did not occur (Prather, 1992), observations at that time showed evidence of greater ozone depletion than that expected in response to the continued growth in stratospheric chlorine abundance (Hofmann et al., 1994; Komhyr et al., 1993). A 6-8 percent loss of ozone in the tropics immediately after the eruption is more likely to have been associated with the vertical lofting that accompanied the strong stratospheric heating by the aerosols (Kinne et al., 1992). Overall, observations by the Total Ozone Mapping Spectrometer (TOMS) showed an additional global ozone deficit of about 2-3 percent by mid-1992 that might be attributed to Mt. Pinatubo (Gleason et al., 1993). The atmosphere had mostly returned to normal a couple of years after the volcanic perturbation, and it is difficult to determine how much of the ozone depletion in these years should be attributed to chlorine increases and how much to volcanic aerosols.

The decrease in polar stratospheric-ozone concentrations that has been documented over the past 30 years is strongly related, at least in part, to the increase in atmospheric chlorine (Figures 5-7 and 5-8). While levels of chlorinated compounds in the atmosphere are still high, their growth rates tend to be decreasing, and in some cases are negative.

Aerosols

Volcanic aerosols can have a significant influence on the radiative balance (defined as the difference between absorbed solar radiation and outgoing longwave radiation) of both the stratosphere and the Earth's climate system. For instance, Labitzke et al. (1983) showed that the aerosols produced by the eruption of El Chichón, which achieved a peak concentration at 24 km between about 10°S and 30°N (for the first six months), warmed this region of the atmosphere by a few degrees. Following the eruption of Mt. Pinatubo, substantial changes to the planetary albedo were observed (Minnis et al., 1993). In addition, substantial heating in the tropical stratosphere was observed immediately after Pinatubo's eruption. This heating was sufficient to cause tropical stratospheric temperatures at 30 hPa to increase as much as three standard deviations above the 26-year mean (Labitzke and McCormick, 1992). On the other hand, global surface temperature was also observed to decrease in the months following the Pinatubo eruption as a result of the increased planetary albedo (see, e.g., Dutton and Christy, 1992), and the temperature remained suppressed through 1993, as predicted (Hansen et al., 1996).

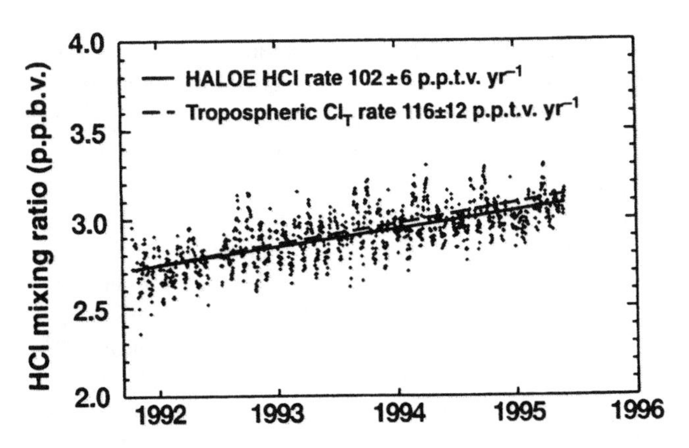

FIGURE 5-8 Stratospheric trend of HCl from 1991 to 1995. HALOE is the Halogen Occultation Experiment. (From Russell et al., 1996; reprinted with permission of Macmillan Magazines, Ltd.)

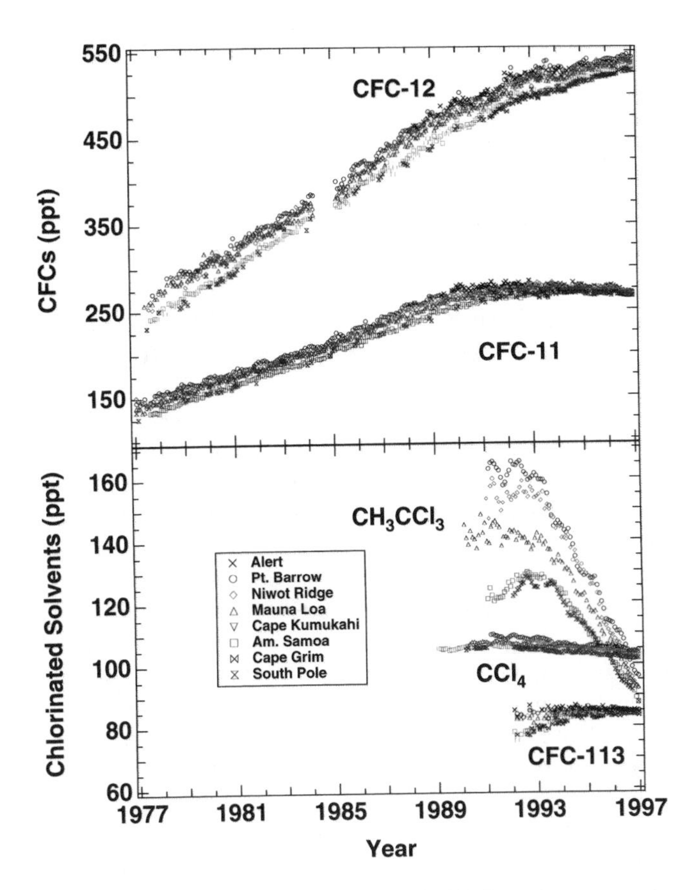

FIGURE 5-7 Atmospheric trends of chlorinated compounds controlled under the Montreal Protocol from 1977 to 1995. The mixing ratios from surface measurements are reported as monthly means in parts per trillion (ppt) in dry air. CFC-11 and CFC-12 data are updated from Elkins et al. (1993); methyl chloroform (CH_3CCl_3), carbon tetrachloride (CCl_4), and CFC-113 (CCl_3F-$CClF_2$) data are updated from Montzka et al. (1996). (Figure courtesy of NOAA/CMDL.)

In addition to these radiative effects of volcanic aerosol, recent work by Solomon et al. (1996) demonstrates that the observed aerosol variability can influence the modeled ozone trends. Periods of peak aerosol loading appear to correlate better with additional ozone depletion than with a trend fitted to the dominant driving force in ozone depletion, stratospheric chlorine levels. It is difficult to interpret this trend in ozone over the 15 years of TOMS data without including the concurrent variations in stratospheric aerosols.

Since the late 1970s, near-global monitoring of stratospheric aerosol distribution has been carried out by in situ (Wilson et al., 1992), ground-based (e.g., Osborn et al., 1995), and satellite-based instrumentation (SAM II and SAGE measurements; see, e.g., Thomason et al., 1997b). Over this period, the primary source of stratospheric aerosol variability has been periodic injections of aerosol, or of gaseous aerosol precursors such as SO_2, by volcanic eruptions. In general, stratospheric aerosols are produced in situ by processes that include the photochemical transformation of gaseous SO_2 into H_2SO_4 aerosol. For example, the composite SAM II/SAGE/SAGE II record of stratospheric-aerosol optical depth shows large effects from the eruptions of El

Chichón in 1982 and Mount Pinatubo in 1991, as well as effects of several smaller eruptions such as Mount St. Helens in 1980, Nevada del Ruiz in 1985, and Kelut in 1990 (McCormick et al., 1993). The eruption of Mt. Pinatubo may have caused the largest perturbation to stratospheric aerosol loading of any eruption since Krakatau in 1883.

The meridional distribution and the residence time of volcanic aerosols are strongly dictated by the latitude of the eruption, the altitude reached by the eruption plume, time of year, and phase of the quasi-biennial oscillation at the time of the initial aerosol injection (Trepte et al., 1993). As a result, any reconstructions of stratospheric aerosol loading resulting from volcanic eruptions are subject to significant uncertainties if they are extended backwards before the start of global measurements in 1978 (see, e.g., Sato et al., 1993), since many of the aforementioned parameters are poorly known. Additional indications of aerosol concentrations over longer time periods can be obtained from the longer records of Sato et al. (1993) (Figure 5-9) and from ice-core analyses (e.g., those of Zielinski et al., 1994). It has been suggested that the non-volcanic background stratospheric aerosol mass has increased by 5 percent annually over the period from 1978 to 1989 (Hofmann, 1990), though SAGE-based evaluations tend to argue against this increase (Thomason et al., 1997a).

Solar Radiation

The sun is the driving force of climate; even small variations in the amount of energy that the Earth receives can apparently have significant impact (for instance, see the last section in this chapter for a discussion of the role of solar variations in the waxing and waning of the great ice ages). By comparison, a doubling of CO_2 in the atmosphere would generate a radiative forcing equivalent to a 1.8 percent increase in solar irradiance. Best estimates derived from solar proxies suggest dec-cen changes in solar irradiance on the order of 0.25 percent over the past 400 years (Nesme-Ribes et al., 1993; Lean et al., 1995; Hoyt and Schatten, 1993). However, the only direct record of solar-irradiance variability we have covers only the last one and one-half solar cycles; as Figure 5-10 illustrates, the recent range of variations is about 0.1 percent. The total-irradiance record shown in Figure 5-10 is based on satellite observations, and involves a modeled reconstruction over this period. (It has long been known from indirect measures of solar radiation that the variability of the sun's UV radiation has an 11-year period.)

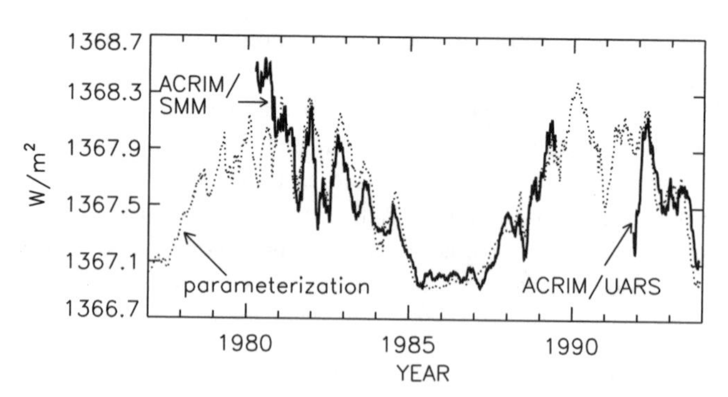

FIGURE 5-10 Total solar irradiance from 1975-1995 measured by the Active Cavity Radiometer Irradiance Monitor/Solar Maximum Mission and Upper Atmosphere Research Satellite (ACRIM/SMM and ACRIM/UARS). The dotted line is a model of the total irradiance variability obtained from a parameterization of the influence of sunspot darkening and facular brightening, which are recognized as the two primary mechanisms of irradiance variability during the 11-year solar cycle. (After Lean et al., 1995; reprinted with permission of the American Geophysical Union.)

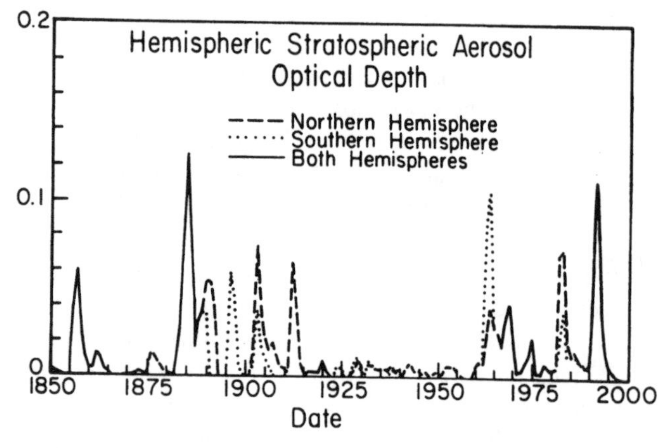

FIGURE 5-9 Stratospheric aerosols as a function of time. For the period 1883-1990, aerosol optical depths are estimated from optical extinction data, whose quality increases with time over that period. For the period 1850-1882, aerosol optical depths are more crudely estimated from volcanological evidence for the volume of ejecta from major known volcanoes. (From Sato et al., 1993; reprinted with permission of the American Geophysical Union.)

Although UV radiation constitutes only a small portion of the total solar irradiance, it is more variable by at least an order of magnitude than the visible-radiation portion, and therefore contributes significantly to total solar variability. This UV variability has special relevance to chemical interactions in the upper atmosphere, where the temperature structure depends partly on the absorption of UV radiation by O_3, O_2, N_2, and other gases. This relationship was highlighted by Hood and McCormack (1992), who showed a strong correlation between O_3 and UV radiation on the 11-year solar cycle.

Additional records of the sun's activity are derived from observations, beginning early in the seventeenth century, of the occurrence of dark spots on the face of the sun; they are not a direct measurement of solar irradiance, but over the period for which we have direct irradiance measures, high sunspot activity correlates strongly with increased irradiance.

Sunspots are associated with bright faculae that surround the dark spot. Although the spots themselves are areas of decreased irradiance, the faculae are longer-lived and more areally extensive, leading to an overall increase in total irradiance at times of sunspot maxima. Sunspot observations indicate that solar activity has varied on an 11-year cycle for the past 300 years (since about AD 1690). But longer-term variation has been inferred from observations of sunspots made over the last several centuries. For instance, during the Maunder Minimum (1650-1690) no sunspots were observed (Lean, 1991). Longer- and shorter-period variance also occurs. The sunspot record exhibits an 80-100 year period known as the Gleissberg cycle, and the apparent alternation of stronger and weaker 11-year cycles produces a concentration of variance with a 22-year period. Over shorter periods, the sun exhibits variations associated with its rotation (which has a 27-day period), and monthly and yearly variations are seen within the envelope of the 11-year activity cycle.

Indirect indicators of solar activity, such as sunspots and the abundance of cosmogenic nuclides (e.g., ^{14}C and ^{10}Be), have considerably longer records than direct observations. Figure 5-11 shows two different indices that are commonly used to infer some measure of solar activity (e.g., solar wind), and are known to correlate with irradiance over the last solar cycle (see, e.g., Wilson and Hudson, 1991). These longer, proxy records show distinct long-term shifts in solar activity over the past several centuries; for example, such shifts can be seen in the record of ^{10}Be found in ice cores. Production of ^{10}Be by galactic cosmic-ray particles in the Earth's atmosphere is modulated by the solar wind; this long-lived radionuclide is removed from the atmosphere by precipitation and preserved in ice cores. Ice-core ^{10}Be abundances were significantly higher in the fifteenth and late seventeenth centuries, implying that the solar wind was much weaker then than it is today. The relationship between solar wind and solar irradiance has been calibrated for the last two solar cycles; the extrapolation for conditions outside the range of direct observations of total solar irradiance—if applicable—implies a dec-cen solar irradiance variation with periods in which irradiance may be lower by as much as 0.25 percent.

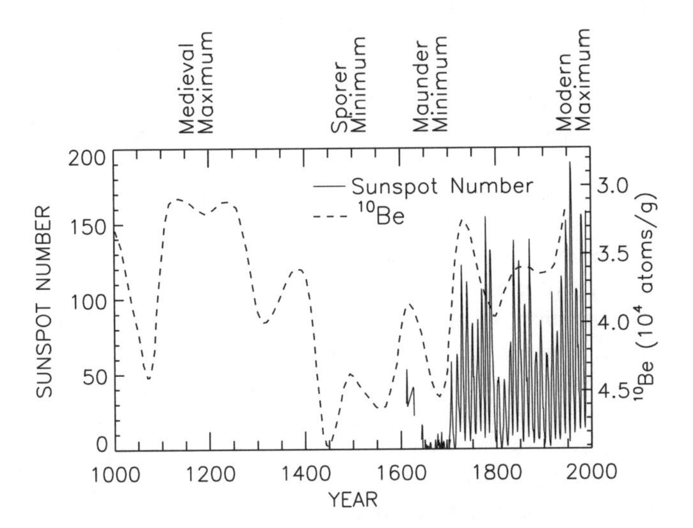

FIGURE 5-11 Time series of sunspot number and ^{10}Be in ice cores, which are both known to correlate with irradiance over the past few solar cycles. (After Beer et al., 1988; reprinted with permission of Macmillan Magazines, Ltd.)

Mechanisms

The sun's radiation, volcanic eruptions, and human emissions of greenhouse gases and aerosols are sources of variability and change that are external to the climate system. Except for the sunspot cycle, they are not predictable at this time. A number of internal and coupled modes of variability within the climate system, however, influence concentrations of trace gases and aerosols in the Earth's atmosphere. Understanding the mechanisms and forcings behind these modes of variability will enhance our ability to predict climate variations.

Greenhouse Gases and the Carbon Cycle

The major externally forced causes of the observed CO_2 increase are the burning of fossil fuels (Marland et al., 1994) and forest destruction (Houghton et al., 1987). Internally, carbon is transferred relatively rapidly among three major "mobile" reservoirs—the oceans, the atmosphere, and the biosphere. About one-seventh of the atmospheric CO_2 enters the oceans each year, and half as much is fixed into organic material by photosynthesis on land. These fluxes are almost balanced by the amounts leaving the oceans or returned to the atmosphere through microbial decay, respectively. We know that more CO_2 is entering these reservoirs than leaving, however, because the rate of atmospheric increase is only about half as large as the global production of CO_2 through the combustion of fossil fuels. From year to year imbalances manifest themselves in interannual variations of the rate at which atmospheric CO_2 increases; the swings of net global CO_2 uptake shown in Figure 5-2 are related to known climate variations such as El Niño. Several studies have found correlations on different time scales between the rate of atmospheric CO_2 increase and global average temperature, as well as ENSO indicators (Elliot et al., 1991; Dai and Fung, 1993; Keeling et al.,1995).

Although carbon dioxide cycles quickly among the mobile reservoirs, it leaves the ocean-atmosphere system only very slowly, through the burial of organic matter and deposition of carbonate rocks. Dissolution of calcite, also a very slow process, adds carbon to the mobile reservoirs, but increases the carbon-holding capacity of the oceans even more by changing their alkalinity. Therefore, the rates at which future anthropogenic CO_2 is removed from the atmosphere will depend mostly on how the additional carbon from fossil-fuel burning is partitioned between the mobile reservoirs, which can essentially be considered to constitute a closed system. Thus, the important factors controlling increases in atmospheric CO_2 concentrations are: first, the rate of fossil-fuel consumption; second, the circulation of the ocean and, to a lesser extent, how the circulation affects marine biological productivity; third, management of the land; and fourth, carbon storage by ecosystems, possibly stimulated by increased CO_2 and anthropogenic deposition of nitrogen. Climate change will affect all of these natural processes.

The magnitude of the ocean's role in the partitioning of CO_2 is dependent on the ocean's chemical capacity to take up CO_2. This capacity is determined by the amounts of carbonate and borate ions, which can be "titrated" by newly dissolved CO_2 into bicarbonate and boric acid, respectively. It takes many centuries for most of this capacity to be accessible to the atmosphere, however, because the ocean turns over very slowly. The ocean's pH could be lowered by a full point if "all" (defined as $400x10^{15}$ mol) fossil-fuel carbon were burned (Tans, 1998). (Since the pre-industrial era, we have consumed about 5 percent of that amount.) These estimates are based on the assumption that the ocean's "biological pump" keeps operating as it does today. This pump represents the photosynthesis in the sunlit surface layer of the ocean and the sinking of organic particles that keeps the carbon content and CO_2 partial pressure lower in surface waters than in the deep oceans. Without any ocean biology, the partial pressure of CO_2 in the atmosphere would be two to three times higher than it is today (Najjar, 1992). At low and temperate latitudes almost all the available phosphate and nitrate are consumed, but this process is only partially effective at high latitudes. Changes in the effectiveness of the biological pump at high latitudes, which result from changes in the balance between the rates of thermohaline overturning and the rates of biological production, have been invoked in attempts to explain the atmospheric CO_2 concentration differences between glacial and interglacial periods (see, e.g., Knox and McElroy, 1984, and Sarmiento and Toggweiler, 1984).

The "solubility pump" is another process by which the ocean maintains a vertical gradient of carbon. Because deep-water formation sites are cold and the solubility of CO_2 is inversely related to temperature, water with a high inorganic carbon content is "pumped" to the deep oceans at deep-water formation sites. A vertical CO_2 gradient is thereby produced between the deep water and the warmer, overlying waters of the remainder of the ocean's surface. The strength of the solubility pump is affected by changes in alkalinity, air-sea gas contrast, and ocean temperature. Without the solubility and biological pumps, the concentration of atmospheric CO_2 would be three to four times higher than it is today (Najjar, 1992).

The carbon delivered to the deep ocean by these pumps is exchanged with that in the atmosphere on time scales of centuries and longer. The strengths of both pumps may change with changes in mixed-layer characteristics and upwelling. The latter will alter both the nutrient supply and the time available for surface phytoplankton to utilize these nutrients, as well as affecting the surface temperature, mixed-layer thickness, and air-sea gas contrast.

Deforestation—which before 1940 or so occurred principally in temperate latitudes, but more recently has been taking place mostly in the tropics—has long been considered a large source of atmospheric CO_2 (Houghton et al., 1987). In the global balance, tropical deforestation is compensated for through increased net uptake by terrestrial ecosystems, principally at temperate latitudes (Tans et al., 1990; Wofsy et al., 1993; Ciais et al., 1995; Battle et al., 1996). Even in the tropics there could be large areas of net CO_2 uptake (Grace et al., 1995). Possible explanations for the observed uptake are fertilization of plants by higher atmospheric CO_2 levels (see, e.g., Mooney et al., 1991) and fertilization by atmospheric nitrogen deposition (see, e.g., Schindler and Bailey, 1993, and Townsend et al., 1996). An additional complication in the internal and coupled mechanisms that produce variations in the carbon cycle is the fact that the balance between source and sink may shift as climate changes (Dai and Fung, 1993), which may account for some increase in terrestrial CO_2 uptake. For example, there is some evidence that the Arctic tundra, once a net sink for atmospheric CO_2, may have turned into a source during the last decades as a result of Arctic warming (Oechel et al., 1993). These results still need to be confirmed by data from additional sampling sites.

Unlike long-lived CO_2, methane has an atmospheric lifetime of about 10 years. The increasing atmospheric methane burden reflects the growth of CH_4 sources in recent decades, with about 60 to 80 percent of this increase attributable to human activities (IPCC, 1996a). The increasing atmospheric concentration of methane directly affects the radiative balance and the chemistry of the troposphere; it accounts for approximately 20 percent of the increase in radiative forcing since the pre-industrial era (IPCC, 1995). In addition, it has an indirect effect on the stratosphere, because, once oxidized, it is an important source of stratospheric water vapor. The most important sources of atmospheric methane are wetlands, rice agriculture, cattle and sheep, biomass burning, fossil fuels, landfills and waste, and termites (see, e.g., Fung et al., 1991).

The atmospheric fate of methane is largely determined by the level of ultraviolet radiation in the troposphere and the concentrations of other key trace gases (e.g., ozone, water vapor, nitrogen oxides, carbon monoxide) responsible for the production and recycling of OH radicals, which in turn initiate the oxidation of methane. Perhaps somewhat surprisingly, no significant decadal trend in OH concentration itself has been detected, in that no change has been seen in the inferred atmospheric lifetime of the synthetic industrial compound chemical methylchloroform, which is also attacked by OH (Prinn et al., 1995). Accurate predictions of future CH_4 levels need to take into account the effects of the relationships between CH_4, CO, and OH (Prather, 1994).

Changes in the concentrations of atmospheric chlorofluorocarbons and similar fully halogenated industrial compounds with no natural sources are controlled primarily by emissions. The lifetimes of the individual gases are determined by transport to and photolysis in the stratosphere. The loss of the equally long-lived nitrous oxide (N_2O) is likewise limited by stratospheric chemistry, but its emissions come from a mix of both natural biogenic sources and anthropogenic perturbations to the nitrogen cycle. Both sets of gases build up in and decay from the atmosphere on a time scale of centuries. Ozone, on the other hand, is also an important greenhouse gas, but its atmospheric concentration is determined primarily by a balance between in situ production (mainly in the stratosphere) and photochemical losses that occur on a time scale of a year or less. The balance is tipped in favor of losses when chlorofluorocarbon (CFC) and N_2O concentrations increase, because they catalyze ozone destruction in the stratosphere.

The source and sinks of anthropogenic radiatively active gases remain a primary concern in determining the extent of the effect of greenhouse gases on global climate. However, an equally important (yet poorly understood) influence on climate is the distribution of increased atmospheric water vapor associated with a speeding up of the hydrologic cycle (discussed in detail later). The atmospheric portion of the hydrologic cycle is complex, and operates on both short and long time scales. The fast processes associated with this mechanism, such as cloud formation and the related intra- and inter-cloud radiative impacts, influence cloud nucleation, longwave radiation, albedo feedbacks, and ultimately the surface energy balance. On dec-cen time scales, the impact of increased water vapor is realized through alterations in large-scale cloud distribution (shown earlier in Figure 2-12), which reflect both the water-vapor distribution and the hydrologic cycle's response. Also, the latent heating of clouds, and its radiative effects, influence the large-scale atmospheric circulation and hydrologic cycle—additional complexities that need to be better understood.

Stratospheric Ozone

Our understanding of the cause of the Antarctic ozone hole has grown considerably thanks to ground-based and aircraft campaigns, and, more recently, satellite missions. The evidence is clear: Observations of high levels of reactive chlorine species coincide with the observations of rapid ozone depletion. The only identified cause of this ozone depletion since the 1970s is the rise in stratospheric chlorine levels, which is driven by the increasing abundance of chlorofluorocarbons and other halocarbons in the lower atmosphere. Laboratory studies have identified and quantified the reactions of chlorine radicals that catalytically destroy ozone, and numerical models of the stratospheric circulation and chemistry predict similar losses. This loss in stratospheric ozone was likely responsible for the recent cooling of the lower stratosphere, and model results indicate that the ozone loss could be expected to have a general cooling effect on the climate (IPCC, 1995). An enhancement of stratospheric ozone destruction would reduce the stratospheric source of tropospheric ozone and increase the UV radiation that drives tropospheric photochemistry. Present models using the best laboratory physics and chemistry can simulate such ozone loss.

CFCs and related halocarbons have no known natural sources, and their atmospheric concentrations in 1950 were negligible. The much higher concentrations currently measured reflect a history in which halocarbons are emitted at the Earth's surface, propagate vertically through the troposphere, and, over the course of five years, ultimately reach the upper levels of the stratosphere. The rise in tropospheric chlorine loading from CFCs and related halocarbons is documented in Figure 5-7. (Note that the concentrations of all of these gases except CFC-12 have begun to fall since 1993, as a result of declining halocarbon emissions.) Upon reaching the stratosphere, these compounds dissociate and release chlorine, which in the upper stratosphere is predominantly in the form of HCl. Figure 5-8 shows the increase in stratospheric HCl observed by satellite during the early 1990s. The total content and magnitude match those of the tropospheric halocarbon sources (allowing for the 5-year lag). The recent decline in tropospheric chlorine should be visible in the HCl record over the next several years.

In addition to their effect on stratospheric ozone, CFCs are potent greenhouse gases; they have contributed to approximately one-quarter of the increase in greenhouse-gas radiative forcing over the past decade (11 percent of the total increase since pre-industrial times). The decline in radiative forcing that CFCs induce through stratospheric ozone depletion is likely somewhat less than their direct radiative contribution (IPCC, 1995).

Aerosols

Since 1978, a series of low-latitude, high-altitude injections of volcanic aerosols has maintained a maximum in aerosol loading in the tropics, centered at altitudes between 20 and 27 km. Although the primary controlling mechanism is external, there are internal mechanisms that serve to limit the spatial distribution and temporal longevity of these injected aerosols. For example, the latitudinal wind gradient in the subtropics impedes transport between the tropics and mid-latitudes. Thus, the maximum in aerosols following a large volcanic eruption in the tropics (e.g., Mt. Pinatubo or El Chichón) remains for a few years in the tropical stratosphere as a long-lived source of aerosol for the middle and high latitudes (Trepte and Hitchman, 1992; Thomason et al., 1997b). Non-volcanic sources of stratospheric aerosols, such as natural organic carbonyl sulfide (OCS) and industrially derived SO_2, also tend to support the presence of a tropical aerosol.

While explosive volcanic eruptions are the most significant source of stratospheric aerosols, a non-volcanic background level of stratospheric aerosols appears to be present. It has been suggested that this may result from the diffusion of tropospheric OCS into the stratosphere (Crutzen, 1976). However, recent research (Chin and Davis, 1995) suggests that OCS has likely produced only negligible amounts of the stratospheric aerosols observed since 1978. Hofmann (1990) has proposed that the possible 5 percent annual increase in the non-volcanic background stratospheric aerosol mass from 1978 to 1989 could be related to the increase in sulfur emissions from commercial aircraft or other anthropogenic sources.

Tropospheric aerosols also influence the overall surface radiative balance through their light-scattering and absorption properties. Tropospheric aerosols, being relatively short-lived in the atmosphere, show regional variability related to the distribution of their sources. Consequently, the non-uniform distribution of these aerosols yields spatially heterogeneous radiative forcing, even given a uniform greenhouse-gas (or natural) forcing. Examples of tropospheric aerosol sources and typical aerosol types are: deserts, which produce mineral dust; vegetation, which produces particulate organic carbon and sulfate aerosols; biomass burning, which produces soot; oceans, which produce sea salt; and industrial centers, which produce dust, soot, and sulfates. The characteristics of the Earth's surface play a role not only in determining the type of aerosols that are emitted, but in determining the dispersal of aerosols. For example, the wind friction caused by surface roughness, which is greatly influenced by the stature and density of the vegetation, affects turbulent mixing of air in the planetary boundary layer.

Depending on the absorption characteristics of a given aerosol, its effect on radiative forcing can be positive or negative. For instance, low-albedo soot aerosols produced from biomass and fossil-fuel burning tend to produce surface warming. However, higher-albedo sulfate and dust aerosols tend to produce surface cooling, and are believed to exceed the radiative influence of darker aerosols on a globally averaged basis by 0 to 1.5 W m^{-2} (IPCC, 1996a).

Predictability

Greenhouse Gases

The primary uncertainty regarding predictions of future warming associated with increased concentrations of greenhouse gases comes from uncertainties in the emission scenarios, as well as tremendous gaps in our understanding of, and ability to represent in models, the myriad feedback processes that may act to enhance or diminish any direct warming. One of the most important feedback processes is the interaction between atmospheric water vapor, clouds, and the surface radiation balance. The details of this complex interaction are still poorly understood. Until this understanding is improved, and better parameterizations constructed, a fundamental question regarding the impact of greenhouse gases cannot be adequately addressed.

A great variety of global carbon-cycle models have been used to estimate future burdens of atmospheric CO_2 on the basis of projected rates of fossil-fuel combustion and associated greenhouse warming. Important model parameters are calibrated such that the past behavior of atmospheric CO_2, radiocarbon (^{14}C), and the $^{13}C/^{12}C$ isotopic ratio are reasonably well reproduced. Representation of the processes responsible for the partitioning of carbon between the mobile reservoirs is often very crude, sometimes speculative, and involves many assumptions. These models are best viewed as extrapolative tools, and their predictive power beyond the next few decades is tenuous.

Prediction of future atmospheric methane concentrations depends primarily on prediction of the CH_4 sources, and secondarily on how the oxidative capability of the atmosphere evolves. The main sources of CH_4 have been identified, but the range of uncertainty in emissions rate on regional and global scales is typically a factor of two or three for each source process.

Stratospheric Ozone

The year-to-year variations of stratospheric ozone at a given location are not yet predictable, and modulation of the depletion on 2- to-3-year periods by major volcanic eruptions (as observed) cannot be predicted. However, the decadal trend in ozone depletion can be predicted from the evolving load of stratospheric chlorine (e.g., Figure 5-7b) and bromine: The tropospheric abundance of their source gases is now declining for the first time, and we can expect stratospheric levels to follow in a few years.

While the behavior of the Antarctic ozone hole on a decadal time scale is fairly straightforward, it is predictable in the long term only if Antarctic meteorology remains more or less the same as today's. If it does, the ozone hole will persist until chlorine levels drop somewhere below about 2 ppb, which is expected to occur around 2050 at best if the phaseout of CFCs and related compounds is successful. This prediction is fairly certain, although the slow decay of CFCs and the degree to which the Montreal Protocol is followed makes the exact date at which we drop below the ozone-hole threshold uncertain within about 20 years. Fortunately, CFC and chlorine increases are very predictable, given the future releases of CFCs and related halocarbons.

Aerosols

Prediction of the long-term influence of tropospheric aerosols may be possible, but a better understanding will be needed of how the spatially heterogeneous, but semi-stationary, distribution of tropospheric aerosols influences the larger-scale climate response. The stratospheric effects of volcanic aerosols are necessarily unpredictable, however, because the occurrence and magnitude of eruptions are not currently predictable. Since the role of volcanic aerosols in ozone trends, and in climate in general, is complex (and aerosols may be more persistent than they have been thought to be), our current ability to produce global depictions of stratospheric aerosol loading (especially in terms of surface-area density) with high vertical resolution is essential. Satellite-based systems such as the EOS/SAGE III instrument will do this well, as long as suitable orbits and frequencies of launch are maintained. Corroborative efforts using ground-based and in situ systems will also be necessary to provide the high-quality data needed to understand the role of stratospheric aerosols in determining climate.

Solar Radiation

The predictability of this phenomenon is limited to empirical extrapolation of solar activity and irradiance. Probably the best prediction regarding solar activity is that the 11-year solar cycle will continue as recently recorded. Although we have observed the sun for only a tiny fraction of its lifetime, the sun is a fairly common type of star and we can derive certain inferences from observations of other similar stars. If observations of many sun-like stars represent comparable stars at different stages in their lifetime, then they can be used to infer the frequency of different behavioral modes. Moreover, these observations provide evidence for linkages among solar-activity proxies and total irradiance. Of sun-like stars with similar mass, radius, rotation rates, activity cycles, and emissions spectra, some 33 percent have no evidence of emission variability associated with sunspot (starspot) variability over a decade of observations. By implication, multidecadal inactive periods such as the seventeenth-century Maunder minimum may be rather common events. Yet our current understanding of solar physics does not enable us to predict between-cycle variability far in advance, much less these larger, multidecade-to-century-scale variations inferred from solar proxies and seen in sun-like stars.

Remaining Issues and Questions

Greenhouse Gases

• *What changes in carbon storage and flux can be expected on dec-cen time scales, including responses to climate change?* Of the carbon emitted by anthropogenic activities (burning of fossil fuels and deforestation) in the past two centuries, just over half has remained in the atmosphere. The carbon that has been removed from atmosphere has gone into the other "mobile reservoirs"—the biosphere and surface ocean. Although initial estimates of carbon sinks failed to account for a large portion of anthropogenic CO_2 emissions, recent measurements suggest that an increase in the amount of carbon stored in the terrestrial biosphere may account for this "missing sink" (for a review of this topic, see Houghton et al., 1998, among others). Uncertainties in estimates of biospheric activity, coupled with a poor knowledge of the surface ocean's uptake and additional uncertainties regarding carbon's pathways and fate within the ocean and terrestrial reservoirs, make it likely that our knowledge of the fate of anthropogenic CO_2 is not yet complete. Atmospheric CO_2 has been increasing steadily since the mid-1800s, but interannual-to-decadal fluctuations in its rate of rise appear in the atmospheric record. These fluctuations may arise from biospheric or from oceanic variability. To predict atmospheric greenhouse-gas concentrations, we must reduce the uncertainties that surround current evaluations of carbon sources, sinks, and fluxes. We must also improve our understanding of how the uptake of each of the mobile reservoirs may vary with changing climate. This means understanding better the relationship between changes in the ocean's mixed layer and its impacts on the biological and solubility uptake of carbon by the ocean, and the response of land vegetation to climate change and its potential for producing methane and carbon dioxide. In the Arctic, the active layer of the soil and the upper permafrost contain approximately 300 gigatons (Gt) of carbon. In recent years, Arctic tundra ecosystems have switched from being a sink of a few tenths of a Gt of carbon per year to a source of a few tenths (Oechel et al., 1993). These recent losses, which clearly change in conjunction with climate, may now exceed 10 percent of the anthropogenic CO_2 emissions. (Further discussions of the ocean and terrestrial uptakes of carbon appear in the "Cryosphere" and the "Land and Vegetation" sections of Chapter 5.)

• *What are the relative contributions of the various sources and sinks to the recent increase in methane?* Although observations document an increase in atmospheric CH_4 since the nineteenth century, the causes of this rise in a potent greenhouse gas remain poorly quantified. Methane sinks appear to be stable over at least the most recent decade, so the total source is well known from the atmospheric rise. However, the contribution of individual components remains poorly quantified. Sources likely to be increasing include agricultural wetlands and livestock herds, biomass burning, fossil-fuel-related industry, and landfills. Future changes in OH will influence the lifetime and atmospheric concentrations not only of methane, but of a number of other radiatively active constituents as well.

• *How does the photochemical breakdown of methane contribute to other chemical and radiative processes in the atmosphere on the dec-cen time scale?* The photochemical breakdown of CH_4 consumes OH and produces H_2O throughout the atmosphere. In the stratosphere, even small amounts of water vapor are extremely efficient contributors to the global greenhouse effect, accounting for nearly half of the radiative impact of water vapor in the atmosphere (the remainder comes from lower-tropospheric water vapor). Dec-cen changes in methane can be expected to alter the vertical distribution of water vapor in the upper troposphere and lower stratosphere, thereby changing radiative forcing.

The actual contribution of water vapor from methane breakdown at the radiatively important altitudes is not known, however. The ice particles that form at these altitudes also provide sites for heterogeneous chemistry. The feedbacks among these chemical processes and atmospheric dynamics are largely unknown.

- *Why is N_2O increasing on dec-cen time scales?* Records of atmospheric N_2O (a greenhouse gas) show a steady rise since the nineteenth century, but a quantitative budget of N_2O sources and sinks explaining this rise cannot be developed from the existing scarce observations and our limited understanding of the processes involved. The global N cycle has been heavily altered by human activities, particularly the widespread application of high-nitrate fertilizers (Melillo, 1995). Atmospheric N_2O concentrations have been affected by anthropogenic changes in the nitrogen cycle (including industrial N_2O production), and oceanic denitrification/nitrification rates have been affected by changes in near-shore productivity caused by increased nitrogen loading. The relative importance of the various components of the nitrogen cycle need to be better understood so that the specific causes of the observed rise in N_2O can be deciphered, and so that our predictions of future N_2O levels can be improved.
- *Why has extratropical tropospheric ozone increased since the last century, and are further increases likely?* Many questions exist in this area. For example, what are the relative importances of the processes known to affect tropospheric ozone? Are precursors, photochemistry, transport, and dilution important? How much is tropospheric ozone influenced by the growing emissions from aircraft and surface sources of NO_x (i.e., nitrogen oxides with an odd number of oxygen molecules) and other species? Although its trend over recent decades is unclear, tropospheric ozone has almost certainly increased substantially since the last century. Because tropospheric ozone is not only a greenhouse gas and a UV shield, but a pollutant with documented health and ecosystem impacts, we need to be able to predict its concentration.
- *What are the cloud-water-vapor-radiation processes and feedbacks that will strongly influence climatic response to increased radiative forcing?* Much of the climate's response to an increase in greenhouse gases will ultimately be associated with the change in atmospheric water vapor, and its vertical distribution. Water vapor influences clouds (e.g., their formation and radiative properties), which in turn influence the surface radiation balance. Our present understanding of how water vapor, clouds, and radiation balance interact is poor; until this fundamental set of feedbacks is better known, one of the most basic questions regarding anthropogenically induced climate change cannot be properly addressed.

Stratospheric Ozone

- *How does the coupling between chemistry, dynamics, and radiation in the lower stratosphere and upper troposphere operate on dec-cen time scales?* For example, what do we need to know in order to predict accurately the timing of stratospheric ozone recovery from anthropogenic halocarbon depletion? How and by what mechanisms do changes in the stratosphere affect atmospheric circulation and the surface radiation balance on longer time scales, recognizing that much of the lower-stratospheric influence is manifested through chemical and transport changes that originate over short time scales? Ozone distribution influences the vertical temperature structure and dynamics of the lower stratosphere because of its absorption of UV radiation. These dynamics determine the distribution of other atmospheric constituents (including ozone), and thus to some degree their chemical interactions, in the upper troposphere as well as the lower stratosphere. Human impacts on these coupled processes include a potential future fleet of high-flying aircraft and changes in the sources of ozone-destroying halogenated compounds. The long-term feedbacks among human influences, chemical interactions, and atmospheric dynamics are largely unknown and must be better understood.

Aerosols

- *How do the spatial distribution, chemical composition, and physical properties of aerosols vary on dec-cen time scales, and how do they relate to climate variability?* For example, how do the impacts of aerosols on the Earth's radiation budget vary by region? Although aerosols appear to constitute a critical radiative-forcing factor that is active on dec-cen time scales, the mechanisms and processes involved remain poorly characterized. Many of them appear to have short-time-scale influences with long-time-scale implications. (This is true for both tropospheric and stratospheric aerosols.) For example, the indirect effects of aerosols on cloud properties and formation, as well as on solar and thermal radiation, remain a major uncertainty—but may represent the primary impact of aerosols on climate. The direct radiative effect of absorbing aerosols (e.g., soot) in the atmosphere over high-albedo surfaces is hypothesized to lead to surface warming. Predicting future aerosol impacts on climate requires an understanding of regional sources and transports, and how those may vary under a given future climate scenario. Human-initiated changes (e.g., in industrial emissions, biomass burning, or land use) will be key to future aerosol variations; natural sources (e.g., volcanic eruptions and natural vegetation emissions) will contribute as well, and may be even less predictable. Our understanding of the influence of tropospheric aerosols on dec-cen climate-change problems is, in general, woefully inadequate, and requires considerable attention to identify the most important aspects.

Solar Radiation

- *How do proxies for solar activity (e.g., sunspots or*

cosmogenic nuclides) relate to total solar irradiance on dec-cen time scales? Direct measurements of the sun's radiative output span only the last solar cycle. Solar radiation appears to correlate well with solar activity as represented by sunspot variations, solar wind intensity, and charged-particle fluxes. Reconstructions of solar activity that extend over nearly the past 400 years can be derived from records of cosmogenic nuclides (e.g., ^{10}Be) and historical sunspot observations. Extrapolated from measured relationships over the past solar cycle, these reconstructions indicate variability in solar radiation of 0.25 percent over decades to centuries. The sensitivity of climate to solar variations has been explored using observational data and climate models, but it is still not clear how well the solar proxies represent irradiance. For example, when sunspot observations indicate no activity during the Maunder minimum (AD 1600-1640), ^{10}Be records continue to indicate relative changes. A better understanding of the sun's influence on dec-cen climate variability requires improving reconstructed solar irradiance.

- *What feedbacks govern climate and ecosystem response to changes in solar UV on dec-cen time scales*? Although the part of the solar spectrum likely to have greatest climatic influence lies in the visible range, the decadal solar variation observed over the last solar cycle occurred primarily in the UV range. Variations in UV have demonstrable effects on stratospheric O_3 levels and on atmospheric electricity, which may influence cloud-droplet nucleation. UV variability may also influence temperature patterns in the middle atmosphere, which affects the surface climate by altering how planetary waves propagate energy. Although primary producers in both the marine and terrestrial realms can be affected, the impact of UV on ecosystems is poorly understood. The impact of UV radiation on the Earth's climate system on dec-cen time scales must be taken into account.
- *To what extent are dec-cen climate changes, as observed in instrumental and paleoclimate records, related to changes in the sun's output, and what mechanisms are involved in the response of climate to changes in solar radiation*? Decadal variations occur in most records of climate with sufficient length and resolution, but the degree to which these fluctuations can be attributed to solar variability is still being debated. Even studies that point to apparently distinct influences of solar variability on climate often indicate highly variable sensitivities for a given change in irradiance. Feedbacks in the climate system, particularly the atmosphere, may account for enhancement or damping of the climatic response to solar forcing. Understanding the sensitivity of the Earth's climate to past changes in solar activity will permit better predictions of future changes in the face of decadally varying solar irradiance. It will also suggest potential responses associated with other sources of change in the radiative forcing (e.g., internal sources, such as changing albedo).

Observations

Resolution of many of the key issues defined here will require data from observing systems that do not yet exist, or to which no long-term commitment has yet been made. For example, current data on solar irradiance come from short-term satellite missions that have no operational (long-term) mandate. Measurements from different missions are significantly offset from one another in terms of accuracy (NRC, 1994). Similarly, planned satellite measurements are capable of detecting the presence of tropospheric aerosols, but they will not assess the properties of aerosols with the accuracy required for full understanding of their potential climatic effects. Addressing issues related to dec-cen solar variability (above) requires a plan for long-term, calibrated solar-irradiance measurements across the solar spectrum. Even the broad-band shortwave radiation at the Earth's surface must be measured with considerably greater precision; the Atmospheric Radiation Measurement (ARM) program measures shortwave radiation to no better than 50-60 W m^{-2}, whereas precisions of 10-20 W m^{-2} are the minimum acceptable.

Resolving carbon-cycle issues will require a CO_2 measurement strategy that accounts for the hierarchy of scales, both temporal and spatial, inherent in ecosystem processes and their controls. We need atmospheric concentration data that allow us to improve our ability to identify and quantify regional sources and sinks, and to assess the response of these sources and sinks to climate fluctuations and human perturbations. These data will provide the information necessary to regionally integrate the carbon fluxes and feedback processes that can be measured, understood, and modeled on smaller spatial and temporal scales. Isotopic data allow distinction between oceanic and biospheric sinks, on regional scales and have provided significant insight into the regional carbon balance (e.g., Tans et al., 1993; Ciais et al., 1995). Measurements of O_2/N_2 ratios in the global atmosphere provide an independent constraint on the balance between net terrestrial and oceanic sinks (Keeling et al., 1996b). The scaling and measurement issues for N_2O and CH_4 are almost identical, and their biogeochemical budgets could be tackled together with a measurement program suitable for CO_2.

An intermediate-scale observation system that would be crucial for estimating the CO_2 budget also holds the key to quantifying the sources of CH_4 on regional scales, especially if full use can be made of isotopic-ratio data. The improved understanding of CH_4 dynamics that would arise from this type of enhanced quantification would probably help predict future CH_4 levels.

Enormous progress in establishing trace-gas budgets could be achieved if a refined method of directly measuring air-sea gas-exchange rates could be developed. Promising candidate methods are air measurements with eddy correlation (e.g., Baldocchi et al., 1996) and/or eddy accumulation (e.g., Businger and Oncley, 1990). Such measurements would eventually yield a more realistic understanding of the processes controlling the rate of air-sea gas exchange, and permit the development of a parameterization that could be applied with confidence worldwide. These improvements would make it possible to utilize the existing climatologies of the partial-pressure differences between air and water for many gases to derive better maps of gas exchange (see, e.g., Takahashi et al., 1997). Oceanic CO_2 partial-pressure data would then become a much more compelling constraint on the atmospheric budget, and the "open top" of surface-ocean gas budgets could be closed.

Processes and Parameterizations

A number of processes that operate predominantly on short time scales need to be better understood and parameterized in order to properly evaluate their role in dec-cen climate variability and change. For example, how do the composition and properties of aerosols determine both their direct and indirect radiative effects? How do tropospheric aerosols contribute to climate change on long time scales? How do aerosols contribute to cloud formation, precipitation, and radiative interaction? Cloud processes in general and their relationship to atmospheric water vapor and the radiation balance, although they occur on time scales far shorter than decadal, remain a major uncertainty in the prediction of future radiation balances; parameterizations need to be improved for cloud formation and distribution as a function of water-vapor distribution, surface boundary conditions, and rate of the hydrologic cycle. These parameterizations must also include the associated radiative impacts.

Probably the most serious uncertainty in our prediction of stratospheric-ozone recovery during the next century is the possibility that significant changes in the chemistry and circulation of the stratosphere will put us well outside the envelope of our experience—for example, continued large increases in CH_4 and stratospheric H_2O, or stratospheric-circulation changes associated with global warming and greenhouse gases. Predictive models of ozone are based in part on first-principle physics and chemistry, and should correctly account for these changes, but observations of atmospheric-chemistry changes in the recent past are used to test and calibrate the models. Observations must continue so that we better understand these processes and their long-term implications, and can develop robust parameterizations that will prove adequate for future scenarios that exceed those we have experienced to date. Our present ability to predict the effects of increased tropospheric ozone on the climate attributes is weak. Global tropospheric-chemistry models, preferably including cloud-chemistry interactions, need to be developed and verified if meaningful prediction of the effects of tropospheric ozone on climate is to be achieved.

Improved parameterizations of carbon (and other gas) uptake by the processes controlling the mobile reservoirs and their gas exchange are needed. Specifically, we must improve our understanding of how changes in mixed-layer processes (e.g., those which occur in response to climate change and variability) interact with biological and solubility-regulating processes to produce changes in ocean carbon storage. Progress in understanding how the biological and solubility CO_2 pumps may vary in the future will require theoretical and modeling studies, as well as a careful study of past changes. The paucity of accurate observations is still a significant obstacle to the quantification of oceanic carbon uptake and its pathways. Current best estimates of the oceanic uptake of anthropogenic CO_2 derive from numerical models. Without better observations and improved model representations, we cannot properly evaluate how the ocean's storage capacities, uptake rates, and uptake efficiencies will change as climate changes.

ATMOSPHERIC CIRCULATION

The atmosphere is the blanket of air surrounding the Earth. It is what we breathe and what we see and hear through. This life-sustaining medium affects almost everything that we depend on, including our water and food supply, our physical activity, and our ability to travel and transport our goods from one place to another. We are, therefore, affected by any variations in the properties of the atmosphere. Variability of these properties on decade-to-century time scales probably arises through external forcing and the atmosphere's interaction with the slower components of the climate system—the sea surface and its temperature, land-surface properties and vegetation, and the snow and ice cover—rather than through interactions internal to the atmosphere. Because the atmosphere is a fast-responding fluid, capable of transporting constituents from one location to a distant one and thus contributing to long-time-scale feedbacks, it is an important component of the climate system on dec-cen time scales.

In examining the characteristics of atmospheric circulation, we must distinguish between its mean behavior and its variations. The climate patterns discussed in Chapter 3 are the most prominent display of the circulation's temporal variability on time scales longer than 10 days or so. These patterns are the vehicle by which atmospheric variations affect the climate attributes of concern to society. Underlying the variations are the permanent mean-circulation features, which are manifested mainly in a three-dimensional asymmetry: In the vertical, gravity and the compressibility of air dictate a rapid drop in pressure with altitude. In the north-south direction from equator to pole, variations in solar radiation and the Earth's rotation give rise to a meridional circulation that produces large-scale belts of different weather and climate. In the east-west direction, topography and land-sea contrast lead to undulations in the atmosphere's most vigorous flow, the zonal wind. Both the mean state and the variability of the atmospheric circulation are seasonally dependent, yet the basic properties of the atmosphere are discernible in all seasons. (For the classical depiction of the mean atmospheric circulation and the theory behind it, see Lorenz (1967).)

The zonal winds are dynamically unstable; they break into "transient" (baroclinic) eddies with horizontal dimensions of a few thousand kilometers, and a time scale of a few days. Between 30 and 60 degrees of latitude (the mid-latitudes) on both sides of the equator, the westerlies reach all the way to the surface; their continuity is broken by both transient eddies and more permanent zonally asymmetric features (stationary eddies) that form in response to topography and air-sea contrasts.

The stationary eddies induce large northward transports of heat and moisture, and set up a zonally uneven distribution of climate regimes, in which east and west sides of continents experience different mean and time-varying conditions. There is a strong link in both location and scale between the climate patterns described in Chapter 3 and the stationary eddies. To some extent, the climate patterns can be thought of as fluctuations in the strength and position of the stationary eddies, as well as variations in zonally symmetric flow. Mid-latitude transient eddies are steered by the stationary eddies, but also affect them. These baroclinic eddies induce poleward movement of warm humid air and equatorward movement of cold dry air, resulting in poleward heat and moisture transport, and also the uplift of air masses. Their inherent two-dimensional asymmetry also produces coherent patterns of momentum transport. Thus, while transient-eddy motions average out, their transport properties have a lasting effect on climate. In fact, baroclinic eddy transports play an extremely important role in the maintenance of the zonally averaged circulation and the stationary waves, and thus should be considered significant contributors to dec-cen variability.

Atmospheric circulation plays an important role in shaping the three-dimensional properties of the world ocean. It determines the amount of solar radiation reaching the surface, the latent and sensible heat exchange with the ocean surface, the amount of freshwater (in the form of precipitation) that falls onto the surface, and the dynamic (wind) forcing of the surface ocean circulation. (For a complete description, see the "Ocean Circulation" section of this chapter.) In turn, it is the ocean that plays the major role in determining the dec-cen variability of the Earth's climate, mainly through the ocean's thermal memory and the delayed effect of its sluggish circulation on the atmosphere (see also Chapter 3).

This section focuses on those aspects of the atmospheric circulation that need to be understood and modeled in order to properly represent the interaction of the atmosphere with the other components of the climate system, and to better identify and understand its unique role in these interactions. Water vapor is treated in the "Hydrologic Cycle" section that follows this one. The contributions of the slower components (the boundary conditions) of the climate system, which are most responsible for driving the dec-cen variability manifested in the atmospheric circulation and the hydrologic cycle, are presented in the last three sections of this chapter—"Ocean Circulation," "Cryospheric Variability," and "Land and Vegetation."

Influence on Attributes

The atmosphere is the fastest-responding component of the climate system, and its circulation, or motion system, the most vigorous. The spectrum of atmospheric motions is extremely broad in time, space, and dynamic range. The atmosphere's turbulent features range in size from centimeters to the circumference of the Earth, and vary on time scales from fractions of an hour to millennia or more. Among the agents that shape and determine the Earth's climate locally and globally, atmospheric circulation plays an important role, even if it is as a slave to, or moderator of, the boundary conditions and forcings. Three of the fundamental climate attributes—precipitation, temperature, and storminess—are directly affected by the atmospheric motion and its vigor. Atmospheric circulation strongly influences the other three climate attributes (solar irradiance, sea level, and ecosystems) as well.

In the most general sense, atmospheric circulation arises in response to the uneven distribution of solar radiation on the Earth's surface, and to variations in surface conditions such as roughness, elevation, reflectivity, and heat and moisture capacity. The result is a wide range of motions that are highly interactive and are affected by variations in water content (in all three phases), in radiatively active gases, and in aerosol concentration. Atmospheric circulation plays a key role in redistributing physical and chemical properties and constituents such as heat, moisture, gases, and aerosols between source and sink regions, thus determining regional variations of climate. Atmospheric circulation directly controls the distribution of temperature, humidity, and rainfall, while affecting the surface radiation via the distribution of aerosols and clouds. The planetary-scale aspects of the circulation regulate the smaller-scale response, in particular the distribution and intensity of storms. The long-term average behavior of the circulation determines the nature and distribution of ecosystems, and by influencing ice melt and growth, it affects sea-level change. Atmospheric circulation also communicates local changes, induced by fluctuating boundary conditions, to other locations, and shapes the response of the other components of the climate system to such changes.

Evidence of Decade-to-Century-Scale Variability and Change

The atmosphere reacts almost instantly to changes in external forcing and to internal interactions. However, because of the nonlinear nature of atmospheric interactions, the time and space scales over which it varies are not directly related to the time and space scales over which it is forced. The atmosphere's circulation thus displays a widely ranging spatial and temporal response.

The globally averaged atmospheric temperature has been the focus of intense study in recent years, particularly because of the interest in anthropogenic climate change. Figure 5-12 shows the change in global-mean land and sea surface temperatures recorded by instruments since the 1850s, and their hemispheric distribution. The overall increase in temperature over time is unequally distributed between land and ocean (IPCC, 1996a). The figure displays considerable low-

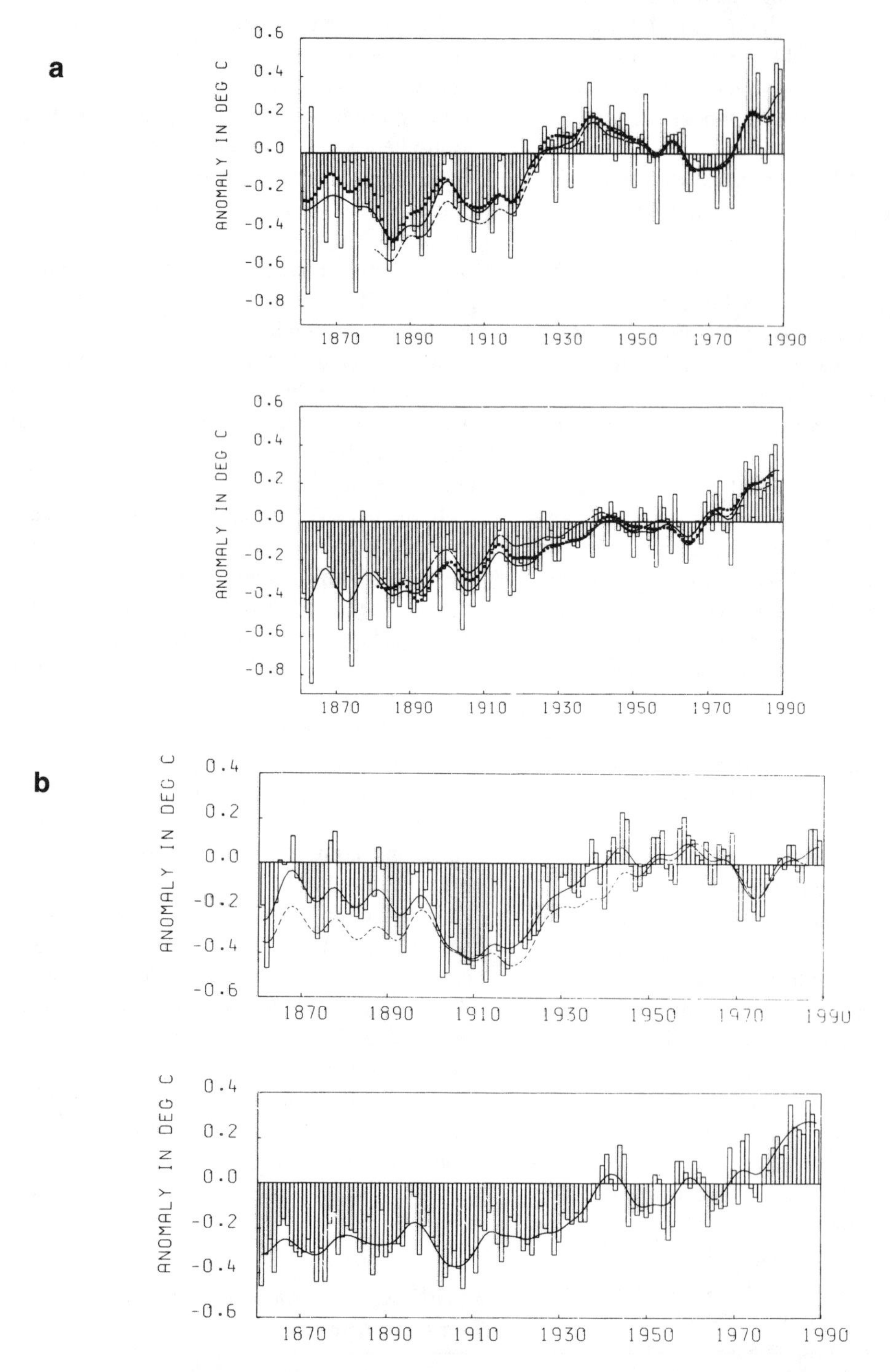

FIGURE 5-12 Change in mean (a) land and (b) sea surface temperatures recorded by instruments since 1861, relative to 1951-1980 mean. The top and bottom panels of each pair represent the Northern and Southern Hemispheres, respectively. (a) Land data: annual values from P.D. Jones; smoothed curves from Jones (1988) (solid), Hansen and Lebedeff (1988) (dashed), and Vinnikov et al. (1990) (dotted). (b) Sea surface data: annual values from U.K. Meteorological Office; smoothed curves from UKMO (solid) and Farmer et al. (1989) (dashed). (All figures from IPCC, 1990; reprinted with permission of the Intergovernmental Panel on Climate Change.)

frequency variability, which has been emphasized through annual averaging and time filtering. The cause of these low-frequency fluctuations is still debatable. Of course, higher-frequency fluctuations exist as well. These fluctuations are due to the internal variability of the climate system, and are governed by, among other things, the atmospheric circulation's response to forcing and to changing boundary conditions.

Interdecadal variability in instrumental temperature records, both global and hemispheric, was contrasted with interannual variability by Ghil and Vautard (1991). The 15- and 25-year near-periodicities detected in the relatively short (135-year-long) records of Jones et al. (1986a,b) and shown in IPCC (1992) have been confirmed with high statistical significance in much longer proxy records having annual resolution (Quinn et al., 1996; Biondi et al., 1997), as well as in the longest available (335 years of local temperature) instrumental record (Plaut et al., 1995). The spatial patterns and physical mechanisms associated with these broad spectral peaks are the focus of intense current research. Thus the 13-to-15-year peak has been associated with variability of the North Atlantic current system (Sutton and Allen, 1997; Moron et al., 1998), while the 20-to-25-year peak has been associated with oscillations in the thermohaline circulation (Delworth et al., 1993; Chen and Ghil, 1996).

A striking example of decadal variability in the atmospheric circulation is evident in Figure 5-13. The decades of 1950-1959 and 1980-1989 were both warm intervals (see the global means in Figure 5-12), yet the geographic distributions of surface-temperature anomalies for the two are significantly different from one another, indicating that the atmospheric circulation had changed considerably. The seasonal distribution of global mean temperature also attests to the importance of atmospheric circulation in climate variability (see Wallace et al., 1993).

In the tropics, ocean-atmosphere interaction leads to a unique ENSO climatic signal (see Chapter 3). While this interaction occurs in the equatorial Pacific, the response of the atmospheric circulation to ENSO is global (Rasmusson and Wallace, 1983). Because of its strength and global influence, and because it has been shown to be predictable up to a year in advance (see, e.g., Barnett et al., 1988, and Barnston et al., 1994), ENSO is arguably the most important mode of climate variability on interannual time scales (see, e.g., Glantz, 1996). In recent years increased attention has been given to the variability of ENSO on decadal and longer time scales (Trenberth and Hurrell, 1994). In particular, the existence of the recent long interval of warm tropical sea surface temperature (SST) in the eastern Pacific and negative Southern Oscillation Index (shown in Figure 5-14) have initiated various speculations on the origin of decadal variability in the tropics and on the nature of the link between these tropical variations and mid-latitude atmospheric circulation (Trenberth, 1995; Latif et al., 1996). Because of the implications of ENSO for global climate, it is imperative that we strive to better understand and predict decadal variability in the tropics. Other examples of tropical variability important to society are the long-term fluctuations of monsoon and Sahel rainfall (see Chapter 4).

In the mid-latitudes, the low-frequency variability of atmospheric circulation—as reflected, for example, by anomalous distribution of sea level pressure and geopotential height—displays the distinct spatial patterns described in Chapter 3. Most clearly discernible during winter, the pressure fluctuations associated with such teleconnection patterns as the PNA and the NAO regulate the meridional distribution of temperature, and moisture advection and convergence, in their region of influence (Trenberth and Hurrell, 1994). In recent years it has become clear that these teleconnection patterns display variability on decadal and longer time scales. For instance, wintertime sea-level pressure (SLP) in the Aleutian low region remained persistently lower than normal for more than a decade between 1977 and 1988 (Trenberth and Hurrell, 1994). The surface pattern during this period and its mid-tropospheric counterpart (see Zhang et al., 1997) resemble the PNA pattern (Figure 5-15). Such a climatic configuration tends to lead to a warmer-than-normal western North America and a somewhat colder-than-normal eastern North America. Changes in baroclinic-eddy activity over the Pacific and North America are also involved in this climatic perturbation.

Over the North Atlantic, SLP has been falling in the Icelandic low region and rising near the center of the Azores high since the 1960s, indicating a remarkably long trend in the strength of the NAO (Hurrell, 1995). Such climate-regime changes are linked to changes in wind distribution and storminess over the North Atlantic, and to the severity of winters (rainfall and temperature) downstream in Europe, Asia, and North Africa (Hurrell and van Loon, 1996).

The NAO and PNA circulation features, which occur predominantly over the ocean basins, are connected with distinct changes in ocean temperature and circulation. They may indicate the existence of a coupled interaction between the atmosphere and ocean that is crucial to understanding and predicting dec-cen variability.

Mechanisms

As noted in Chapter 4, three types of interaction govern the variability of the atmospheric circulation:

1. External forcing (solar, volcanic, and anthropogenic);
2. Interaction with the other components of the climate system (ocean, land, cryosphere, and biosphere); and
3. Internal atmospheric interactions (transient-transient and transient-mean interactions, and other nonlinearities).

The relationship between atmospheric-circulation variability and external radiative forcing has not been clearly resolved. Many studies have attempted to identify periodic behavior in a number of atmospheric variables that could be attributed to periodic changes in solar forcing. One of the foci of this search has been ascertaining the climatic response to orbital fluctuations on millennial time scales (Imbrie et al., 1992)—a topic beyond the scope of this document. Many attempts have also been made to detect a climate response to shorter-scale variations in solar forcing. One prominent example is the search for effects of the periodic 11-year fluctuations in solar luminosity and its 22-year modulation (see the "Atmospheric Composition and Radiative Forcing" section of this chapter). Satellite observations (available for only one and a half sunspot cycles so far) indicate that during a period of sunspot maximum, solar irradiance is higher by more than 1 W m^{-2} than it is during a sunspot minimum. Since the lower atmosphere absorbs only a small portion of incoming solar radiation, it is hard to see how such a weak signal can affect climate unless a positive feedback exists in the climate system. Such mechanisms have been proposed

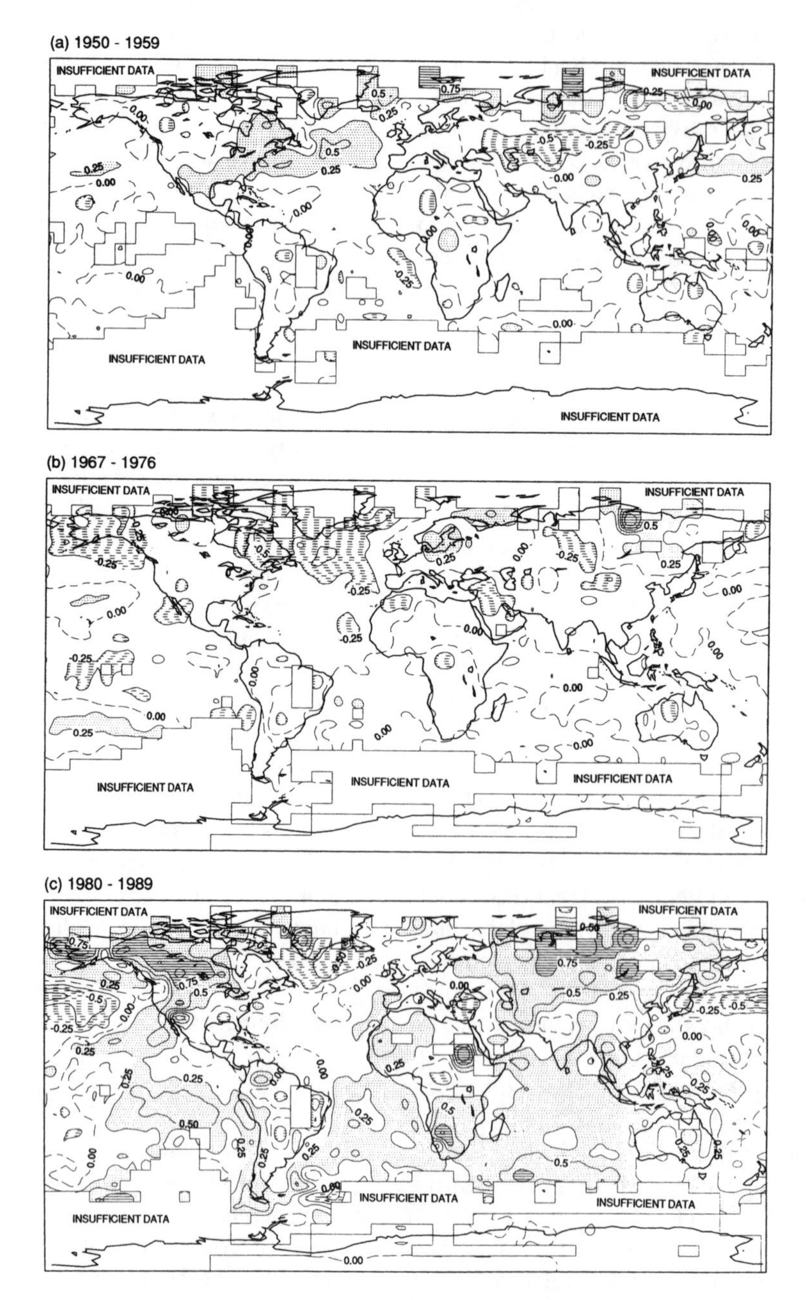

FIGURE 5-13 Decadal surface-temperature anomalies relative to 1951-1980. The contour interval is 0.25°C. The 0°C contour is dashed. Dash-filled areas have negative anomalies less than -0.25°C; white areas have anomalies between -0.25°C and +0.25°C; dot-filled areas have positive anomalies between 0.25 and 0.75°C; striped areas have positive anomalies greater than 0.75°C. (From IPCC, 1990; reprinted with permission of the Intergovernmental Panel on Climate Change.)

but not authenticated (see, e.g., Hansen and Lacis, 1990). Nonetheless, evidence of just such an effect in the climate record (see Mitchell et al., 1979; Cook et al., 1997; White et al., 1997b), while compelling, has generally been received with skepticism (e.g., Pittock, 1978, 1983). The main reason for such skepticism is the weak impact these fluctuations have on the solar constant (~0.1 percent, see Sarachik et al., 1996). The subject has recently received renewed attention, both because of popular concern about anthropogenic global climate change and because of the apparent changes in sunspot cycles over long (centennial) time scales (Kelly and Wigley, 1992; Schlesinger and Ramankutty, 1992; Lean et al., 1995). (The atmospheric response to changes in anthropogenic radiative forcing was discussed more fully in the "Atmospheric Composition and Radiative Forcing" section of this chapter.)

Volcanic eruptions may release large amounts of aerosols and gases into the troposphere and the lower stratosphere, as noted earlier. In the stratosphere the residence time of volcanic aerosols is longer, allowing them to be distributed around the globe by stratospheric winds. These stratospheric aerosols are responsible for a complex radiative interaction, involving the absorption of both solar and terrestrial radiation, that can have a significant influence on climate (Lacis et al., 1992). The effect of Mount Pinatubo's eruption in the summer of 1991 on the climate of the following winter has been much discussed (e.g., Hansen et al., 1992). However, it appears that individual strong volcanic eruptions affect the atmosphere for only short periods of time (1-2 years) before the aerosols are flushed out of the lower stratosphere (Sear et al., 1987; Bradley, 1988; Mass and Portman, 1989). Thus, a series of major explosive volcanic events would be needed to produce a long-term effect on climate.

The issue of volcanic influence on long-term climate variability has been the subject of speculation and debate over the last several decades (Budyko, 1969; Zielinski et al., 1994). Attempts are being made to construct a record of past explosive volcanic eruptions so that their effects on climate can be explored (Bradley and Jones, 1992; Robock and Mao, 1995). It is interesting to note that several reconstructions of volcanic activity indicate a long span, from 1920 to 1960, of low stratospheric-dust levels—a period during which global atmospheric temperature generally increased and peaked (Robock, 1991). Increased volcanic activity between 1960 and 1980 parallels a period of global cooling that began in the early 1960s.

Myriad intricate interactions within the atmosphere, and between it and each of the other components of the climate system, give rise to feedbacks that can either amplify or weaken the effect of external forcing. For example, atmospheric circulation determines the vertical and horizontal distribution of moisture, radiatively active gases, clouds, and aerosols. Because the vertical and latitudinal distributions of these constituents influence the radiative forcing of the planet, changes in atmospheric circulation directly affect the global mean temperature and its horizontal and vertical distribution. Changes in temperature in turn affect the distribution of winds and their convergence patterns, leading to changes in evaporation, in the atmospheric moisture content and its distribution, and ultimately to changes in radiative

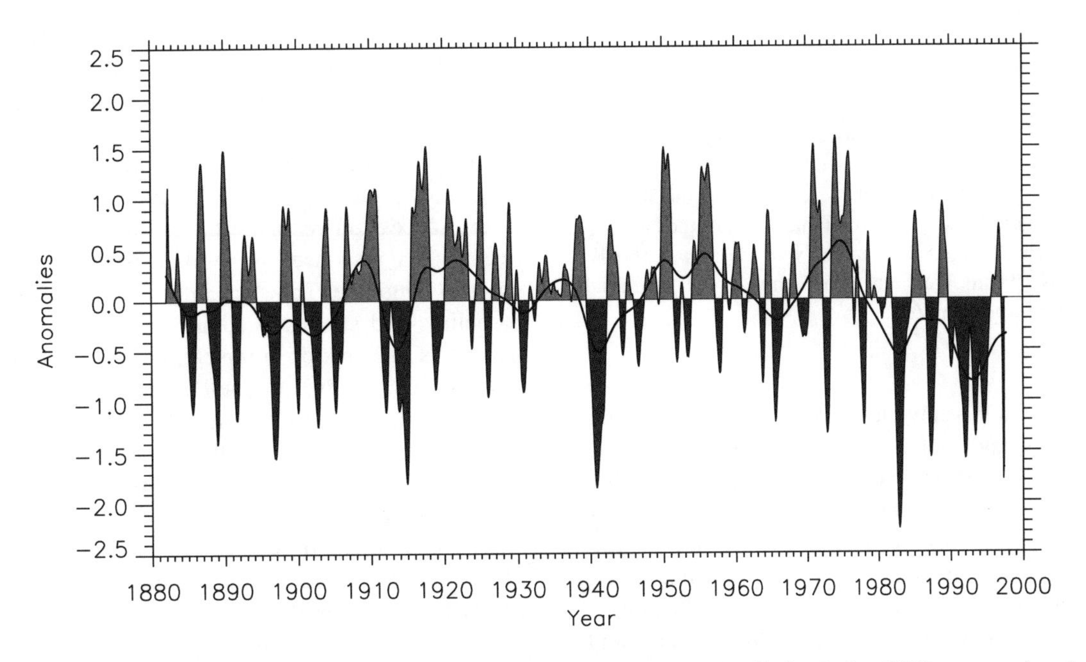

FIGURE 5-14 Standardized monthly and low-pass filtered time series of a Southern Oscillation Index (SOI) computed as the negative of sea-level pressure at Darwin, Australia. It is one of the key indicators used to assess tropical atmospheric activity associated with El Niño. (After Trenberth, 1984; reprinted with permission of the American Meteorological Society.)

a

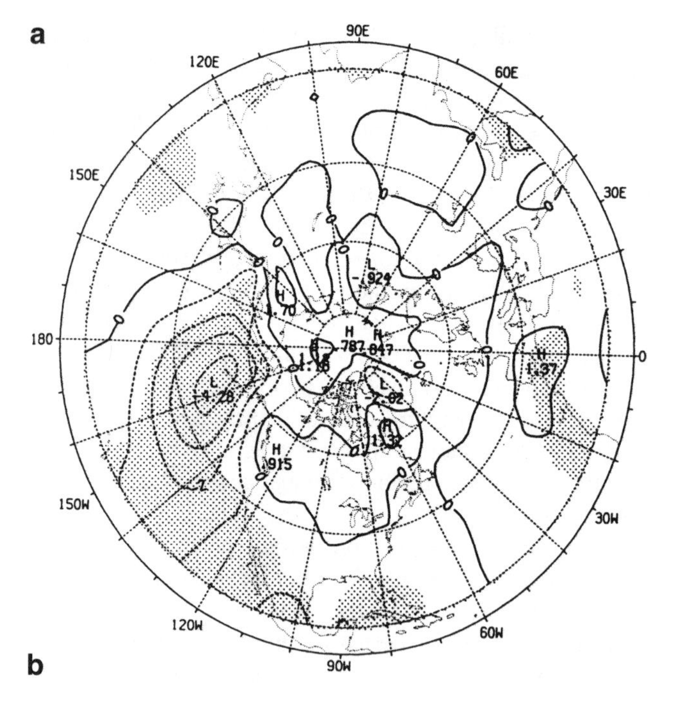

b

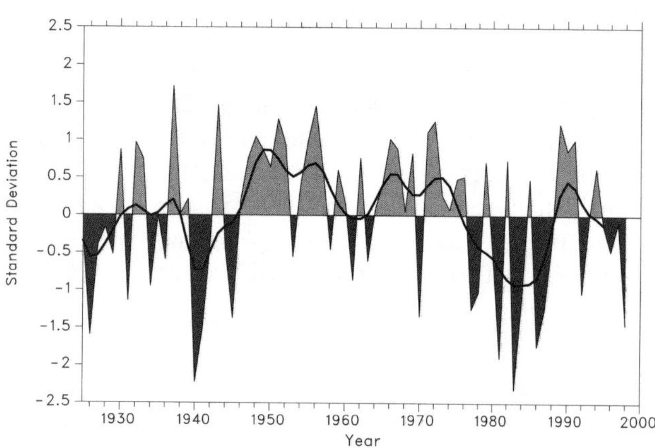

FIGURE 5-15 (a) The anomalous November-through-March averaged SLP differences during the period 1977-1988 with reference to the period 1924-1976. Note the resemblance to the 500 hPa PNA pattern shown in Figure 3-2a. (b) Time series of the November-through-March averaged SLP anomaly in the region 160°E-140°W, 30-65°N. (From Trenberth and Hurrell, 1994; reprinted with permission of Springer-Verlag.)

forcing. It is crucial to study these interactions and to better understand the behavior of climate on both short and long time scales. Several recent studies have examined whether such interactions should lead to a stable or an unstable tropical climate in the presence of external forcing (Pierrehumbert, 1995). Similarly, interactions among the atmosphere, cryosphere, and ocean explain the poleward amplification of the warming signal observed in the recent climate record and in climate models forced with increased CO_2 (see, e.g., Meehl and Washington, 1995).

It is commonly assumed that if the atmosphere were not interacting with the other components of the climate system, the spectral energy of its variability would not increase on time scales longer than that of the annual cycle. While this assertion may be disputed (James and James, 1989; Lorenz, 1990), fluctuations in the atmospheric circulation on the deccen time scale that do not arise from changes in external forcing are most likely influenced by atmospheric interactions with the more slowly varying parts of the climate system. For instance, ocean-atmosphere interaction gives rise to ENSO variability in the tropics, and may, through teleconnections, explain decadal-scale variability in the mid-latitude Pacific (Graham, 1994). The possibility of direct atmosphere-ocean interactions in the mid-latitudes themselves has also been noted, most recently by Latif and Barnett (1994, 1996) and Barsugli and Battisti (1998). Latif (1998) classifies interdecadal atmosphere-ocean variability into four classes: tropical interdecadal variability; interdecadal variability that involves both the tropics and the extratropics as active regions; mid-latitude interdecadal variability involving the wind-driven ocean gyres; and mid-latitude variability involving the thermohaline circulation. Long-term changes in precipitation patterns in the subtropics (e.g., the Sahel) have been attributed both to changes in global SST (Folland et al., 1986) and to positive feedback from land-surface processes (Charney, 1975). Whichever the cause of the variability is, and wherever it originates, the atmospheric circulation can distribute the outcome over wide regions, and communicate its effects from ocean to land and vice versa.

The way the atmospheric circulation responds to external forcing and to interactions in the climate system is determined by dynamic laws governing the atmospheric flow. Several specific dynamic and thermodynamic mechanisms have been proposed to explain the atmospheric response in terms of patterns and teleconnections. It has been suggested that the global effects of ENSO, particularly in the extratropics, are the result of planetary stationary waves, forced at the tropics by deep tropospheric heating, that distribute energy into the mid-latitudes (Hoskins and Karoly, 1981). This theory successfully explains the seasonal dependence of the extratropical response, and to some extent its phase with respect to longitude. Transient interactions with mid-latitude baroclinic and stationary eddies must also be included to make this explanation complete, however (Hoskins, 1983; Simmons et al., 1983). In the mid-latitudes, similar dynamics explain the shape and location of atmospheric teleconnection patterns, but the response to surface forcing is less well understood (see Kushnir and Held, 1996, and references therein).

The appearance of similar spatial patterns in the atmosphere's intraseasonal, interannual, and interdecadal variability (see Chapter 3) indicates that at least some of these patterns correspond to basic modes of internal variability of the atmosphere. It has been suggested (Legras and Ghil, 1985; Palmer, 1993) that the bridging of time scales that is apparent in these common spatial patterns might be due to the fact that changes in the atmosphere's surface boundary conditions affect the frequency of occurrence of one or more weather regimes (Reinhold and Pierrehumbert, 1982; Ghil and Childress, 1987). Supporting evidence for this idea has been provided by Horel and Mechoso (1988) and Kimoto (1989). Further discussion of possible mechanisms governing decadal climate variability can be found in WCRP (1995), Latif et al. (1996), and Sarachik et al. (1996). A review of coupled ocean-atmosphere modeling studies of decadal-scale variability is given by Latif (1998).

Predictability

The pioneering work of Lorenz (1963, 1969), and the research it spurred over the years, established that the atmospheric flow, on the scale of individual weather disturbances, is predictable only a week or so in advance. Climate prediction is based on the expectation that part of the atmospheric circulation is evolving more slowly, either because of its internal dynamics or because of slowly changing boundary conditions (land and ocean surface properties). In moving beyond the limits of weather prediction, we hope to be able to specify future mean or variance conditions in terms of their relation to normal conditions, and to define the degree of uncertainty in that prediction. Because of its strategic societal importance, climate prediction has been actively studied for many years (Namias, 1968). In the tropics these studies have born fruit in the form of the discovery that ENSO is predictable up to a year or more in advance (Cane and Zebiak, 1987; Barnett et al., 1988; Barnston et al., 1994). It has proven more difficult to infer the global response of the atmospheric circulation from knowledge of upcoming conditions in the equatorial Pacific (Barnett, 1995). Attempts at seasonal-scale prediction of other atmospheric features, such as tropical cyclone activity and terrestrial rainfall in semi-arid regions, show skill and promise (Palmer and Anderson, 1993; Hastenrath, 1995; see also the Center for Ocean-Land-Atmosphere Studies quarterly Experimental Long-Lead Forecast Bulletin (http://www.iges.org/ellfb)). The implications of these capabilities for our ability to make longer-term forecasts or predictions are not yet clear, however.

Chapter 4 distinguished between two types of prediction: non-initialized, an example of which is the prediction of the response to the anthropogenic addition of greenhouse constituents; and initialized, exemplified by detailed prediction of the boundary conditions that determine the atmospheric statistics, which requires the specifications of initial conditions tied to a given point in time. The importance of climate variability on decadal and longer time scales has been demonstrated by the drop in skill of ENSO-prediction methods earlier in this decade, followed by a recent increase in skill. It is hypothesized that decadal variability in the Pacific ocean-atmosphere system is responsible for the recent change in the characteristics of ENSO (Latif et al., 1996). Little is known about the predictability of climate on the dec-cen scale, though recent modeling results (Latif and Barnett, 1996) show the possibility of 5-to-6-year prediction of North Pacific SLP patterns, which are directly related to atmospheric circulation and the PNA pattern. Clearly this question cannot be addressed until it has been established that climate models are capable of representing the range of dec-cen variability displayed in the instrumental and proxy records.

Remaining Issues and Questions

Many of the remaining issues and questions about the role of atmospheric circulation in dec-cen variability are related to climate patterns, interaction with other climate-system components, and the role of changing boundary conditions in triggering and maintaining long-term circulation anomalies (see Chapter 3 and other sections of Chapter 5). Here we highlight the issues especially important for atmospheric circulation.

- *How much of the dec-cen variability in atmospheric circulation is unforced?* For example, to what extent are dec-cen-scale circulation variations related to PNA, NAO, and other climate patterns driven by natural atmospheric variations that are produced through nonlinear interactions in the atmosphere? What role do coupled interactions with the other components of the climate system play in the atmosphere's dec-cen natural variability? Are some or all of these variations driven or shaped by changes in the radiative forcing resulting from an anthropogenic increase of greenhouse gases, changes in the levels of natural (e.g., volcanic and surface dust) or anthropogenic aerosols, or variations in solar radiance?
- *How do global-scale, dec-cen circulation changes affect regional-scale, higher-frequency climate variability and severe weather?* We need to better document, understand, and predict the local and higher-frequency outcomes of dec-cen-scale changes in atmospheric circulation. The societal impacts of dec-cen variability are determined primarily by changes in regional climate states, storm tracks (both mid-latitude and tropical), and rainfall, which typically vary over shorter time scales. In other words, how do variations in the mean climate state influence the spatio-temporal distribution of the higher-frequency variance, and how might this relationship be used to exploit knowledge of long-time-scale variations to make shorter-time-scale climate predictions?
- *What are the magnitudes, spatio-temporal patterns, and mechanisms of the mid-latitude atmospheric response to both mid-latitude and tropical SSTs?* The hypothesis that the ocean plays a key role in dec-cen climate variability hinges on the ability of the atmosphere to respond to changing SST conditions, and the patterns of its responses. While we now have a solid understanding of how this response occurs in the tropics, we do not yet fully understand how the tropical response is communicated to the mid-latitudes. Even

less clear is the nature of the atmospheric response to midlatitude SST, for which theory, observations, and modeling are not yet consistent. Several mechanisms have been hypothesized to explain observed teleconnections between the tropical and extratropical latitudes, and the contributions they might make to dec-cen variability—for example, slower propagation of anomalies back to the equator via ocean processes and faster propagation from the tropics via atmospheric processes. These processes must be evaluated to determine their importance in explaining climate variability, the dominant mechanisms by which they occur, and how they maintain and propagate anomalies. The exploration of the spatial extent of such teleconnections should also be continued to determine more fully the scales over which local and regional anomalies and influences can be communicated. Moreover, links to tropical SST and the decadal variability of ENSO should also be investigated more fully, as well as links to variations in the annual cycle of the Southern Hemisphere. Further investigation into the origin and maintenance of these phenomena, and their mutual relationships, is warranted.

- *What are the mechanisms of interaction between the atmosphere and land-surface processes on dec-cen time scales?* Land-surface characteristics, such as snow and ice cover and soil moisture content, are known to affect short-term (interannual) climate variability. Longer-term fluctuations may also be related to variations in land-surface characteristics. Changes in the lower boundary conditions may initiate regime changes in the atmosphere, which may emerge as dec-cen-scale climate variations. The land surface's ability to extend the memory of the climate system, even though smaller than the ocean's, can be important in such regime shifts.

- *What are the mechanisms of region-to-region and basin-to-basin interaction on the dec-cen time scale?* Given our short instrumental records and a limited number of long coupled atmosphere-ocean GCM simulations, the issue of cross-basin and remote interaction of climate anomalies remains a largely unresolved problem. The most dramatic climatic events of the last 20 to 30 years appear to be related to almost synchronous long-term climate-regime changes over the North Pacific and North Atlantic regions, as well as in the tropics. It is not clear whether this coherence is coincidental or whether these changes arose through dynamic linkages between these regions. Furthermore, it is not clear to what extent these synchronous changes were driven by anthropogenic effects. On geological time scales, other dramatic climatic events in the Earth's history (e.g., deglaciation; see Bender et al., 1994) were nearly global in extent. While many of these changes happened on longer time scales, there is little doubt that understanding them is helpful for understanding dec-cen variability, because some of the underlying mechanisms driving past changes might be responsible for changes occurring in the future. In addition, some climatic events documented in ice cores and elsewhere indicate that major climate-system reorganizations can take place in a matter of decades (e.g., the Younger Dryas termination; see Alley et al., 1993).

- *How do dec-cen-scale changes in atmospheric trace gases, aerosols, and cloud cover affect radiative balance and thus atmospheric circulation, and vice versa?* The delicate feedback mechanisms associated with the interaction between radiation and dynamics are not well understood. Subgrid-scale parameterizations are involved in modeling these feedback mechanisms, and need to be put on a more robust physical basis. There has been considerable controversy recently about the sufficiencies of model simulations of upper-tropospheric moisture transport and phase changes, about the role of anthropogenic aerosols in climate change, and about the role of clouds in stabilizing the global climate. These important issues need to be resolved if we are to simulate and predict dec-cen variability.

Observations

The study of dec-cen variability of the atmospheric circulation faces many challenges. The widespread current interest in anthropogenic climate change has motivated the climate-research communities to reorganize the available instrumental and proxy data and to increase the volume of their archives through data "archeology" and additions of new data. These efforts need to proceed in tandem with the establishment of clear guidelines for future atmospheric observations, and with long-range, coordinated planning of the observational networks so that adequacy, continuity, and homogeneity of the observational records are assured. Observations should focus on describing both the state variables—wind, pressure, temperature, humidity, and rainfall—and the forcings or related variables—solar radiation, clouds, aerosols and chemical composition.

Processes and Parameterizations

Models of the climate system are a powerful tool for the study of climate variability, as Chapter 6 will make clear. Such models must be developed further until they are comprehensive enough to allow the coupled simulation of ocean, atmosphere, cryosphere, and changes in continental surface conditions. The processes controlling the evolution of all these important components must be improved in atmospheric-circulation models so that the dominant interactions driving long-term climate change in the atmosphere can be properly evaluated. In particular, the processes controlling the feedbacks and interactions between water vapor, cloud-formation processes, and atmospheric circulation must be better understood and parameterized. These processes are particularly important for tropical regions, where convective activity and cumulus formation exert the dominant control on atmospheric dynamics. Also, the processes controlling the boundary-layer physics, including all interactions (e.g., exchange of heat, moisture, and momentum) and feedbacks, need to be better understood. Much of the dec-cen variability in the atmospheric circulation arises through interactions and coupling with the boundaries, yet this complex interface region is poorly understood, and poorly resolved in models. Because of climate models' relatively coarse vertical resolution, the dynamics of this interaction cannot be resolved in models, so they must be parameterized in terms of large-scale circulation quantities. Our ability to correctly model the atmospheric response to greenhouse-gas increases and to natural surface-temperature and moisture anomalies, as well as the ability to correctly force ocean models, depend on boundary-layer processes. Finally, the processes controlling the large-scale wave features and their interaction with the major climate patterns described in Chapter 3, including such shorter-time-scale features as ENSO, must be better understood to be correctly represented in models.

HYDROLOGIC CYCLE

The Earth's hydrologic cycle involves the movement of the planet's water through its many phases in the atmosphere, ocean, and land. Water evaporates from the surface of both the land and the oceans, condenses in the atmosphere, and precipitates back onto both the land and ocean surfaces. Ultimately it finds its way back into the atmosphere as vapor, sometimes residing in surface or subsurface reservoirs for hundreds or thousands of years, or longer, before completing the cycle. (For convenience, snow and ice are dealt with in detail in the "Cryosphere" section of this chapter, rather than directly below, even though they are an important part of the cycle.) In the course of this cycling, water influences all of the climate attributes discussed in Chapter 2. The hydrologic cycle involves complex interactions with factors such as vegetation and clouds, as well as atmospheric and oceanic circulation. For instance, atmospheric circulation helps determine the convergences and divergences of vapor (and thus the ratio of precipitation to evaporation for a given locale), but the condensation processes in turn help determine the circulation of the atmosphere.

Influence on Attributes

The hydrologic cycle is obviously the dominant control of the climate attribute of precipitation and water availability. It also directly influences sea level. If more precipitation falls as snow than is returned by evaporation (or sublimation, the direct vaporization of the snow cover), the snow will accumulate. Eventually it will form glaciers, which can account for considerable changes in sea level. In fact, aside from the local sea-level changes associated with the coastland response to the weight of the waxing and waning continental ice sheets, the largest changes in sea level arise from the addition or subtraction of freshwater as those ice sheets change in size. During the last ice age, enough freshwater was removed from the oceans and stored in continental ice to lower the sea level by more than 120 m. In the relatively moderate climate conditions of the past century, the return of freshwater to the oceans from melting alpine glaciers, and possibly from the existing continental ice sheets, is thought to be responsible for approximately half of the 20 cm sea-level rise that has been observed.

The amount of water involved in these sea-level changes is small relative to the total amount of water involved in the hydrologic cycle (see Figure 5-16). About 16 million metric tons of water falls on the Earth every second. In the mean,

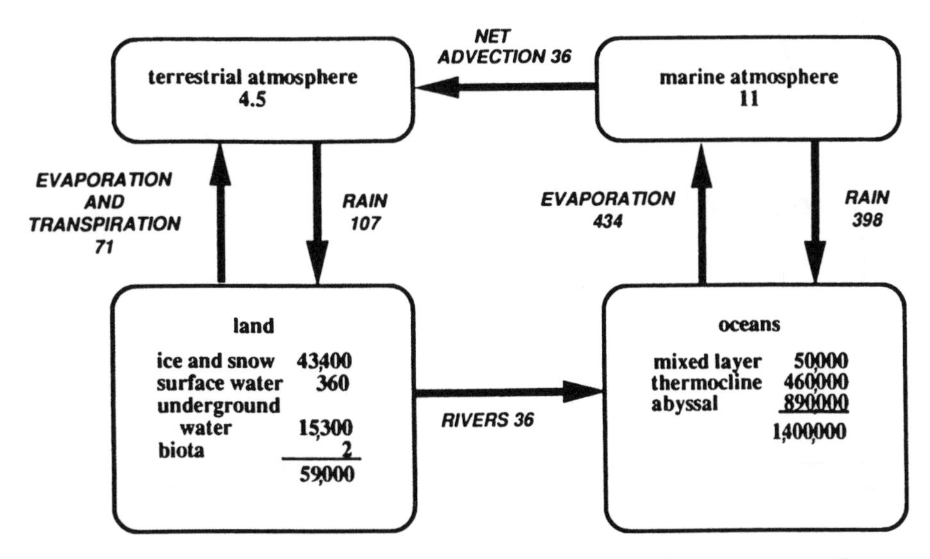

FIGURE 5-16 Gross budget of the mean global water cycle: Reservoir figures are in 10^{15} kg, fluxes in 10^{15} kg per year. (From Chahine, 1992; reprinted with permission of Macmillan Magazines, Ltd.)

assuming a nearly steady state, roughly that same amount must leave the surface. Note that 1 million metric tons per second corresponds to 1 sverdrup in oceanographers' units and to 32,000 km^3 per year in the units more commonly used by hydrologists. It should also be noted that the figures for global fluxes given in Figure 5-16 are uncertain to about 10 percent.

About 80 percent of the total precipitation falls on the surface of the ocean, where it reduces the local salinity, or, if it falls as snow on sea ice, raises the local albedo. Likewise, where precipitation falls as snow on land, it raises the local albedo and influences the regional radiation budget. Of the total water leaving the surface of the Earth, 85 percent evaporates from the surface of the ocean. Some of the water vapor evaporated from the oceans (36,000 km^3 per year, or 8 percent) must be diverted from ocean areas to over land, where it condenses as precipitation, in order to balance the water that flows as runoff from land to ocean (Chahine, 1992). The total rainfall over land, 107,000 km^3 per year, arises from local evaporation and from the aforementioned importation of marine vapor.

Climate variability at decadal to centennial time scales translates into significant impacts on surface and subsurface hydrology and, thus, on water-resource management. Regional flood potential, water-quality trends, hydropower and recreation potential, and irrigation and municipal-water demands and supply are modulated at these time scales as a function of climatic variability. Dec-cen climate variability is especially important because decisions on project sizing, as well as policies governing water-project operation, are based on an analysis of 10 to 30 years of data, and on the assumption that the underlying processes are stationary. These decisions undergo severe tests in the subsequent period. Look, for example, at the Colorado River compact that allocates the water of the Colorado River between California and other Western states. The water was initially allocated on the basis of flows over a prior 30-year wet period. (Decadal-scale variability in the Colorado River flow can be seen on the lower curve of Figure 5-17.) When Glen Canyon Dam was built on the Colorado, it was sized on the basis of data from a similar period. Not for 30 years, until the record El Niño event of 1982-1983, was the reservoir behind Glen Canyon Dam filled. It is now felt that the Colorado River is over-appropriated. A related situation involving changing flood frequency has occurred on the American River above Sacramento, California. The Folsom Dam, built in 1945, provides flood protection for Sacramento. However, the flood frequency in this basin has changed since its construction; eight floods greater than the largest flood in the 1905-1945 period have occurred since 1945. This has led to concern about the level of flood protection actually provided by the dam, and, more important, how flood risk should be analyzed. A better understanding of dec-cen variability at regional scales will clearly have a significant impact on the water-resources sector.

An adequate water supply is critical to the health of natural ecosystems, as well as to human habitation. Low-flow periods in streams lead to higher pollutant concentrations

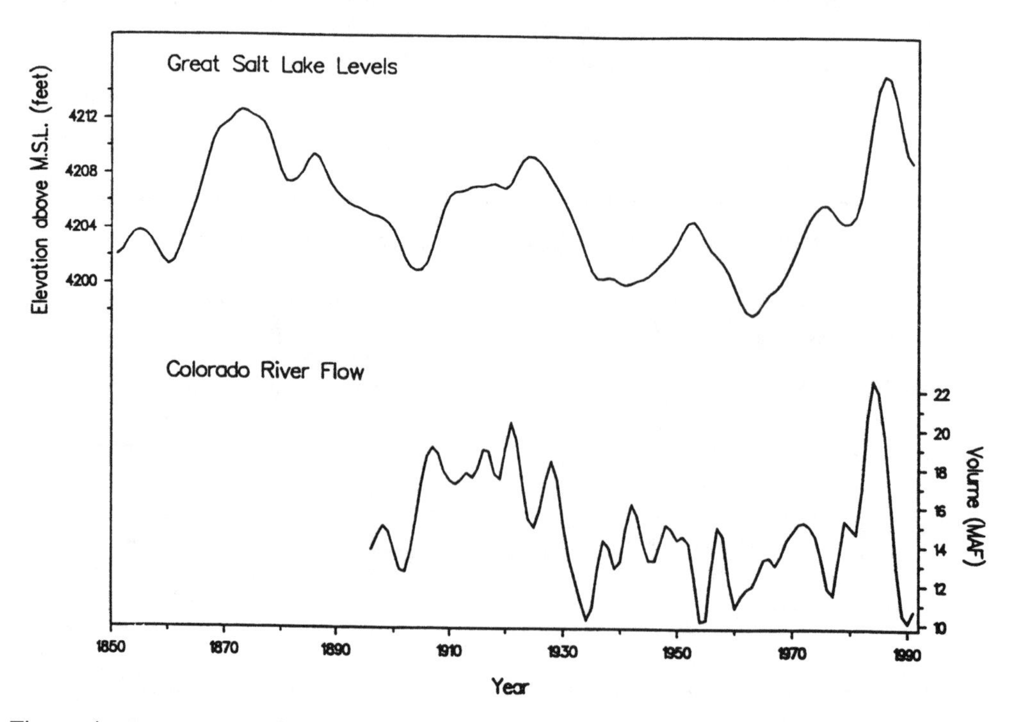

FIGURE 5-17 Time-series measurements of the level of the Great Salt Lake and the flow of the Colorado River. (From Diaz and Anderson, 1995; reprinted with permission of the American Geophysical Union.)

and to an increased potential for damage to aquatic habitat. This situation is exacerbated by human activity during protracted drought. Increased pumping of groundwater may further reduce baseflow in streams. This groundwater may be used for irrigation, in which case the surface or subsurface agricultural runoff will carry nutrients at high concentrations into streams. The threat to water quality from such conditions is the most critical in the last 200 years, given the changes in agricultural practices, the marginal water quality in many rivers, and the precarious balance between supply and demand, particularly in the midwestern and western states. Catastrophic impacts on the ecological, agricultural, hydropower, recreation, and municipal-use sectors would take place if a drought such as that of the 1930s were to recur.

The hydrologic cycle also influences ecosystems, because the supply of freshwater is central to terrestrial life. Because much of the evaporation from the land's surface is in fact water transpired from the soil to plants through their root systems and then evaporated from the stomata in their leaves, the total vapor leaving the land surface is referred to as evapotranspiration. In this respect, ecosystems influence the hydrologic cycle. Since the evapotranspiration over land is 71,000 km^3 per year, the 36,000 km^3 per year that remains from precipitation is available as surface water, or finds its way into the soils. Ultimately this surface water is captured in drainage basins (see Figure 5-18 for the major basins), and rivers transport it back to the oceans. Only about 10,000 km^3 per year of this surface water is available as global renewable freshwater resources serving all human, agricultural, and industrial purposes for the entire world's population (Cohen, 1995; Postel et al., 1996). It is of interest to note that if the entire world used the same amount of renewable freshwater per capita as the United States (2,100 m^3 per year per person), the globe could not support more than 5 billion people, which is less than its current population (Cohen, 1995).

Rind et al. (1997) show that under a doubled CO_2 scenario, vegetation changes would constitute only a moderate feedback to climate. However, such a warming's impact on ecosystems, particularly vegetation itself, could be dramatic because of hydrologic stress. In their GCM simulation with interactive vegetation, the impact of increased hydrologic activity is particularly enhanced over land, driving considerable evaporation from the vegetation through transpiration. The vegetation attempts to limit this drying by closing stomata. This tactic enables vegetation to survive short-term, drought-like conditions, but in the long run reduces productivity and eventually destroys the vegetation (particularly in lower latitude zones). This result is not revealed in simple GCM studies that do not include a treatment of vegetation strain response, such as those used in the first IPCC assessment (IPCC, 1990), though it is indicated in impact studies such as those used in the second IPCC assessment (IPCC, 1996a). Likewise, Sellers et al. (1996) reveal that changes in stomatal resistance under doubled-CO_2 scenarios feed back

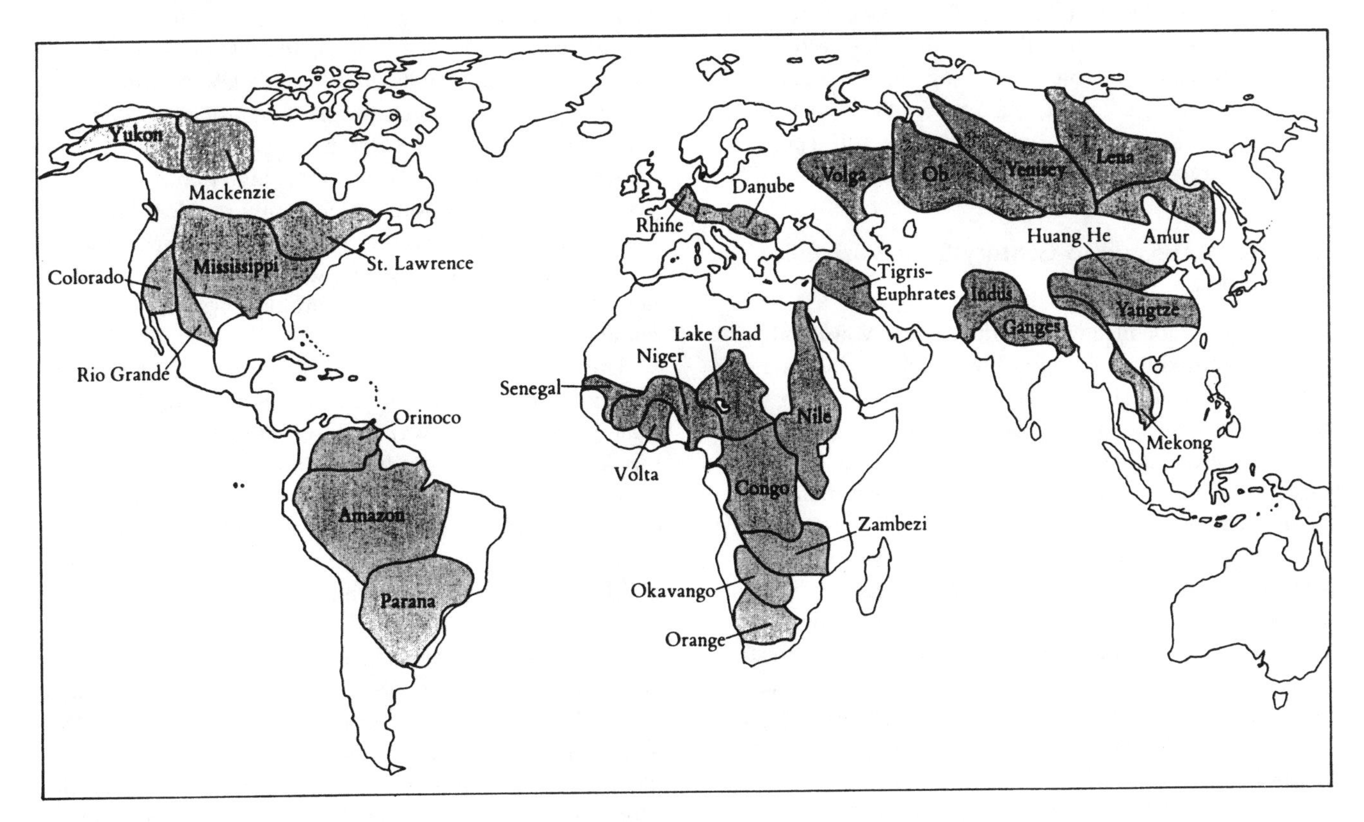

FIGURE 5-18 The world's major drainage basins. (From WRI, 1996; reprinted with permission of the World Resources Institute.)

into the hydrologic cycle. Betts et al. (1997) find that changes in the type and extent of vegetation canopy resulting from CO_2 fertilization and CO_2-induced climate changes would also have significant effects on the hydrologic cycle.

Finally, the hydrologic cycle influences global temperature, since water vapor is the Earth's primary greenhouse gas. That is, while carbon dioxide is probably the most notorious of the greenhouse gases, the real danger from increased carbon dioxide is the potential of altering the hydrologic cycle and increasing the amount of water vapor residing in the atmosphere beyond what is typically present. This increase could significantly amplify the relatively small direct warming associated with carbon dioxide.

The issue of upper-level water-vapor feedback was raised by Lindzen (1996). He argues that, in the tropics at least, upper-level water vapor will decrease, not increase, with the addition of greenhouse gases. Thus, in the tropics, the net water-vapor feedback from the entire atmospheric column could be weak or negative, rather than strongly positive (which would be the case if upper-level water-vapor concentrations were to increase). The total magnitude of the response to added greenhouse gases without a strong positive water-vapor feedback would be relatively small. It is difficult to make direct measurements of upper-level water vapor, however, and the results to date are contradictory (see Chapter 4 of the second IPCC assessment (IPCC, 1996a) for a summary of observations).

Consequently, the mean hydrologic cycle is inseparably intertwined with the mean climate of the globe. Since each of the processes composing the hydrologic cycle involves the other processes, the role of hydrology in the climate system is exceedingly complex, and a thorough understanding of its influences, feedbacks, and sensitivities still represents one of the largest challenges facing the climate community.

Evidence of Decade-to-Century-Scale Variability and Change

Estimates of the mean annual fluxes, reservoirs, and processes associated with the global hydrologic cycle are hard won and still relatively uncertain, and the year-to-year variability of the global cycle is poorly known. Not surprisingly, there is a paucity of information documenting the long-term variability of the hydrologic cycle. Variability is more easily documented locally than at larger scales; it shows up as variability of precipitation, but also as variability of water storage and runoff. While there is everywhere a fast component of rainfall variability because of the short time scales of the atmospheric hydrologic cycle, here we are more interested in the slower changes. Proxies for precipitation (such as outgoing longwave radiation and microwave emissions) can be used to give a global view of precipitation processes and variability (Rasmusson and Arkin, 1993). For very long time records (e.g., 100 years), however, station data for precipitation must be used, because most relevant satellite-based observations extend back little more than a decade into the past. An investigation of the adequacy of the U.S. and southern Canadian station networks by Groisman and Easterling (1994) has shown that useful climatological information and statistics can be gathered from the existing station data in those regions. Dai et al. (1997) show evidence of decadal precipitation variability in their global gridded dataset of monthly precipitation anomalies.

There is some evidence (Keables, 1989; Dettinger and Cayan, 1995; Rajagopalan and Lall, 1995; Mann and Park, 1996; Baldwin and Lall, in press; Jain et al., in press) that the expression of temperature, precipitation, and streamflow "seasonality" in the mid-latitudes has undergone significant changes at dec-cen time scales. While some of these changes in the intensity and timing of the annual cycles of warming and moistening may be related to ENSO activity, there are pronounced decade-to-century-scale trends that are unexplained. Such changes in seasonality are of considerable importance to water supply and demand and to water-resource management. The surface hydrologic system may amplify such changes in the underlying climatic state. One example is the advance in the date of the peak annual flood in the American River in California that has accompanied the previously mentioned recent increase in flood magnitudes (Dettinger and Cayan, 1995). Warmer temperatures earlier in the year have apparently led to warmer, moister air masses flowing into the area, and also to more precipitation falling in the form of rain rather than snow. This heavy rain, falling on the snow that generally exists early in the year, has exacerbated the flood problem. Interestingly, years with early peak annual floods tend to have less rain in the late spring and early summer, posing a significant problem for reservoir operators and water managers. Reservoirs may be emptied in response to concern about high floods early in the year, leading to a serious water-supply problem later. The limited capacity of the reservoir system and the complicated institutional rules that govern reservoir operation expose the lack of resilience of managed hydrologic systems to climatic variability.

The terrestrial regions in which decadal-scale hydrologic variability is most evident are the Sahel and the United States. (Precipitation variability is extremely important for the Asian monsoon, but that variability is primarily interannual and is not discussed here.) In the United States, hydrologic variability is manifested not only as precipitation variability, but also as variability of streamflow and of storage. Figure 5-17 shows the temporal variation in the level of the Great Salt Lake as well as the flow rate of the Colorado River since the nineteenth century. The dramatic rise of the Great Salt Lake in 1983-1987, and the even more dramatic rise of Devil's Lake, North Dakota, in 1982-1998, reflect the ability of the surface and subsurface hydrology to integrate the dec-cen climate signal. The American River floods also show a change in the seasonality of floods at dec-cen time scales, and both the timing and the change in flood frequency can be explained in terms of the large-scale climate fluctuations at dec-cen time scales (Lall et al., 1998). Even average precipitation years in such a run of events can lead to dramatic effects in the surface hydrologic system. Such hydrologic changes can be large enough to have significant impacts on many systems (e.g., agriculture, natural ecosystems, and hydroelectric power generation) that are important to society.

Not only do these direct observations indicate decadal-scale variability, but paleoclimatic records provide ample evidence of prolonged episodes of extreme hydrologic conditions. Dune fields (now vegetated) that were active during the past 1,000 to 10,000 years (Madole, 1994), tree stumps in modern natural lakes (Stine, 1994), long periods of slow growth revealed in drought-sensitive tree-ring records (Cook et al., in press), and more localized records (see, e.g., Laird et al., 1996; Madole, 1994; Muhs and Maat, 1993) all indicate that droughts more intense and prolonged than the U.S. Dust Bowl period have recurred throughout the past few millennia. Ancient dune fields in the Great Plains, now covered by vegetation, imply that the ability of this region to maintain vegetative cover without human intervention is marginal (Madole, 1994). There are indications of droughts in California at approximately the same time as those in the Great Plains (Stine, 1994). Evidence supports the existence of extreme droughts in Patagonia during these common periods, suggesting a common global cause for all these droughts. Tree-ring records from California spanning several millennia indicate periods of below-average rainfall lasting on the order of a millennium (Hughes and Graumlich, 1996). It is currently believed that discharges from the melting ice pack that covered much of the mid- to high latitudes of North America during the last glacial maximum significantly influenced thermohaline circulation, and thus the temperature, of the North Atlantic. Modulations by iceberg discharges (Heinrich events) are thought to have caused significant punctuations in the climate record of this region throughout the glacial period (see, e.g., Broecker, 1994, and references therein).

Mechanisms

The hydrologic cycle varies as a function of internal, coupled, and external mechanisms. Regarding the latter, it is generally agreed that the addition of radiatively active gases to the atmosphere will warm the Earth's surface (the degree of warming is in dispute) and significantly speed up the hydrologic cycle, so that the total global evaporation and precipitation should both increase. This is one of the most consistent predictions of change in models simulating anthropogenic greenhouse warming. Rind et al. (1997) suggest that a doubling of atmospheric CO_2 would increase the precipitation and evaporation rate by approximately 10 percent, resulting in an increase in total atmospheric water vapor of 30 percent. Unfortunately, models probably cannot adequately estimate the net change and distribution of water vapor in the atmosphere, and not enough global data exist to confirm or refute their predictions. Trenberth (in press) does find evidence of increases in water vapor and in extreme precipitation events coincident with the recent global warming. Presumably, a change in the surface-radiation budget associated with a reduction of surface irradiation because of volcanic eruptions could drive a similar response in the hydrologic cycle.

Coupled mechanisms affecting the hydrologic cycle are easily illustrated by the well-known examples of variability mentioned previously, such as the precipitation variability in the Sahel and United States. Such coupled mechanisms have also been observed in several other regions, including the Brazilian Nordeste and Australia (see, e.g., Ropelewski and Halpert, 1987). Rainfall variability in both the Sahel and the United States has been correlated with SST variability in the oceans. For instance, Sahel rainfall has been correlated with SST variability in the tropical Atlantic (Folland et al., 1986). Much of this SST variability appears to be correlated with the Atlantic dipole's oscillation about the mean location of the intertropical convergence zone (ITCZ) in the Atlantic (Houghton and Tourre, 1992), influencing the location of atmospheric convection and the trade winds. However, SST in other regions of the globe has also been related to precipitation variability in the Sahel (Folland et al., 1991). Many of the SST-related processes in the Atlantic that seem to affect the location of the ITCZ on interannual time scales also seem to affect it on decadal time scales. Even less clear is why something in the ocean that varies primarily interannually should affect rainfall in the Sahel on longer time scales. At least two possible explanations have been offered: First, the rainfall and vegetation in the Sahel may be connected through a positive feedback, allowing them to produce longer-term variability than would exist without this feedback. Second, SST in other parts of the globe, especially the Indian Ocean, may vary on longer time scales than in the Atlantic, producing longer, decadal-scale modulations of Sahel rainfall (Ward, 1992).

Changes in the rate of the hydrologic cycle are expected to influence the budgets of precipitation and evaporation, and the linkages between these two processes can produce internal modes of variability. More complicated internal mechanisms undoubtedly exist, but have yet to be fully described or documented. River runoff fluxes may play a fundamental role in Arctic climate and the thermohaline circulation; it has also been suggested that complex, large-scale feedback loops (see, e.g., Slonosky, et al., 1997) link atmospheric convection and circulation to the hydrologic cycle and Arctic river runoff, to sea-ice formation and freshwater transport, and to the thermohaline circulation. Such relationships need to be more thoroughly investigated.

Rainfall variability in the United States correlates with SST variations in the Pacific (Figure 5-19). Much of the U.S. precipitation variability is connected with SST in the North

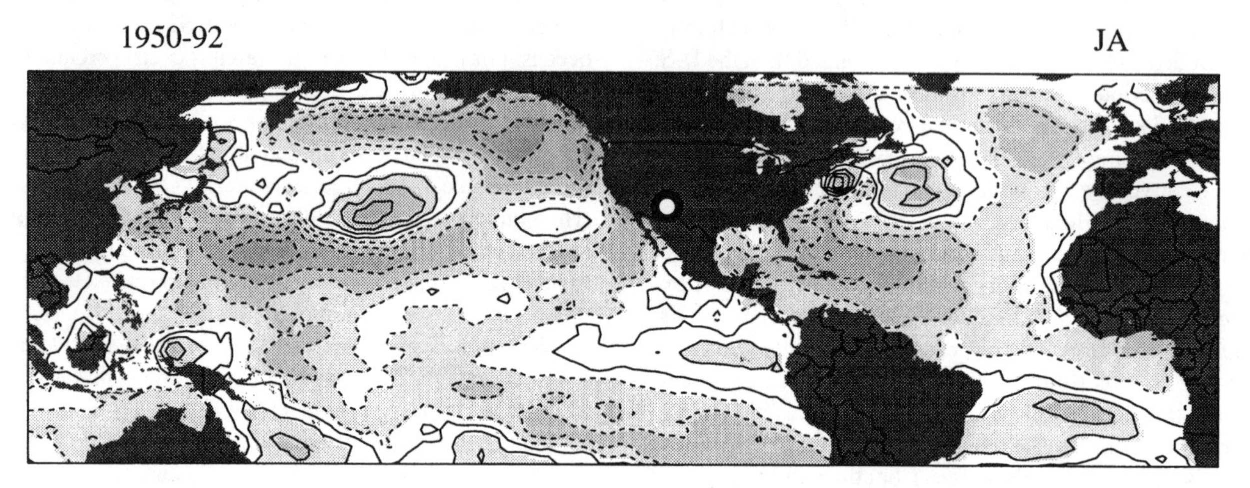

FIGURE 5-19 Correlation between U.S. rainfall at two sites, U.S. Great Plains (top panel) and Arizona (bottom panel), and SST over the period 1950-1992. Rainy-season months included are noted above each panel. Positive and negative correlation coefficients are indicated by solid and dashed isopleths, respectively (contour interval is 0.1). The SST-precipitation correlations tend to be strongest (both positive and negative) in the Pacific Ocean, with some exceptions. (Figure courtesy of T. Mitchell, JISAO/University of Washington.)

Pacific (Ting and Wang, 1997) and in the eastern and central equatorial Pacific (see, e.g., Ropelewski and Halpert, 1987). The ENSO phenomenon plays the most important role in producing this SST variability, and as a result of extensive developments in our understanding of this process and ability to model it, SST in the eastern Pacific can now be predicted with considerable skill a year in advance. Recently it has become more certain that ENSO itself is modulated by decadal processes of unknown origin, as discussed in Chapter 3. Since the teleconnections between the tropical (or subtropical) Pacific and the United States take place on relatively rapid (e.g., interannual) time scales, it seems likely that decadal modulations of SST will produce decadal modulations of precipitation in the teleconnected regions. As of this writing, it is clear that the decadal variability of ENSO has a broader meridional scale than its interannual variability (Tanimoto et al., 1993; Zhang et al., 1997), but it is not clear whether its dynamics are at all related to the interannual dynamics. Understanding the decadal variability of SST in the Pacific offers the best hope for skillful predictions of U.S. precipitation over medium and long time scales.

In most cases, it is not obvious how teleconnections influence drought. The mechanisms that initiate multidecade or multicentury droughts and enable them to persist. Feedbacks from vegetation and long-term SST anomalies that influence the supply and delivery of moisture are probably involved. However, no obvious mechanistic explanation is apparent even for the interannual-to-decadal droughts documented by instrumental records (e.g., the Dust Bowl of the 1930s and the killing droughts of the Sahel in more recent decades).

In contrast to droughts, floods are often thought of as high-frequency events caused by precipitation outbreaks with a duration of hours to days. However, flood (and the associated erosion and sediment yield) potential depends strongly on antecedent moisture conditions, on the baseflow in the stream, and on the state of the vegetation in the basin.

Each of these factors will depend on the long-term state of the climatic forcing. A condition that leads to a wet period characterized by a sequence of moderate storms may result in a higher flood potential than a period with an intense storm. Likewise, anthropogenic changes in land use in a watershed over dec-cen time scales can significantly alter its response, and its water and sediment production potential. The interaction between anthropogenic and climatic factors at these time scales will be key to better flood-plain management.

The deeper groundwater reservoir responds at time scales of tens to thousands of years to climatic perturbations—rather like the response time of the ocean. Baseflow in streams represents a discharge of groundwater to the surface-water system that, because of its slow response time, is inherently characterized by low-frequency dynamics. Human influences also significantly modulate regional groundwater resources. Groundwater pumping, which tends to be greatest in dry periods, has led to significant declines (>100 m) in regional aquifers (often covering greater than 106 km^2), such as the Ogallala aquifer in the central United States. These human and natural processes affect stream-aquifer interactions on dec-cen time scales. Moreover, the temporal distribution of the global hydrologic budget between the atmospheric, oceanic, surface, and subsurface reservoirs is likely being significantly affected as large amounts of groundwater are withdrawn and deposited in the other reservoirs. The resultant impacts on atmospheric water vapor, cloudiness, and regional precipitation are largely unknown.

Predictability

Prediction of the global hydrologic cycle is critical to our ability to predict the magnitude and distribution of anthropogenic greenhouse warming, because the hydrologic cycle influences greenhouse warming through a number of feedbacks, including those involving the distribution and amount of atmospheric water vapor. Successful prediction will require improved understanding of the intricate processes and atmospheric feedback mechanisms controlling water-vapor content, as well as better observations against which representations of our understanding can be tested. At present we cannot say how successful we will be in predicting the response of the hydrologic cycle to externally induced change, because our understanding of the present functioning of the hydrologic cycle is incomplete.

The directions of the feedbacks among the several elements of the interacting ecological, hydrologic, and climatic systems remains the source of some debate. The multi-scale nature of the dynamics serves to render the problem complex. However, to the extent that precipitation variability over land depends on SST variability, there is some hope of predicting precipitation several years in advance. SST predictions with this type of lead time might be possible if the ocean can be initialized accurately in models and if atmospheric "noise" does not overly contaminate the "signal" that exists in the depths of the ocean and eventually works its way to the surface. The atmospheric "noise" in the tropical Pacific is small, and SST in that region has been shown to affect rainfall and temperature over the United States. Fore-

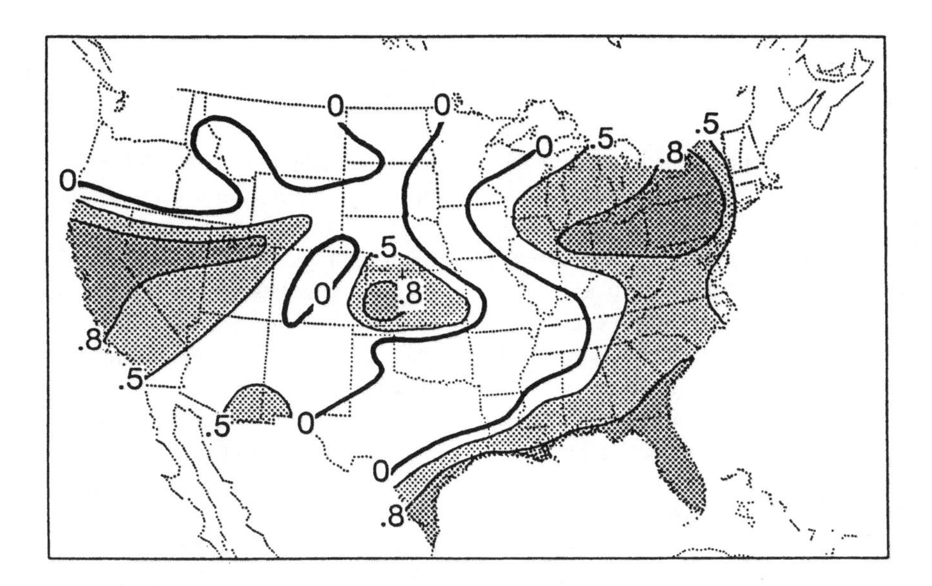

FIGURE 5-20 Correlation skill for 7-month forecasts of U.S. precipitation for January-February for seven strong ENSO events made with the Scripps-Max-Planck-Institut coupled model. Skill is expressed as a correlation between model forecasts and observations. (From Barnston et al., 1994; reprinted with permission of the American Meteorological Society.)

casting methodologies being developed on the basis of predicted SST in the tropical Pacific (Ji et al., 1994; Barnston et al., 1994) are already showing reasonable skill in forecasting U.S. precipitation two seasons in advance (Figure 5-20). More recently, Ting and Wang (1997) have found a relationship between mid-latitude North Pacific SST and U.S. precipitation. Dettinger et al. (1993) have shown a correlation of streamflow in the northwestern and southwestern United States with extreme ENSO events that may improve our ability to forecast precipitation and runoff, under certain conditions. Lall et al. (1998) have provided evidence for the strong correlation between dec-cen variations of ENSO and like variations in the streamflow of the San Juan and other rivers in the western United States. In addition, Lall and Mann (1995), Mann et al. (1995a), and Moon and Lall (1996) have connected fluctuations in the level of the Great Salt Lake to quasi-periodic variability organized at ENSO, decadal, and interdecadal time scales in the hemispheric circulation. Lall et al. (1996) and Moon (1995) have also provided remarkably successful multi-year (2-4 years ahead) forecasts of the level of the Great Salt Lake, using nonlinear time-series analysis methods on the historical time-series observations of the Great Salt Lake together with selected climate indicators. To take full advantage of such advances and to extend the range of hydrologic predictions over the United States will require the ability to predict SST further ahead, using comprehensive coupled models accurately initialized with ocean data.

For the Sahel, some skill seems to exist at ranges of a year or so, in the regular precipitation forecasts issued by the U.K. Meteorological Office, which utilize existing SSTs as predictors and assume that SSTs do not change during the period (Folland et al., 1991; Hulme et al., 1992). Again, since the prediction of precipitation changes is intrinsically a coupled atmosphere-ocean problem, greater skill in predicting Sahel rainfall will be realized through the use of more comprehensive dynamic coupled models that include SST predictions.

Additional factors that are important in predicting the hydrologic cycle are cloud physics and cloud-radiative feedbacks, which encompass a variety of fundamental processes that control or influence the formation, distribution, and evolution of clouds, in both the liquid-water and ice phases. The response of clouds to a change in the hydrologic cycle, and the subsequent impacts of that response, cannot be accurately predicted until the aforementioned cloud processes are understood well enough to be more realistically represented in models. Although cloud processes (e.g., cloud formation and changes in structure) operate on very short time scales, systematic changes in their character can have long-time-scale implications. Because clouds are thought to represent a sensitive component of the climate system, understanding them is critical to the goal of climate prediction on dec-cen scales. We need to better comprehend their temporal and three-dimensional spatial distribution and internal properties, including multi-layering and the cloud characteristics that affect the radiation balance.

Remaining Issues and Questions

To understand the hydrologic cycle better, we need to improve both our knowledge and the model representations of the processes controlling the rates, pathways, storage, and redistribution of water in all its forms in the hydrologic cycle. One of the dimensions of the WCRP's Global Energy and Water Cycle Experiment (GEWEX) program is an investigation of the detailed land-surface hydrology in major drainage basins. (The Mississippi basin is the primary U.S. focus, but concurrent land experiments are proposed for other parts of the world.) The basic issue is the parameterization of aspects of multi-scale land-surface properties and processes that are reliable across a wide range of climatic conditions and are appropriate for use in relatively coarse-scale models (e.g., NRC, 1998b). Among the most urgent questions are the following:

- *What are the patterns of and mechanisms causing prolonged drought on dec-cen time scales*? Paleoclimatic records reveal ample evidence of climate extremes that persisted for many decades or centuries in regions that experience quite different conditions today. Drastic consequences would ensue to such vulnerable and populous regions as California if such droughts were to recur during present times. An understanding of the conditions that could lead to such a situation, and its expected severity and duration, would be very valuable. The hydrologic aspects of an analysis to assess the initiation and persistence of droughts would include a better accounting of the global water balance, of changes in land-surface conditions, and the human use and natural drawdown of the near-surface and deeper reservoirs. A thorough examination of the patterns of past hydrologic variability, and the testing of plausible mechanisms with focused observational and simulation strategies, should lead to a better predictive understanding of this critical climate feature. An interesting aspect of this research would be the reconciliation of natural and human dynamics as they affect the hydrologic cycle. At present no research programs are investigating these issues at dec-cen time scales.
- *How do the distribution of water vapor, precipitation, and clouds respond to and interact with surface boundary conditions and changes in forcings on dec-cen time scales*? The hydrologic cycle plays a poorly defined role in producing and responding to the patterns of climate variability we have identified in Chapter 3. These patterns and large-scale precipitation seem to co-vary, however, so efforts to use knowledge of climate patterns to predict regional precipitation should be enhanced. There is clearly a need for better documentation and understanding of the nature and sensitivity of these co-varying relationships, and for establishing and refining the mechanisms responsible for driving them.

In addition to improving model representations of the hydrologic cycle, we must also better document its temporal variability. The distribution of water vapor in the atmosphere, particularly the vertical distribution, needs to be better understood, because considerable controversy surrounds it. One theory about anthropogenic enhancement of the greenhouse effect suggests that increases in moisture in lower levels of the atmosphere will be offset by decreases in the upper troposphere, reducing surface warming (and its moisture increase) while cooling upper layers. If this offset does indeed occur, it would greatly reduce the net warming that would otherwise be anticipated to result from an increase in total atmospheric water vapor (see, e.g., Lindzen, 1996). As mentioned previously, direct measurements of water vapor in the upper levels of the atmosphere are difficult to make, and the results have been contradictory. Clearly, accurate treatment of upper-level water vapor is essential to realistic modeling of the climate system's radiative response to anthropogenic increases in greenhouse gases (and other external forcing factors), and to reliably estimating the greenhouse-warming response.

• W*hat combination of remote and in situ observations can be used to measure the large-scale distribution of precipitation and evaporation on dec-cen time scales*? Precipitation is vital to nearly all of society's activities, and thus to the global economy. It is also a significant expression of dec-cen variability. While observed covariations suggest that teleconnections, such as those associated with the NAO and PNA, directly influence land precipitation, measurements that would confirm these relationships are poor, and are being made only sporadically. Existing global observation climatologies significantly disagree among themselves, so that not even a baseline of large-scale evaporation has been established. (A recent GEWEX initiative, the Global Precipitation Climatology Project (GPCP), now provides a baseline for precipitation.) Satellite measurements have now provided a global view of radiative proxies of precipitation, but the calibration needed to translate the proxy fields into precipitation is just starting to become credible, partly because of the GPCP results. In situ measurements are helping to quantify the radiation measurements made by satellites, which will permit absolute calibration of satellite measurements, as well as providing spatial-gradient information. Both the in situ and satellite measurements need to be continued in order to provide optimal estimates of large-scale fields of precipitation. Measuring evaporation is an even more difficult problem, though over the oceans a metric for precipitation-minus-evaporation (P-E) may ultimately be realized through monitoring of surface salinity fields by autonomous platforms. Knowledge of these fields is also critical for determining the ocean thermohaline circulation, as discussed in the next section. For measurements on dec-cen time scales, it is clear that a combination of remote and in situ systems must be designed to ensure long-term consistency and accuracy—which implies a commitment to long-term intercalibrated measurements.

• *What spatial/temporal changes occur in the storage and pathways of land water, including the flux of water to the oceans, over decades and centuries*? The buoyancy state of the ocean (defined by temperature and salinity) determines the water-transformation properties of the ocean, the rates of deep- and bottom-water production, and therefore ultimately the transport properties of the ocean and the stability of its internal oscillations (see the next section, "Ocean Circulation"). Crucial to these processes is the geographic distribution and amount of freshwater inputs—directly, by precipitation over the ocean, and indirectly, by input by runoff, groundwater discharge, and discharge of glaciers and other forms of land snow and ice, either through rivers or through ground discharge. The inputs' geographic distribution depends on the locations of rivers and their respective flows, and on the amount and distribution of groundwater discharge. River and groundwater flows are determined by the relative amounts of precipitation and evaporation over time in drainage basins. (The detailed paths of water on land, and how to model these pathways on the catchment level, are major themes of the GEWEX program.) When freshwater reaches the ocean, it affects sea level, regional ocean temperature, and chemistry, including salinity and alkalinity. Freshwater and thermal inputs alter the ocean's buoyancy state and water-mass transformation properties, as these inputs are mixed and advected throughout the oceans.

Research is needed that integrates our evolving understanding of the spatial aspects of runoff generation in a watershed with the temporal aspects of climatic variability and watershed urbanization. Ecological aspects related to changes in the vegetation at these time scales (see the "Land and Vegetation" section of this chapter for more discussion of this topic) and the resultant change in sediment yield also need to be investigated. A related and important area of research is the development of procedures for decision and risk analysis that explicitly recognize the nonstationarity in the flood process, as opposed to the historical institutional framework, which has been predicated on the assumption of a "steady state" or static risk of floods.

Observations

As noted earlier, precipitation is the key variable for hydrology. For most studies of dec-cen variability and its effects, we need global fields of precipitation over those same time scales. So far we have only regional instrumental records of that length. In the future, if we are to relate precipitation to global boundary conditions, simultaneous measurements of sea-surface salinity and temperature, vegetative ground cover and soil moisture, sea and land ice, and snow must be made. Climate models suffer from a lack of precipitation data for evaluation purposes. Information on precipitation, both large- and small-scale, and the fate of hydrometeors ejected from clouds in the upper atmosphere (which evaporate to become vapor rather than fall as precipitation) is crucial if we are to understand the processes governing the hydrologic cycle and its role in determining climate. The regions in which the surface hydrology is most sensitive to anthropogenic and natural climatic variability need to be identified. The paucity of long time-scale records is a primary difficulty in conducting all these analyses. The development of proxy records using tree rings and other surrogates will be very valuable in this regard.

Processes and Parameterizations

In order to develop robust model parameterizations, additional understanding is needed of evapotranspiration, the dynamics of vegetative cover, and the dynamics of soil moisture. Accurate simulation of terrestrial water flow and storage is a difficult modeling problem, particularly because water pathways depend on conditions far more local than climate models can currently resolve or are likely to be able to resolve in the foreseeable future.

Better parameterizations of the relationships of water vapor, precipitation, and clouds are required in order to improve their representation in climate models, because their responses and feedbacks are critical to the long-term climate response to changes in forcing and concomitant changes in the fundamental climatic state. Process and larger-scale studies are needed to help us better understand the coupled relationship between the ocean surface's temperature and salinity fields and the overlying P-E fields. Likewise, the interdependencies of soil moisture, soil and vegetative cover characteristics, and P-E must be better understood and represented in models. Improvements in precipitation prediction can likely be made through the use of more comprehensive dynamic coupled models (e.g., that of Stockdale et al., 1998), because precipitation variability and change are so tightly interwoven with the entire climate system. In addition, low-order, statistical-dynamic models will be useful in improving our mechanistic understanding of the non-linearities of climate-hydrology interactions on decadal time scales.

Empirical assessments are needed of the sensitivity of regional hydrologic extremes (high-frequency floods and persistent droughts) to decadal and longer climate variations identified in specific oceans and the atmosphere. Improvements are needed in statistical methods used to analyze space-time data for low-frequency components, using limited records. Fingerprinting moisture sources and transport mechanisms that correspond to different phases of quasi-decadal climatic oscillations may be useful for understanding regional hydrologic responses at dec-cen time scales. An interesting aspect of these analyses is that there is strong evidence that high correlations at decadal time scales can exist between distant locations (e.g., precipitation in the Sahel and the American Southwest), but not necessarily between areas that are geographically closer (but perhaps "distant" in terms of the operative climatic mechanisms).

The question of how regional flood potential varies in response to the dec-cen climate variation patterns is of great interest. Many hydrologic parameters, including the flood potential, reflect an interaction of slow and fast time scales of large-scale climate and regional hydrologic evolution. To advance our knowledge of how regional flood potential may change, a better understanding is needed of how the potential for storms is changed in response to the larger-scale spatial-flow configuration and the local energetics. In addition, a better understanding is required of how the evolving antecedent moisture and baseflow conditions and vegetation on the land surface alter the dynamics of flood generation in a watershed.

Long-term hydrologic forecasts need to be predicated on the state of the large-scale climate system. This state may be represented through regional or local projections of the phase and amplitude of large-scale space-time oscillatory phenomena. Nonlinear, multivariate time-series modeling approaches accounting for the relationships between atmospheric pressure and precipitation, and between precipitation and flow, may be useful in this regard. However, there is also a question whether purely statistical forecasting approaches will be adequate even if they recognize nonstationary behavior. Forecasts are meaningless without an accompanying understanding of the dynamic processes. Once again, a conceptual framework for the multi-scale evolution of the hydroclimatic state is needed to understand what variables are useful for forecasting and how they should enter into forecasting models, as well as to identify the limits and nature of predictability. Some of the key long-term hydrologic forecasting questions include: Are there regimes that are inherently predictable or unpredictable? How should ensembles of forecasts be constructed? Can it be assumed that hydrologic predictability depends solely on conditions at a given time, not on how they evolved? How should one interpret probabilities—as measures of uncertainty, or as a quantification of the relative local frequency of a climate phenomenon?

OCEAN CIRCULATION

The Earth's present climate is intrinsically affected by the ocean. The ocean covers 70 percent of the Earth's surface to an average depth of about 4 km, and without this ocean, the climate would be different in many essential ways: Without the evaporation of water from the sea surface, the hydrologic cycle would be different; without the ocean's heat transport and uptake, the temperature distribution of the globe and thus the atmospheric circulation would be different; and without the biota in the ocean, the total amount of carbon in the atmosphere would be many times its current value. Yet, while we may appreciate the role of the ocean in climate, the difficulty and expense of making measurements below the ocean's surface has rendered the vast volume of the ocean a sort of mare incognita.

The ocean circulation is usually divided into a wind-driven component, which is confined predominantly to the upper ocean (Color plate 2), and a component that results from variations in density (i.e., temperature and salinity) called the thermohaline circulation, or THC (Color plate 3). While this division cannot be made rigorously (each circulation influences the other in essential ways), it is a convenient one. The THC originates as dense surface waters plunge deep into the sea carrying surface properties, especially heat, salt, and gases (such as carbon dioxide), along with them. Water gets dense enough to sink into the deep sea in only in a few, unique regions of the world, predominantly in the high latitudes where cold water is produced by contact with the frigid atmosphere. Its density increases further as it grows salty by sea-ice formation. Despite the tiny areas of the world's oceans that are thermohaline source regions, the transports involved are considerable, on the of order 20 million m^3 sec^{-1}—more than the rate of rainfall onto the entire surface of the Earth.

The thermohaline circulation situates dense water under light water, and is therefore responsible for maintaining the stable stratification of the world's oceans. The water that sinks must be replaced by water at the surface levels of the ocean. This combined circulation—sinking of cold water that works its way deep into the ocean, and its replacement at the surface by water that makes its way to the sinking regions—leads to large oceanic heat transports. A glance at Color plate 3 shows two important characteristics about the THC: It reaches to various depths (intermediate, deep, bottom), and the Antarctic Circumpolar Current blends the outputs of the THC from the world's major basins, redistributing their waters to other basins.

Variations in density play a strong role in the surface circulation as well, since the density gradients determine the volume of water contained in the surface waters, and thus the mass of water that is moved by the surface winds. This volume determines how much temperature change will result in surface waters for a given flux of heat across the air-sea boundary. Therefore, the circulation systems and property distributions of the deep and surface waters are intimately linked, and constantly interact. This link is important, since it implies that the surface properties of the ocean, which play a direct and immediate role in climate, are controlled by both surface and deep-ocean processes.

Ultimately, processes in the ocean affect the atmosphere by exchanges through the air-sea interface. Therefore, ultimately, we have to understand how the surface and thermohaline circulation systems operate separately and together to influence the surface, and how properties from the ocean's interior affect the ocean's surface. In general, as we consider longer time scales, the lateral distances over which the ocean can transport heat, and the extent to which interactions between the surface and deep waters can occur, we must begin to consider the ocean's various processes through more and more of its volume. For example, the deeper the point at which a parcel of water begins to rise, the longer it probably takes to reach the surface. Thus, as the time scale of interest lengthens, the region of the ocean that can affect the surface expands downward. For the decadal-to-centennial time scales of interest in this document, the region from which the water can affect the surface is not at all well known, although it is probably not the entire ocean.

It should be noted that water descending from the surface to the interior can proceed into deep regions on relatively rapid time scales, at least in water-mass transformation regions of the ocean (see, e.g., Smethie et al., 1993), and this water can produce variability at deeper levels of the ocean. Since our interest in this report is how the ocean affects climate variability, we concentrate mainly on those processes and those regions of the ocean that can affect the atmosphere on dec-cen time scales.

Influence on Attributes

The ocean circulation participates in the climate system through three primary agents:

- exchange of heat, water vapor, and carbon dioxide with the atmosphere and cryosphere;
- sequestering of heat, freshwater, and carbon dioxide at depth in the ocean and sediments for long times before possible return to the atmosphere;
- redistribution of heat, freshwater, and carbon dioxide through the action of large-scale ocean currents (surface and deep), which subsequently affects the distribution of these constituents in the atmosphere.

The ocean influences the climate attributes discussed in Chapter 2 either directly, through these primary agents, or indirectly, through the primary agents' altering some aspect of the climate system that affects one of the attributes. For example, the heat capacity of the ocean is such that the upper 5 to 10 m of water contains as much heat (enthalpy relative to freezing) as the entire column of air overlying it. Therefore, any exchange of heat between the ocean and a stationary column of air must result in a significant change of air temperature relative to a very minor change in ocean temperature (unless the ocean's stability is such that this exchange occurs only at its "skin"). Consequently, the substantial involvement of the ocean in absorbing and storing heat from the atmosphere, moving the heat great distances, and subsequently returning it to the atmosphere is central to determining the regional and global distribution of temperature in both the atmosphere and ocean. This involvement also implies that SST is most easily changed not by altering the rate of exchange with the atmosphere (ignoring the radiative transfers), but by changing the volume of surface water involved in the exchange. This can happen through alteration of the ocean's vertical stratification through any number of dynamic or thermodynamic means.

The ocean provides the primary source of moisture for the Earth's precipitation. Also, the ocean's circulation moves salinity anomalies, making it instrumental in adjusting the regional imbalances between precipitation and evaporation. In other words, the oceans prevent the positive precipitation-minus-evaporation (P-E) values over the mid-latitude oceans from making this region's surface water increasingly fresh, and also prevent the negative P-E balance over the subtropical oceans from making that region increasingly saline.

Storms are a critical coastal issue. The position and intensity of atmospheric storms over the oceans depend on atmospheric circulation and ocean temperature distribution (among other things). The high population densities in coastal regions are frequently exposed to tropical and large-scale frontal storms, which receive much of their energy from heat at the sea surface. The combination of wind, precipitation, and sea-level fluctuations during the passage of coastal storms can inflict large economic and human losses, such as the $3 billion to $6 billion in damages and 270 deaths caused by the blizzard of 1993 (Lott, 1993), which intensified as a result of air-sea interactions along the eastern coast of the United States. In addition to these vulnerabilities on land, there are others at sea. The distribution of severe maritime weather, sea ice, and anomalous ocean currents can be critical to the economic vitality and even the safety of maritime activities, such as transport, fishing (see, e.g., Kawasaki et al., 1991), and oil extraction (Epps, 1997). Also, the ocean's volume, dictated by its temperature and mass (which is dominated on dec-cen time scales by the quantity of water storage on land), is the primary agent in sea-level rise.

In addition to having these physical influences, the characteristics of the oceans obviously play a vital role in determining the nature and distributions of marine ecosystems. Changes in upper-ocean conditions (e.g., upwelling rate, temperature, salinity) may alter the food chain, and thus the locations and stability of marine biological communities.

Evidence of Decade-to-Century-Scale Variability and Change

Examples of dec-cen variability in the oceans are available from relatively short instrumental records at single stations; from diverse measurements collected at different locations over many decades (see, e.g., the salinity and temperature atlases of Levitus et al. (1994) and Levitus and Boyer (1994), respectively); and from proxy evidence recorded in ocean corals, ocean sediments, and ice cores and trees in areas adjoining ocean regions. Unfortunately, few ocean observation stations have been maintained over the time scales of interest, so our understanding of the behavior of the ocean on dec-cen time scales is fragmentary and incomplete. The three major oceans are discussed separately in the sections below. The Atlantic Ocean has a longer history and greater density of measurements (especially the North Atlantic) than any other ocean basin, so some examples drawn from this basin may have analogs in other basins that have not yet been detected.

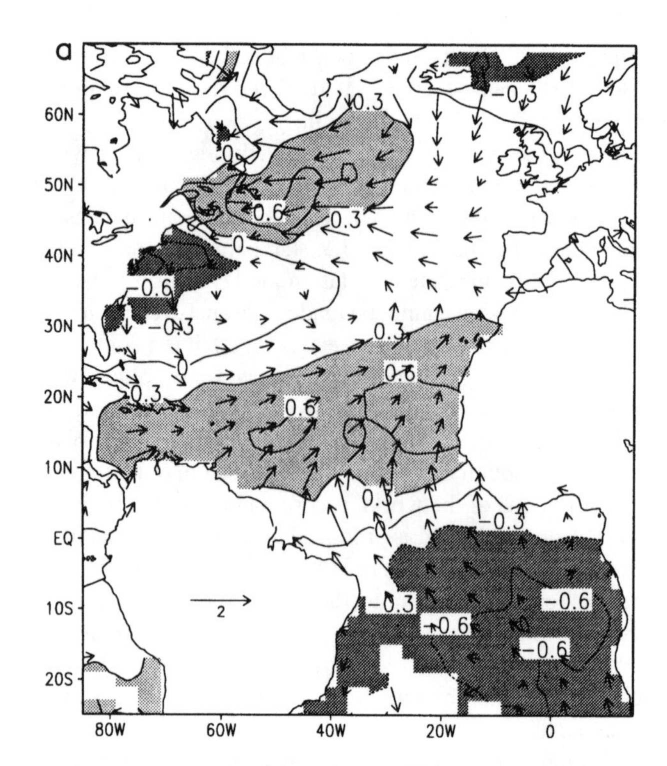

FIGURE 5-21 Basin-wide annual-average SST (°C) and wind-velocity (m s^{-1}) anomalies for six years of high (1969,1970, 1978, 1980, 1981, 1982) and six of low (1972, 1973, 1974, 1984, 1985, 1986) Tropical Decadal Oscillation index (defined as the difference in SST between 10-20°N and 15-5°S). (From Xie and Tanimoto, 1998; reprinted with permission of the American Geophysical Union.)

The Atlantic Ocean

Upper Ocean

SST is known to vary coherently on dec-cen time scales in the Atlantic. In the North Atlantic, the time scale of the cold and warm epochs that coincide with longer-term variations of the North Atlantic Oscillation (NAO) is relatively long, on the order of multiple decades (see, for example, the pattern shown earlier in Figure 3-5). In the subtropical Atlantic, the north-south dipole mode has a time scale closer to a decade (see, e.g., Mehta and Delworth, 1995). It has recently become clear that the variability in the North Atlantic co-varies with the subtropical dipole, and an Atlantic-wide oscillation may be present; Figure 5-21 shows the basin-wide connection between these modes (Xie and Tanimoto, 1998). This basin-wide banded pattern also correlates strongly with rainfall in northeast Brazil. Hansen and Bezdek's (1996)

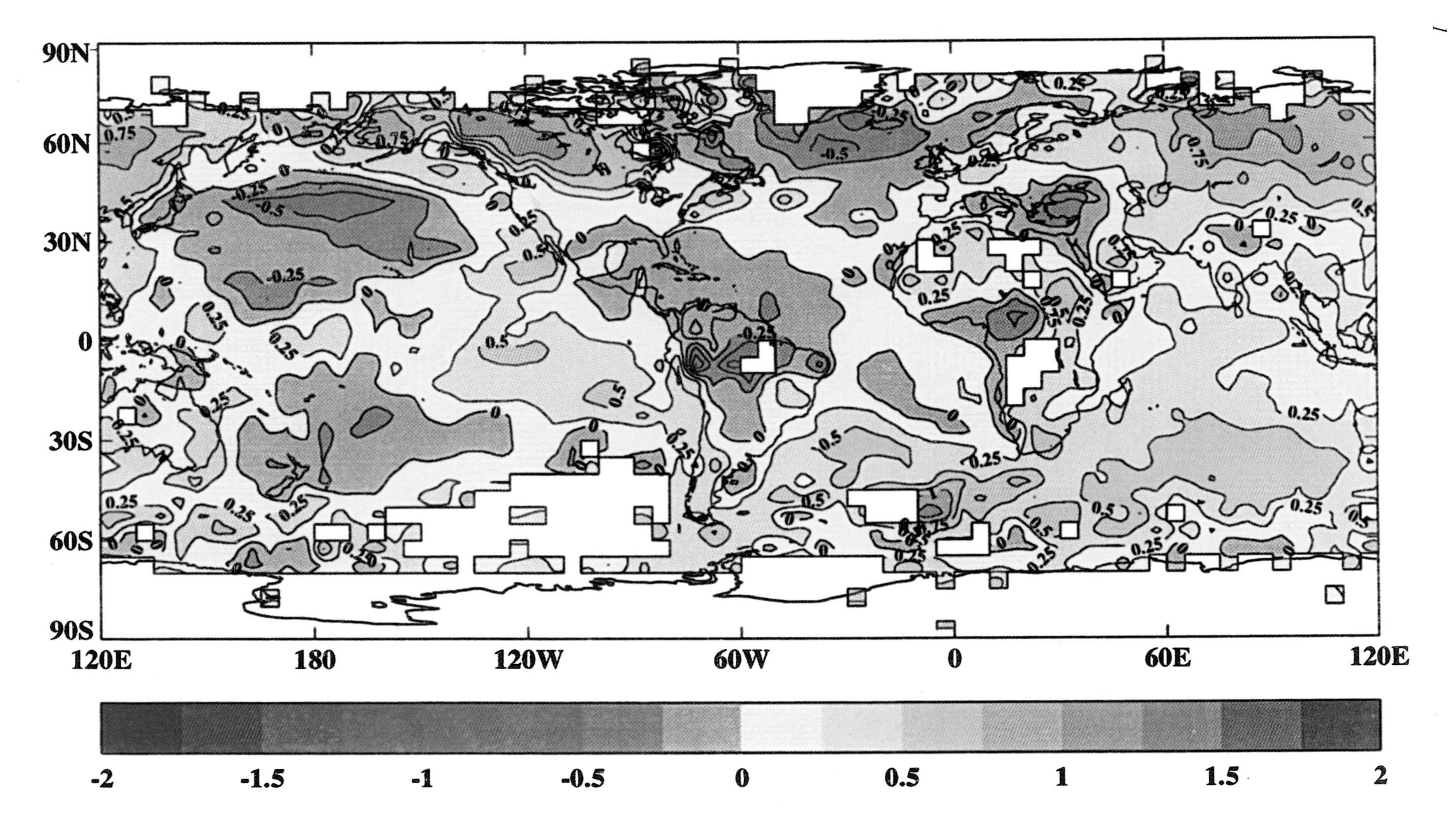

PLATE 1 Annual surface temperature in °C for 1975-1994 relative to 1955-1974. (From IPCC, 1996a; reprinted with permission of Intergovernmental Panel on Climate Change.)

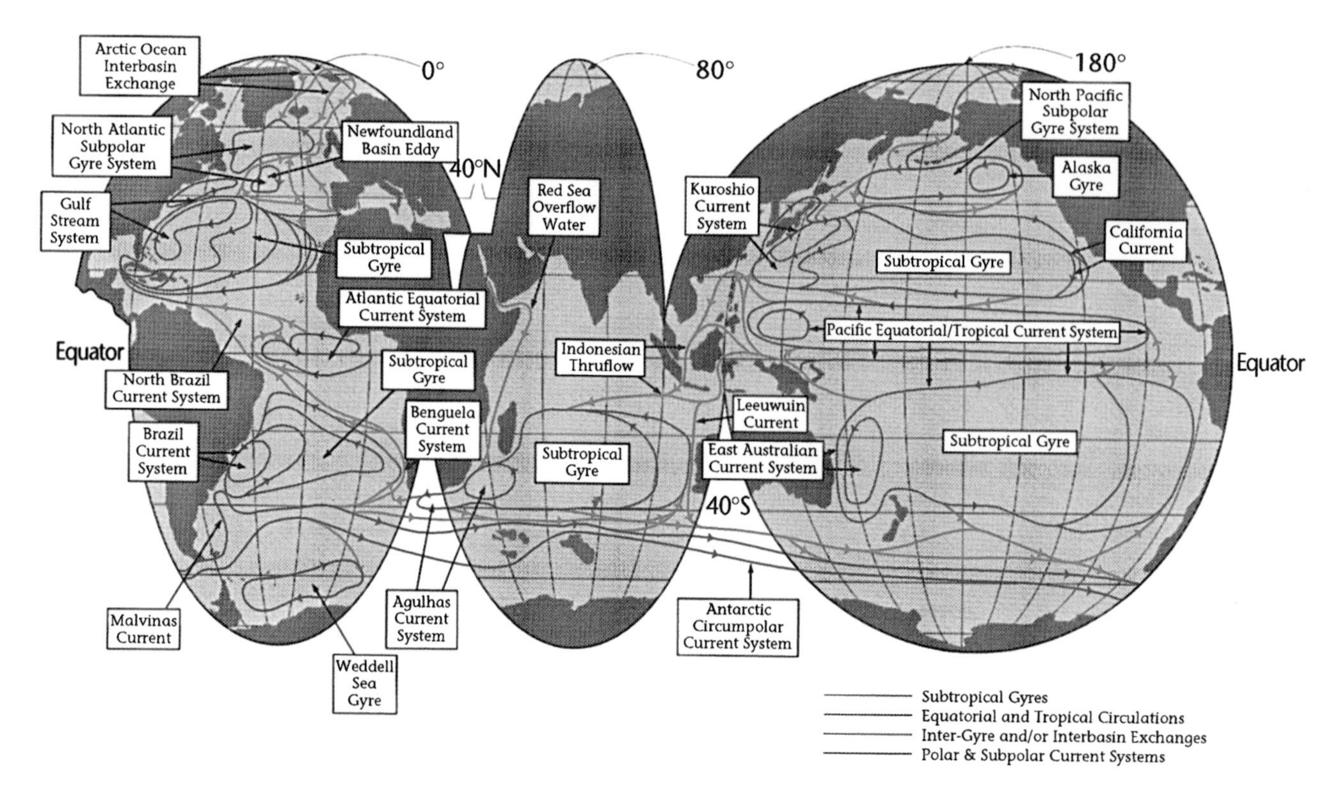

PLATE 2 The global distribution of upper-ocean flow, which is primarily wind-driven. (From Schmitz, 1996; reprinted with permission of the Woods Holes Oceanographic Institution.)

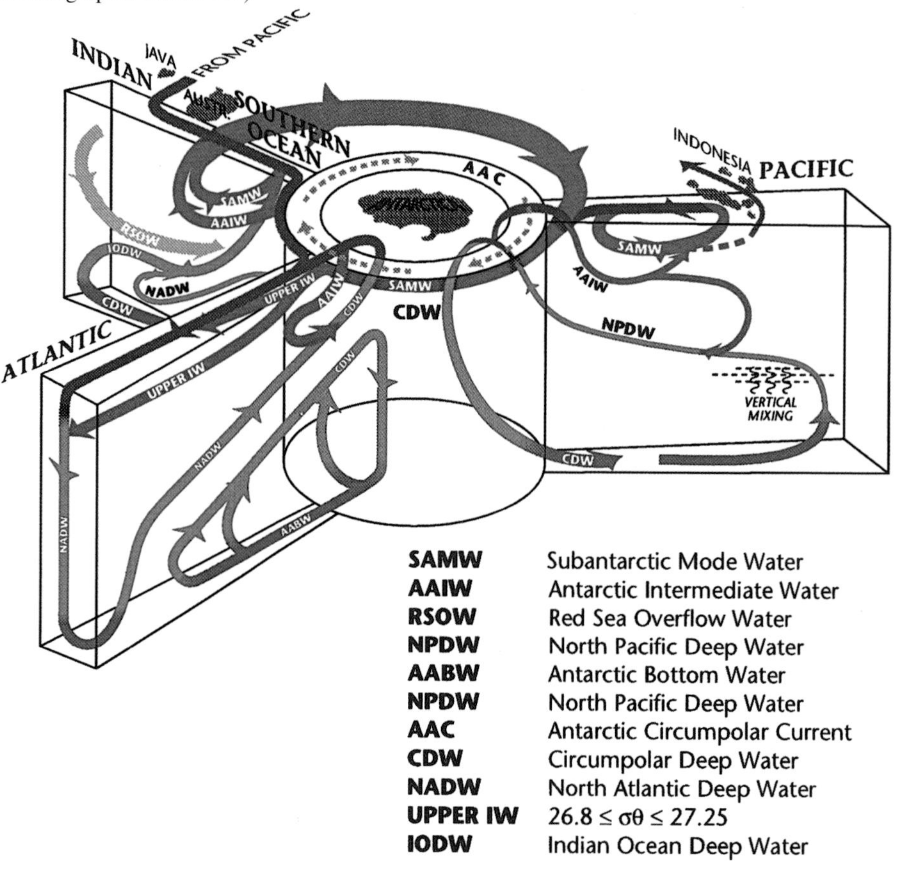

PLATE 3 Schematic of the global thermohaline circulation. (From Schmitz, 1996; reprinted with permission of the Woods Hole Oceanographic Institution.)

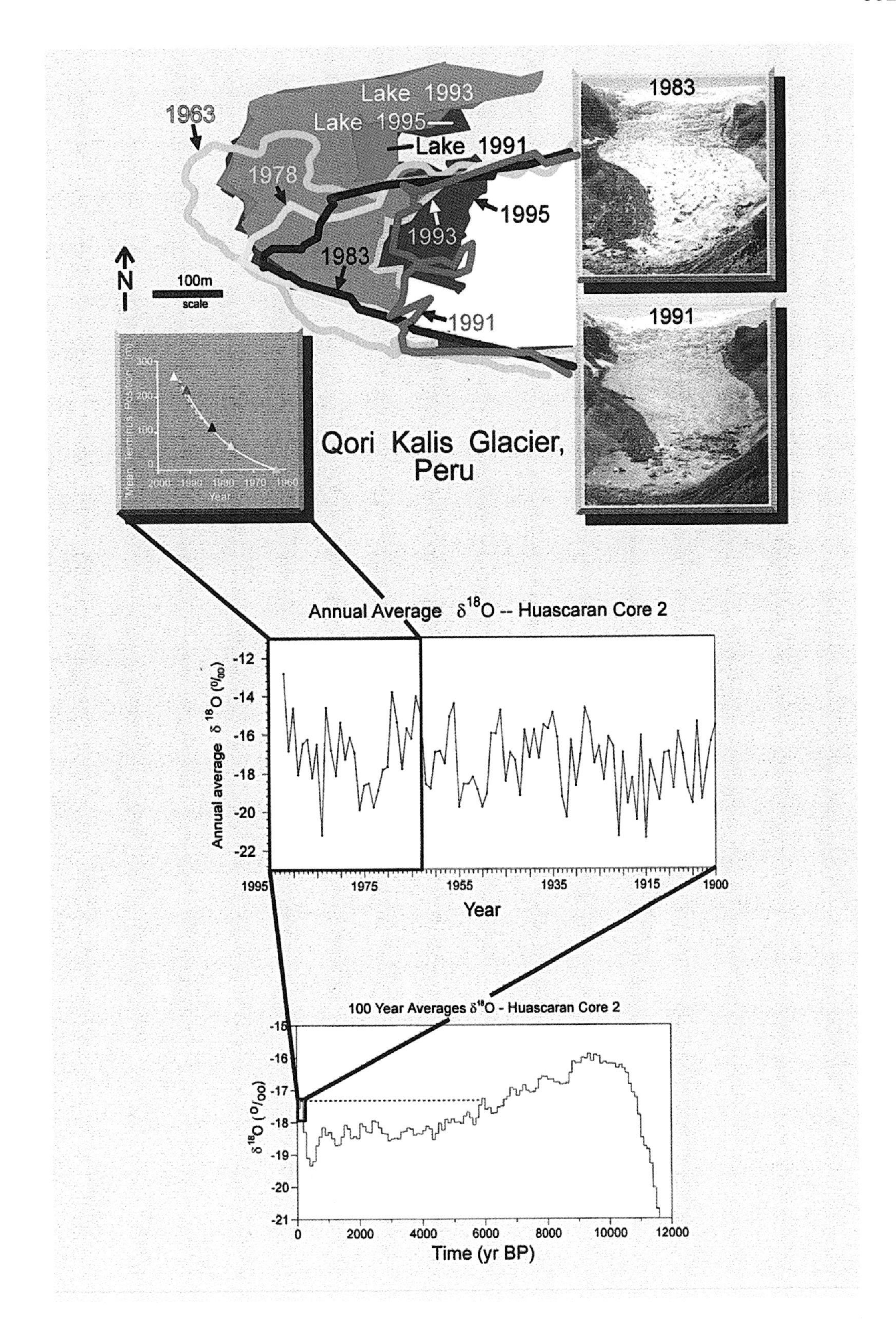

PLATE 4 Climate change in the Peruvian Andes. Upper part, a map of the changes in extent of the tropical glacier Qori Kalis from 1963 to 1995, with photographs of the change over 8 years. The left-hand graph relates the shrinkage of Qori Kalis to the rising temperatures indicated by the ice-core data below it. Lower part, oxygen-isotope values from a Huascarán ice core. Bottom panel, century averages; top panel, annual averages since 1900. (After Thompson et al., 1995; reprinted with permission of the American Association for the Advancement of Science.) (Figure courtesy of Lonnie G. Thompson, Byrd Polar Research Center.)

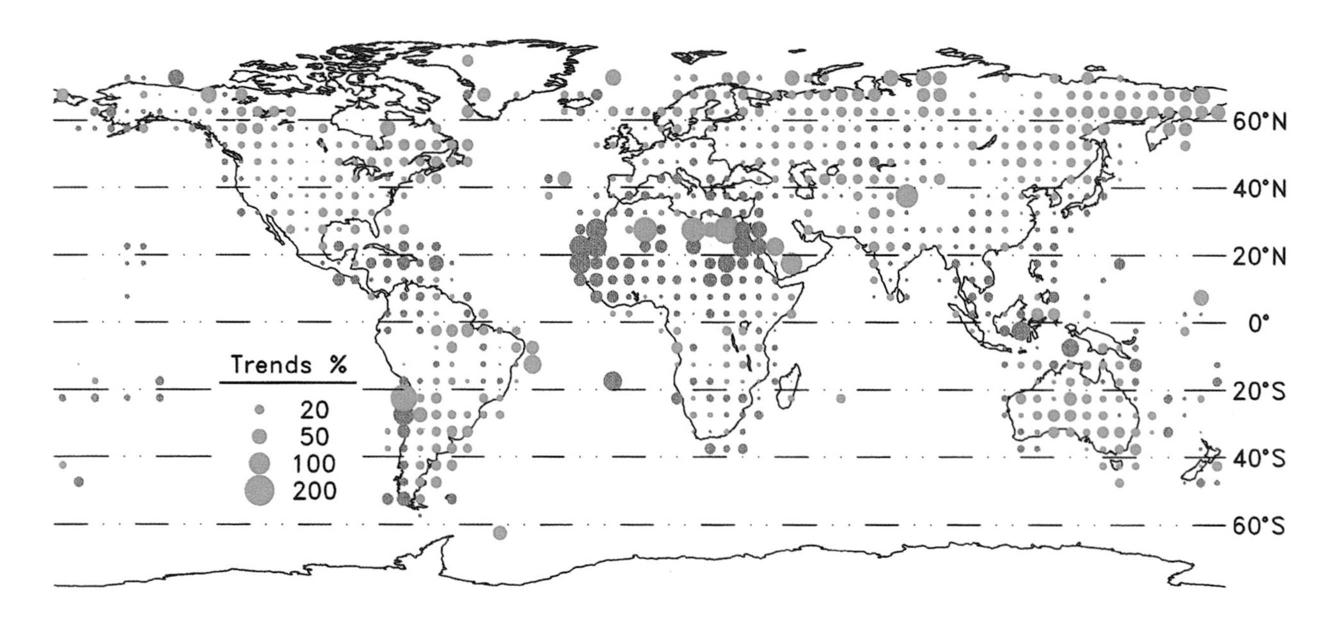

PLATE 5 Precipitation trends over land, 1900-1995. The trend is expressed in percent per century relative to the mean precipitation from 1961-1990. The magnitude of the trend at each location is reflected in the size of the circle. Green circles represent increases in precipitation, and brown circles represent decreases. (From IPCC, 1998; reprinted with permission of the Intergovernmental Panel on Climate Change.)

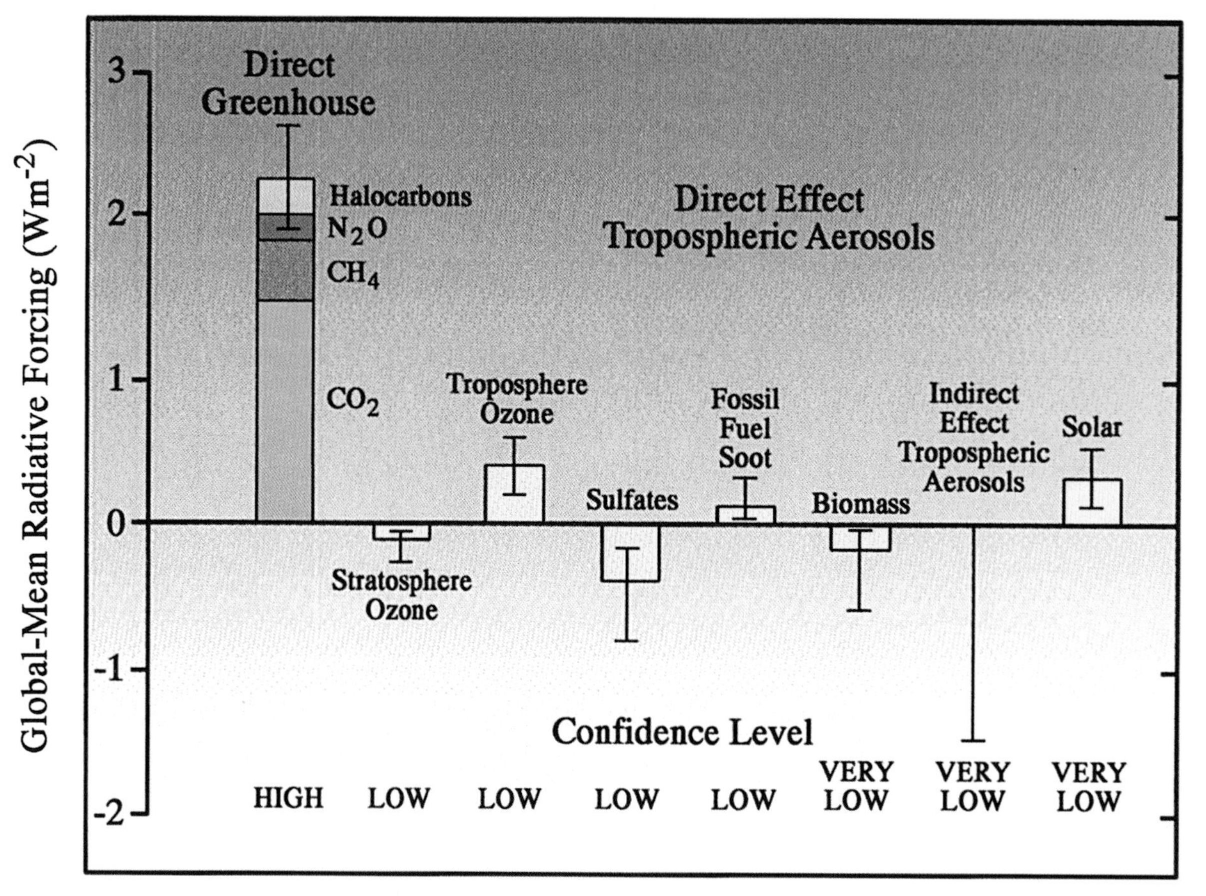

PLATE 6 Estimates of the globally and annually averaged anthropogenic radiative forcing (in W m^{-2}) attributable to changes in concentrations of greenhouse gases and aerosols from pre-industrial times to the present day, and to natural changes in solar output from 1850 to the present day. Error bars are shown for all forcings. The indirect effect of sulfate aerosols through their interaction with clouds is so uncertain that no central estimate of radiative forcing is provided. (From IPCC, 1996a; reprinted with permission of the Intergovernmental Panel on Climate Change.)

and Sutton and Allen's (1997) analyses of SST indicate that the warm/cold epochs of Kushnir (1994) involve systematic interannual-to-decadal propagation of winter SST anomalies along the primary circulation pathways in the North Atlantic, which suggests an interplay between ocean dynamics and sequestering of temperature information (making it unavailable to the atmosphere) in the upper ocean.

Because salinity is not routinely measured, it is much harder to assess the decadal variability of sea surface salinity than SST. There is, however, a well-known low-salinity event that was observed over much of its 20-year lifetime. Referred to as the Great Salinity Anomaly (GSA), it was first observed in the Labrador Sea in 1968, and peaked in 1971 (Dickson et al., 1988). The GSA migrated out of the Labrador Basin and flowed eastward in the southern subpolar gyre and merged with the cold subtropical gyre, before reappearing again around Greenland and the Labrador Basin in the early 1980s. The origin or the GSA is thought to lie in the increased discharge of ice through Fram Strait and its subsequent melting, although this theory is still a source of debate.

Temperature variations have also been documented below the surface. Figure 5-22 shows the 500 m temperature difference between two 5-year averages (pentads) 15 years apart (Levitus, 1989). Differences of more that half a degree are clearly seen. Similar results have been found in two deeper surveys (Roemmich and Wunsch, 1984; Parilla et al., 1994). Also, the heat content of the upper ocean near Bermuda appears to have increased since the 1970s, and the thermocline to have moved correspondingly (Joyce and Robbins, 1996). These observations reflect interdecadal changes in both the properties and the volume of the upper water mass, in response to changes in the dynamics (wind forcing and associated thermocline response, or "heave") and thermodynamics (surface buoyancy flux and associated convection). For example, there is a cooling of the dominant regional convective water mass, known as 18°C water, from the mid-1950s to about 1970, and then a warming till the mid-1990s. The cooling and warming are in phase with those in the Greenland-Iceland-Norwegian (GIN) Seas, and opposite in phase to those in the Labrador Sea and the NAO (Dickson et al., 1996).

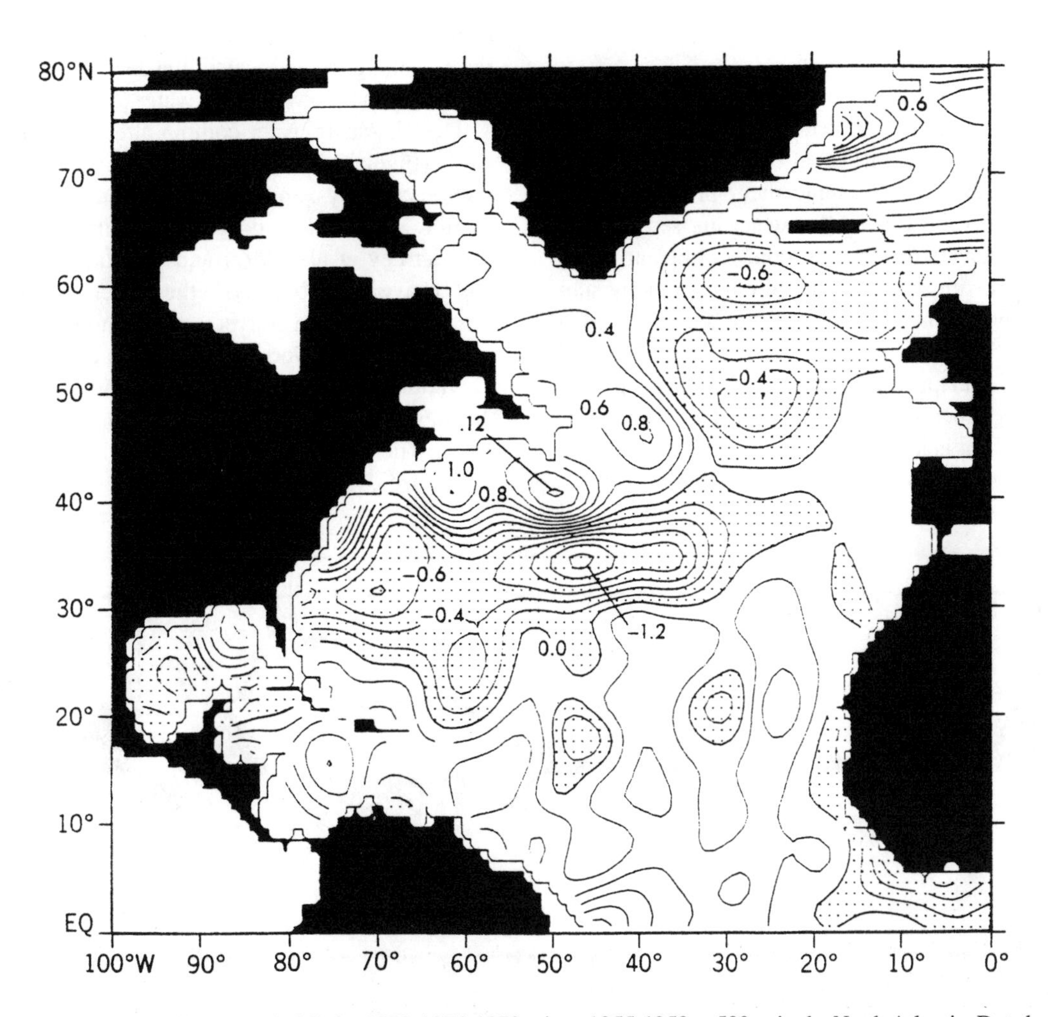

FIGURE 5-22 Temperature differences (in °C) for 1970-1974-1979 minus 1955-1959 at 500 m in the North Atlantic. Dot shading indicates negative values. (From Levitus, 1989; reprinted with permission of the American Geophysical Union.)

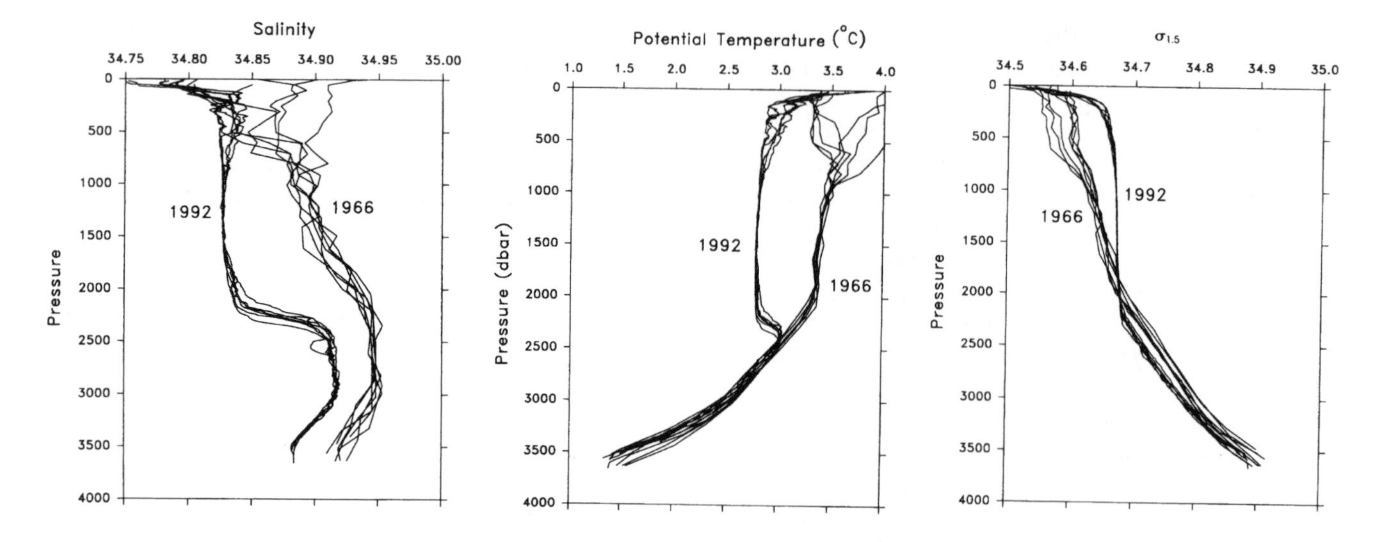

FIGURE 5-23 Temperature and salinity profiles obtained at six stations in the central Labrador Sea in March 1966 and in June 1992. Left, profiles of salinity; center, profiles of potential temperatures; right, profiles of density referenced to 1500 db. (From Lazier, 1995; reprinted with permission of the National Academy Press.).

Thermohaline Circulation

Warm, salty water proceeds northward in the Gulf Stream, cooling as it crosses the Atlantic in a northeastward direction as the North Atlantic Current. Upon reaching the subpolar regions, it enters the GIN Seas and then spreads into the Arctic, where it mixes with resident water masses and further increases in density due to additional cooling and salinization associated with sea-ice formation. It eventually subducts and/or convects and flows back out of the GIN Seas at intermediate depths, whereupon it sinks to depths of several thousand meters and proceeds southward as a deep cold boundary current beneath the Gulf Stream. This deep western boundary current is joined in its upper parts by convecting water from the Labrador Sea. Any variation in convection in the source regions of the THC will be reflected in changes in transport, and may therefore affect subsurface regions along the pathway of the deep boundary current, and the sea surface to which the water eventually returns.

Decadal-scale water-column alterations in the Labrador Basin (Figure 5-23) have been documented through repeated sections, surveys, and time-series station observations (e.g., Lazier, 1995). The downstream effects on circulation (McCartney et al., 1997) and ventilation (Molinari et al., 1997) have been observed in the western North Atlantic. Figure 5-23 shows that convection was more intense and deeper in 1992 than in 1966. A time series of the temperature of Labrador Sea Water (LSW), shown in Figure 5-24, reveals the interdecadal nature of its history (McCartney et al., 1997). The temporal evolution of the NAO is also illustrated in Figure 5-24; generally it varies inversely with the

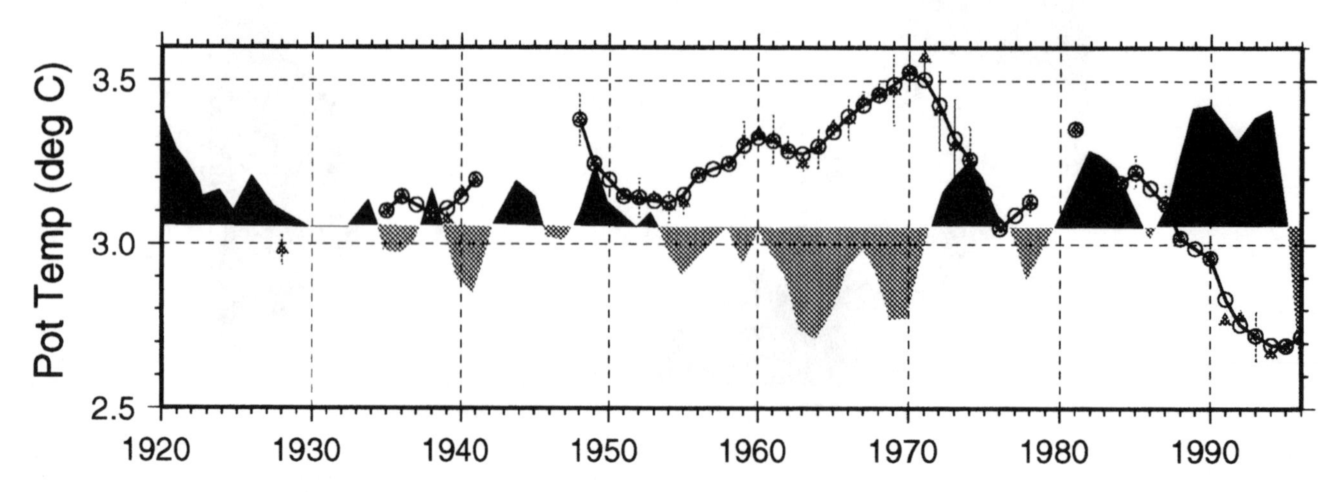

FIGURE 5-24 LSW temperature time series with NAO overlay. (From McCartney et al., 1997 .)

interdecadal progression of the LSW. It has been suggested (Dickson et al, 1996) that the NAO has large-scale effects on convection in the Greenland Sea, which showed a remarkable reduction in the 1980s (Schlosser et al., 1991). There is also evidence that the Labrador Sea was affected by the passage of the GSA in the late 1960s, undergoing reduced convection and subsequent warming.

A few measurements document interdecadal variability in the Southern Hemisphere deep and bottom water-formation regions. While the changes observed are small, they are significant in that they involve vast quantities of water, and therefore must be related to large-scale persistent changes in the environment at the air-sea interface (see, e.g., Coles et al., 1996; Zenk and Hogg, 1996). An abyssal warming and a decline in bottom-water transport were observed in 1992-1994 at the equator at the northern end of the Brazil Basin; the temperatures involved were all lower than measurements made in the same location in 1972, 1983, and 1989, however, suggesting recovery from some intervening cold event (Hall et al., 1997). Hall et al. also found a pronounced seasonal cycle in the deep circulation. This rapid cycling suggests that the deep limb of the THC is less isolated from higher-frequency forcing than might be expected, and that perhaps in the tropics the behaviors of the deep/cold and shallow/warm limbs of the THC are dynamically coupled.

Abrupt Climate Change

A variety of evidence from Greenland ice cores and ocean sediments has been used to suggest that the patterns and rates of deep oceanic circulation have undergone large changes over the past several hundred thousand years (see, e.g., Broecker and Denton, 1989). These changes have been interpreted to be the result of the recurrent cessation and initiation of the Atlantic thermohaline circulation. While the changes associated with the glacial stages reflect millennial-scale change, recent analyses indicate that some of the more abrupt events in these ice and sediment cores reflect changes that may have taken place in a matter of decades or less (see, e.g., Severinghaus et al., 1998; Grootes, 1995).

It is unclear what mechanisms are responsible for initially triggering these ocean-atmosphere reorganizations. Substantial increases in iceberg calving, inferred from deep-sea sediment cores, have recurred at intervals of 2,000 to 10,000 years. These phenomena, known as Heinrich events, have been related to circulation changes in both high-latitude and equatorial regions (McIntyre and Molfino, 1996). They have been correlated with warm/cold climate oscillations known as Dansgaard-Oeschger events (Bond and Lotti, 1995). Analyses of deep-sea cores from the North Atlantic suggest that abrupt shifts in climate occur independently of the glacial-interglacial fluctuations on a millennial-scale cycle; they have been inferred to exist from at least 500,000 years ago through the Holocene (Bond et al., 1997; Oppo et al., 1998). Evidence from both poles indicates a high level of global synchronicity in some of these abrupt climate changes (Bender et al., 1997), though the degree of synchroneity is still controversial for some of the shorter time-scale events. Date derived from ice cores indicate that Antarctic climate changes often precede those in Greenland by more than 1,000 years (Blunier et al., 1997, 1998).

While attempts have been made to reconstruct the SSTs that existed during the last glacial maximum (CLIMAP, 1981), much of our understanding of the ocean-circulation patterns that may have existed during periods when ocean circulation was different from the present (e.g., the last glacial maximum) comes from modeling studies (e.g., Mikolajewicz et al., 1997; Bush and Philander, 1998). For instance, the study by Mikolajewicz et al. (1997) using a coupled ocean-atmosphere GCM suggests that a shutdown of North Atlantic Deep Water formation during the Younger Dryas (approximately 12,000 BP) may have also brought about a cooling in the North Pacific. Paleo-proxy data support such a link between the Atlantic and Pacific (Broecker and Denton, 1989). Among the other mechanisms that have been proposed to account for such large-scale changes in climate regime are the onset of a large-amplitude internal mode of oscillation in thermohaline circulation (Tziperman, 1997; Weaver and Hughes, 1994), and a non-linear response to gradual changes in solar insolation resulting from changes in the Earth's orbital parameters (Herbert, 1997). It has also been suggested that the warming from increased atmospheric greenhouse-gas levels could lead to a significant reduction in THC (Manabe and Stouffer, 1994). Stocker and Schmittner (1997) found that the degree to which the thermohaline-circulation rate can decrease depends on the rate of increase in atmospheric greenhouse gases. Their coupled ocean-atmosphere model indicates that if the warming is rapid enough, deep-water formation in the North Atlantic may effectively cease.

The study of abrupt climate change began relatively recently, and it is not yet fully understood, but the observational evidence is tantalizing in that it suggests the possibility that major climate shifts may occur over relatively short time scales. This possibility underscores the need to improve our understanding of the sensitivities of climate to changes in the thermohaline circulation, and the dependence of the thermohaline circulation on changes in its source areas.

The Pacific Ocean

Upper Ocean

Examination of the zonally averaged circulation in the Pacific reveals a shallow meridional overturning cell that connects the tropics and subtropics (Hirst et al., 1996). This overturning involves equatorial upwelling, poleward flow near the surface, subduction in the subtropics, and equatorward return flow in the upper part of the thermocline (the band where there is a strong thermal gradient). Forced by both Ekman (wind-driven) pumping and thermocline fluxes (see, e.g., McCreary and Lu, 1994), this overturning cell helps determine equatorial SSTs, the amount of tropical heat exported to the subtropics, and thermocline stratification throughout the region. Variability of each of these processes is likely on all time scales.

The dominant SST variations in the Pacific Basin were discussed in Chapter 3 (see Figures 3-6 through 3-9). The dec-cen pattern shown in Figure 3-8 has a broader meridional SST-anomaly pattern in the tropics than the ENSO pattern shown in Figure 3-6, and has clearly stronger values of mid-latitude SST relative to the tropical (ENSO) values. Thus, the analysis representative of longer time scales (Figure 3-8) indicates that the equatorial and mid-latitude regions contribute comparably strong anomalies, and that the regional contributions to climate are greater in geographical extent than the contributions from interannual-scale variability. The time coefficient of the interdecadal pattern in Figure 3-8 exhibits "regime shifts" in 1976 (Trenberth, 1990). Since the mid-latitude manifestation of this mode is so strong, it can be identified with a specifically North Pacific oscillation now called the Pacific Decadal Oscillation (PDO). The PDO can be identified from the pattern illustrated in Figure 3-8 or obtained by direct EOF analysis of North Pacific SST (see Mantua et al., 1997).

The PDO index, shown in Figure 5-25, co-varies with an index of pressure over the North Pacific. When the water over the North Pacific (Figure 2-18) is cold (as it has been since 1976), the Aleutian low is deep and the PNA pattern is

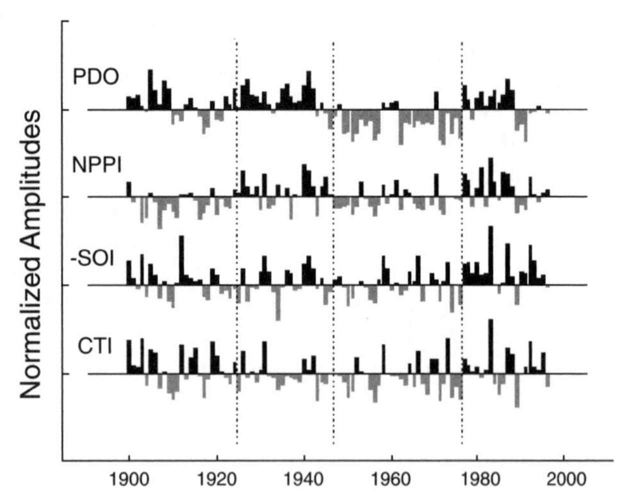

FIGURE 5-25 Normalized winter mean (November-to-March) time histories of Pacific climate indices. Dotted vertical lines are drawn to mark the PDO polarity reversal times in 1925, 1947, and 1977. Positive values of the NPPI (North Pacific SLP) correspond to years with a deepened Aleutian low. The negative SOI is plotted so that it is in phase with the equatorial SST anomalies captured by the cold-tongue index (CTI). Bars with positive values are filled with black, those with negative values with grey shading. (From Mantua et al., 1997; reprinted with permission of the American Meteorological Society.)

strong. There is a definite correlation between the PDO and the salmon catch in the mid-latitude North Pacific (Mantua et al., 1997) and, in particular, between the 1976 Pacific climate-regime change and fish migration patterns (Mysak, 1986). As is seen in Figure 5-25, the regime shifts in mid-1976 also manifest themselves in several other ways: as an increase in tropical SST indicated by the Southern Oscillation Index (see also Graham, 1994); as a change in subsurface conditions in the tropical (Guilderson and Schrag, 1998) and mid-latitude Pacific (Zhang and Levitus, 1996); and as an increase in the intensity and regularity of warm phases of ENSO. While the temporal character of interdecadal variations cannot clearly be defined from the record shown in Figure 5-24 (it begins only in this century), Minobe (1997) has extended the PDO record using tree-ring records from North America; he deduces a time scale of 50-70 years for the variability. Because EOF methods do not cleanly separate time scales, studies based on such analyses have reported characteristic time scales ranging from interannual to 70 years. Until the record is extended, this limitation will probably persist.

While changes in surface temperature are obviously important, it is the combination of temperature and salinity that dictates the density profile and thus much of the vertical redistribution of enthalpy, freshwater, and momentum. These factors influence not only the magnitude of surface-temperature anomalies, but also the speed and structure of the ocean currents, and thus the lateral redistribution of any anomalies. The vertical structure of anomalies and currents will vary according to the relative phasing of the temperature and salinity signals, with implications for the evolution of other ocean-atmosphere phenomena over a variety of time scales. The importance of the relative phasing of temperature and salinity implies that the regional response to an El Niño event may differ from year to year.

Deser et al. (1996) have related decadal-length SST records to upper-ocean temperature changes and to zonal-wind anomalies (Figure 5-26). The subsurface temperature data provide clear evidence of variability across a range of interannual to decadal time scales; it becomes increasingly decadal with depth. Also, Deser et al. (1996) and Watanabe and Mizuno (1994) suggest that water subducts from the high-latitude cold pool into the subtropical gyre of the North Pacific on time scales of a decade or more. Analysis of tracer-derived ages shows that water subducted into the eastern North Pacific subtropical gyre can be transported on decadal time scales into the equatorial Pacific (Fine et al., 1987), and that those waters can be found beneath the warm and cold pools associated with ENSO. Furthermore, these tracer data can be used to infer that the loop back to the surface for the subducted water is closed by upwelling at the equator (Jenkins, 1998). In an analysis of temperature observations, Zhang et al. (1998) find that the Pacific upper-ocean warming and decadal changes in ENSO after 1976 may have their origin in changes in higher latitudes that have propagated

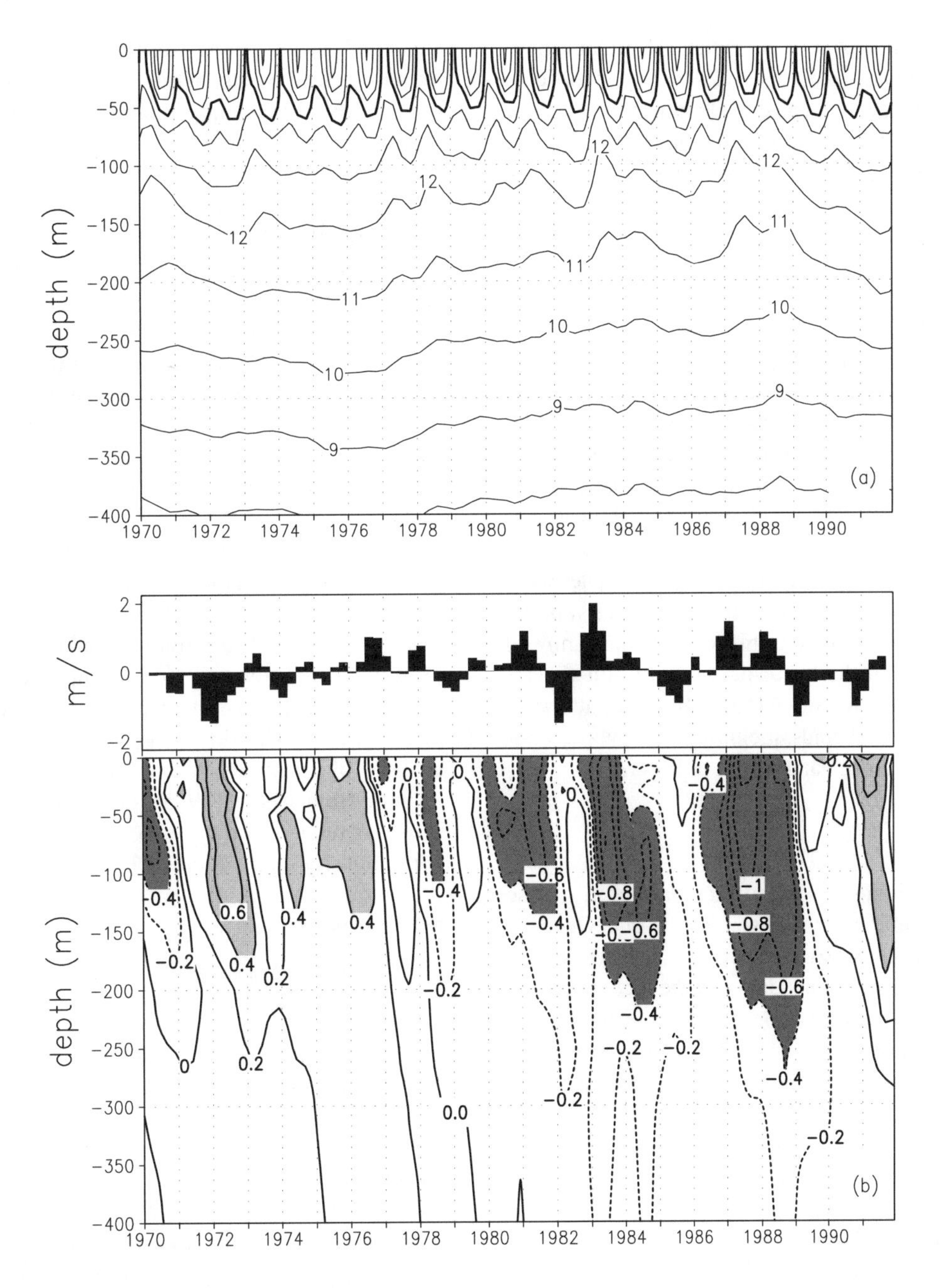

FIGURE 5-26 Time-depth plot of seasonal (a) total and (b) anomalous temperature in the central North Pacific region. In (a) the contour interval is 1°C and the 15°C contour is thickened. In (b) the contour interval is 0.2°C, negative anomalies are dashed, and anomalies greater than 0.4° (less than -0.4°C) are indicated by light (dark) shading. The bar graph directly above (b) shows seasonal zonal wind anomalies; positive (negative) values denote stronger (weaker) than normal winds from the west. (From Deser et al., 1996; reprinted with permission of the American Meteorological Society.)

equatorward via subsurface transport. These dynamics have been illustrated in various ocean-only models (e.g., Liu et al., 1994), and demonstrated in an ocean assimilation model, where the assimilation of observed temperature data at high latitudes generates decadal variability in equatorial regions (Lysne et al., 1998).

The Indonesian Seas are the only warm-water pathway between ocean basins. By transporting anomalies between basins, the Indonesian throughflow plays an important role in climate variability. Gordon and Fine (1996) find seasonal (monsoonal) variations in the ratio of North and South Pacific water contributing to the throughflow of water to the Indian Ocean. They conclude that the ratio changes in response to the phase of ENSO, and may provide feedback to ENSO by influencing the extent of the warm pool in the western equatorial Pacific. The tropical biennial oscillation and ENSO have been linked to the Asian-Australian monsoon system. In addition, Allan et al. (1995) find significant interdecadal SST anomalies in post-1900 Indian Ocean data. Reason et al. (1996a,b) suggest that the principal explanation for the mid-latitude decadal SST anomalies is found in equatorial wind anomalies from the Pacific, which drive variations in the Indonesian throughflow, resulting in variations in the Agulhas Current.

Circulation Changes

Other changes in the North Pacific subtropical gyre have been deduced from brief records. Qiu and Joyce (1992) found a gradual increase in the Kuroshio and North Equatorial Current transport of about 5 sverdrups (1 Sv = 10^6 m^3/s) per decade, starting in 1970 and continuing into the 1980s, which implies greater wind forcing. Bingham (1992) showed that during the period 1978-1982 the subtropical gyre was stronger than during 1938-1942, which again implies greater wind forcing. Deser et al. (in press) found an intensification of the winds associated with the regime shift of 1976, and showed that the enhancement of the Kuroshio was part of a general North Pacific response to this wind shift. The pattern of decadal variability of the winds, including those winds connected with the 1976 shift, can be seen in Figure 3-8.

The Southern Ocean

Constructing an image of subsurface variability in the South Atlantic, South Pacific, and southern Indian Ocean is more difficult than in the northern oceans, because of the relative paucity of hydrographic data and the lack of meteorological data in the high-latitude Southern Hemisphere. Despite this scarcity, several studies have been able to glean evidence of decadal-scale variability from what data are available.

Changes in temperature and salinity of similar magnitude to those reported in the deep North Atlantic Ocean (Bryden et al., 1996) are also seen in Subantarctic Mode Water and Antarctic Intermediate Water in the Tasman Sea (Bindoff and Church, 1992), the South Pacific (Johnson and Orsi, 1997), and the southern Indian Ocean (Bindoff and McDougall, in press). The trend towards cooling and freshening on isopycnals and warming on depth surfaces is coherent over a large spatial scale between the late 1960s and early 1990s, but the long time elapsed between surveys may mean that the measurements are not strictly comparable. The changes are consistent with surface warming at high southern latitudes and surface freshening in higher southern latitudes at some time during the two decades that elapsed between the repeat hydrographic sections (Bindoff and McDougall, 1994). There is also evidence of changes in the Weddell Deep Water (Fahrbach et al., in press), though the records are not yet long enough to fully characterize the time scales of the variability.

The Antarctic Circumpolar Wave (ACW), described in Chapter 3, is a traveling-wave disturbance in SST, SLP, sea ice and surface winds (White and Peterson, 1996). It is predominantly interannual in nature, but it displays a longer-scale component in the relatively short record documenting it. While distinct in the Southern Ocean circumpolar belt, this phenomenon may be global in scope and involve linkage to ENSO (Barnett, 1985; White and Peterson, 1996; Tourre and White, 1997; Yuan et al., 1996), though some argue that it is primarily a regional phenomenon (e.g., Qiu and Jin, 1997). The ACW signal is large in the higher-latitude sectors of the Southern Hemisphere; its impact may influence intermediate- and deep-water formation, which may in turn alter the thermocline characteristics of the subtropical gyres, influencing lower-latitude properties and spreading the ACW influence to broader spatial and temporal scales.

Even less is known about low-frequency atmosphere-ocean interactions in the South Atlantic. Kushnir et al. (1998) show correlated NAO, SST, and SLP signals in the tropical Atlantic and parts of the South Atlantic. Barnett's 1985 study of global SLP indicated NAO-related SLP signals in these South Atlantic areas as well, but with a time delay: The high (low) SLP there follows the high (low) NAO states. An analysis of transports in the upper waters of the southeast Atlantic from 1993 to 1995 suggests that the observed variability is related to gyre wobble, rather than to changes in gyre intensity (Garzoli et al., 1997).

Mechanisms

Global Ocean Processes

Climatically relevant oceanic variability is manifested in many ways, among them anomalies of SST, salinity, sea ice, and the underlying internal distribution of heat and salt content, as well as changes in the patterns and intensities of oceanic circulation. The ocean's participation in climate-change phenomena can be essentially passive—the ocean merely responds to forcing by an evolving atmosphere—or it can be active—variability in air-sea exchange and subsequent sequestration and redistribution of heat, freshwater, and carbon dioxide feed back to affect the evolution of the overlying atmosphere. Active participation can be seen as primarily a cause-and-effect relationship, in which some internal natural mode of variability in the ocean forces an atmospheric response, or as a coupling, in which changes in ocean and changes in atmosphere mutually reinforce or oppose each other.

The vast thermal, kinetic-energy, and chemical capacities of the ocean, by comparison with those of the atmosphere, generally make the ocean a much more slowly changing element of the system, and thus a potential modulator of the fast-responding atmospheric-circulation system. But that slow modulating role may be only partially realized at any given time, because the strong vertical stratification of the water insulates much of the ocean's volume from direct interaction with the atmosphere. In many regions, the impact of air-sea exchanges on the ocean is restricted to a relatively thin near-surface layer whose smaller storage capacity allows a faster response. Indeed, on seasonal and interannual time scales the oceanic response to atmospheric change is essentially in phase, reflecting a more local balance of the heat budget on these short time scales. For example, the observed strong correlation between SST tendency and the amplitude and sign of local heat-flux anomalies is to a first approximation a "passive ocean" response of the local mixed layer to atmospheric forcing. This is not the case on longer time scales, where the ocean transport of anomalies alters the heat balance and thus the phasing (and even direction) of the atmospheric change.

The large mass of the ocean compared to that of the atmosphere allows the relatively sluggish ocean to be comparably effective in its redistribution of properties, especially heat. Poleward transport of heat within an ocean basin is achieved when the aggregate of poleward flows at a given latitude in a basin is warmer than the aggregate of equatorward flows. The slow-moving THC involves large surface-to-deep-water temperature differences, so the amount of heat it transports meridionally is similar to that transported by the faster-moving wind-driven circulation.

In addition to the THC's storage and redistribution of heat, and the global movement of heat from the equator to the poles, there is also a basin-scale influence associated with the lateral redistribution of heat that is important on longer time scales. On short time scales, such as the seasonal-to-interannual, the local storage of surface heat is balanced predominantly by vertical diffusion; on longer time scales, however, the rate of change of surface temperature represents a balance between lateral advection and diffusion (Moisan and Niiler, 1998). That is, on longer time scales, surface anomalies are displaced laterally by the slow but steady movement of the ocean's surface currents. This lateral redistribution of heat typically swamps any local heat-storage imbalance.

This power of the ocean's heat transport is particularly evident in the climate system. For example, the oceanic meridional transport of heat northward across 24°N (the aerial midpoint between the equator and pole) is estimated directly from oceanic measurements to be about 2 petawatts (PW; Bryden et al., 1991). This figure represents about half of the net meridional heat transport required by the Earth's radiation budget at this latitude, suggesting that the contributions of atmospheric and oceanic circulation to the overall budget are nearly equal (Trenberth and Solomon, 1994). At 24°N, the North Pacific and North Atlantic together span only a bit more than half the Earth's circumference, so the ocean is about twice as effective (per unit of longitude) as the atmosphere in transporting heat meridionally at that latitude. The mean currents are responsible for most of the heat transport in the ocean with the transient eddies contributing comparatively little.

Of the ocean's contribution to meridional heat transport, 1.2 PW occurs in the North Atlantic and 0.8 PW in the North Pacific (Bryden et al., 1991). The North Atlantic is only half the width of the North Pacific, and spans about a fifth of the Earth's circumference, so it is three times as effective (per unit longitude) as the North Pacific (and more than three times as effective as the atmosphere) in transporting heat meridionally at 24°N. This contribution is so strong because the North Atlantic is the dominant Northern Hemisphere site for conversion of warm water to cold water within the THC. The North Pacific also converts warm water to cold, but achieves less poleward heat transport because the warm-to-cold-water conversion occurs primarily above the thermocline and undergoes a smaller effective temperature change than the deep-reaching overturning in the North Atlantic. The Indian Ocean also supports large poleward heat transport through a deep-reaching overturning circulation with a large effective temperature change.

Freshwater Transport

The ocean's redistribution of freshwater is a dominant agent in the hydrologic cycle. It compensates almost completely for the meridional transport of water vapor in the atmosphere. The remainder of the contribution from river flow is approximately an order of magnitude smaller. There are significant differences in the patterns of net water flux into the various ocean basins. For instance, the North Pacific receives much more precipitation than the North Atlantic. As a result, the North Pacific is a fresher ocean that does not support deep-water formation or a significant thermohaline circulation, as previously mentioned. This limits the amount of heat it can carry through its large volume of deep water, although its large basin is involved in considerable movement of surface anomalies over broad distances. The more saline Atlantic, in contrast, has sites of both deep- and intermediate-water formation and maintains a vigorous thermohaline circulation, permitting greater meridional heat transport.

Understanding the linkages between heat and freshwater transport in the ocean is a fundamental requirement for solving many of the puzzles of climate variability. In addition to modulating oceanic heat transport and density distribution (which influences currents and the volume of surface water with which surface fluxes interact), the differences in net water flux into the various ocean basins also lead to interbasin flows, such as that through the Bering Strait. Thus, the net southward flow of 0.3 Sv at 24°N is the difference between a large southward flux in the North Atlantic and a smaller northward flux in the North Pacific (Schmitt and Wijffels, 1992). This net southward flow compensates for the net atmospheric transport of water vapor from the Atlantic to the Pacific (Broecker, 1991). Therefore, the latent heat flux associated with this atmospheric water-vapor transport can be indirectly estimated from oceanic mass flux, which permits the magnitude of two major components of meridional heat transport (i.e., oceanic advective heat flux and atmospheric latent heat flux between the Atlantic and Pacific) to be determined from ocean data alone.

The *net* meridional transports of water, discussed above, are dwarfed by the large evaporation and precipitation fluxes over the ocean; some 12.6 Sv of water is thought to be precipitating on to the ocean, nearly 4 times the water that is falling on land. The net excess of terrestrial precipitation over evaporation is returned to the sea by rivers at a rate of more than 1 Sv (see, e.g., Berner and Berner, 1987). Thus, the largest portion of the global water cycle occurs over the oceans, where it is poorly known, but appears to have significant impact on oceanic climate processes (e.g., the GSA) and on surface freshening and stabilization. Small redirections of the water transports from ocean to land could have dramatic consequences. For example, an increase in precipitation over the Mississippi River basin equivalent to a mere 1 percent of Atlantic precipitation would more than double the discharge of the Mississippi River. Expanded measurements of oceanic salinity fields are required to better constrain our estimates of the oceanic water cycle's fluxes and reservoirs, as well as its role in climate fluctuations.

Carbon Transport

The ocean sequesters a great amount of carbon dioxide, and thus is a major consideration in estimating the ultimate fate (and buildup) of anthropogenic carbon dioxide in the atmosphere, and the resulting increases in greenhouse warming. The oceanic inventory of carbon is estimated to be 39,000 gigatons (Gt), with the deep ocean containing over 95 percent (IPCC, 1995). This quantity is more than 50 times the atmosphere's inventory, and about 18 times the amount contained in terrestrial biota and soils (IPCC, 1995). The ocean's inventory is growing at a rate of approximately 2 Gt per year, which corresponds to roughly one-quarter to one-third of the rate of release of carbon to the atmosphere by all current anthropogenic sources of carbon dioxide. The ocean is thus a major carbon reservoir, capable of moderating the growth of carbon dioxide in the atmosphere and redistributing that carbon globally. Therefore, by influencing the concentration of atmospheric CO_2, the oceans indirectly affect the Earth's radiation budget and the overall energy balance of the climate system.

The components of the global carbon budget are not as firmly established as those of the heat and freshwater budgets. The IPCC (1995) places error limits (with 90 percent confidence) of 0.8 Gt per year on the 2.0 Gt per year best estimate of net oceanic carbon uptake. The ocean absorbs CO_2 by two mechanisms: the solubility pump (effectively the ability to hold more or less CO_2 as water cools or warms), and the biological pump (the biological incorporation of carbon into living cells and its ultimate transfer to deep water after cell death). These two pumps are discussed in greater detail in the "Atmospheric Composition and Radiative Forcing" section earlier in this chapter. Feedbacks associated with changes in ocean dynamics and thermal composition over local, regional, and global scales increase the uncertainty in predicting future oceanic uptake. These feedbacks involve complex interactions between physical and biogeochemical processes operating over a broad range of scales.

For example, the study by Sarmiento et al. (1998) investigating the role of the ocean in modulating the anthropogenic increase of atmospheric CO_2, identifies the Southern Ocean's circumpolar region as the region showing the most sensitivity to different model formulations while having the largest global impact on the net atmospheric CO_2 concentration. Sarmiento et al. (1998) suggest that the effectiveness of the circumpolar region in CO_2 uptake, which is a function of the biological and solubility pumps, is enhanced by strong isopycnal mixing that tends to efficiently transport (subduct) the freshly absorbed CO_2 from the surface layer into the deep waters to the north, strengthening the solubility pump. As atmospheric CO_2 increases, their model freshens the circumpolar region, which stabilizes the surface layer. This stabilization reduces the subduction, decreasing the effectiveness of the CO_2 uptake via the solubility pump. The stabilization also reduces the upward mixing of deep nutrients into the surface layer. This might in turn alter the biological productivity and effectiveness of the biological pump; however, the biological formulations in the model used by Sarmiento et al. were not designed to quantify the likely changes in productivity. In contrast to indications in the Sarmiento et al. study that the biological pump might be significantly affected by climate-induced changes in the ocean, the modeling study by Maier-Reimer et al. (1996) suggests that feedbacks involving changes in the biological pump might be relatively minor in importance.

Keeling and Peng (1995) estimate that the North Atlantic Ocean takes up and transports about 0.4 Gt of carbon per year across the equator into the Southern Hemisphere. Keeling et al. (1996b) show that atmospheric oxygen concentrations in the Southern Hemisphere are currently higher than in the Northern Hemisphere. A consistent explanation for this interhemispheric gradient can be given by a combination of factors: More fossil fuel is burned in the Northern Hemisphere (releasing CO_2 and consuming O_2); terrestrial ecosystems in the Northern Hemisphere are likely a net CO_2 sink (and source of O_2); and net transport by the Atlantic Ocean is southward. Thus, the Keeling et al. result is broadly consistent with the Keeling and Peng (1995) result for oceanic carbon transport in the Atlantic. Little is known about oceanic carbon transports in the Indian or Pacific Oceans.

Observations of oceanic uptake and transport of anthropogenic CO_2 (see, e.g., Gruber, 1998) can be used to assess the veracity of ocean GCMs. A recent study by Stephens et al. (1998) using this approach concludes that the models they examined tend to take up too much anthropogenic CO_2 in the Southern Ocean, and therefore do not accurately reproduce the interhemispheric transport of carbon and oxygen that is seen from or implied by atmospheric observations.

Local Ocean Processes

For longer time scales, for some regions, and for some climatic phenomena, the ocean is not simply a well-mixed layer passively responding to flux anomalies forced by the atmosphere. The ocean may also have active modes of internal variability that force an atmospheric response, or participate actively in coupled ocean-atmosphere modes of variability on the dec-cen time scale. The ocean's active role in dec-cen climate variability involves several local mechanisms.

First, the ocean acts as a recorder and integrator of the forcing history—the farther downstream along the paths of surface or thermohaline circulation, the older the record. By subducting parcels of water containing climate information reflecting a particular time, and later returning them to the surface, the THC temporally translates climate conditions and influences the evolution of future climatic changes. (This record of past climate, manifested in the temperature and chemistry of the water, is also a valuable scientific tool that can provide insight into the history of climatic processes on the time scale of years to many centuries.) For example, the relatively deep-reaching convection in the northern North Atlantic, which typically extends down to 500 m but often attains depths exceeding 1000 m, provides an advective memory. A heat anomaly spanning the thermocline—for example, warm/cold 18°C water atop an unusually deep/shallow thermocline—can "remember" and preserve a residual influence for great distances, not just through a seasonal sequestering cycle, but through many years of recurrent sequestering and re-exposure. McCartney et al. (1997) show that the anomalies that arrive in the area west of Ireland 10 years after their initiation in the western subtropical gyre are not dissipated there. They not only propagate northeastward in the Norwegian Current, but move west across the subpolar gyre to the Labrador Basin, contributing a 20-year-old "memory" to the warming and cooling cycles of the LSW.

A second mechanism important to dec-cen variability is the subduction or detraining of mixed-layer water conditions into the thermocline in some regions. This process reorganizes the vertical stratification on which the wind-driving of the circulation and the surface thermodynamic forcing act. Subduction alters both the advection fields and the distribution of the properties on which advection acts, thus modifying the ocean's redistribution of heat, freshwater, and carbon dioxide. This sequestering of mixed-layer water is central to the ocean's active role in climate variability. Subsurface alteration continues at the same time that the advecting waters are isolated from the mixed layer through stirring by eddies and mixing. Also, all these subsurface changes are reflected in the near-surface field of currents in which the surface mixed layer and Ekman layers (i.e., layers affected by wind stress) reside. These deeper conditions, along with dynamic thermocline adjustments, therefore contribute to local SST variability through advection. While the "flushing time" by subduction in the subtropical thermocline appears to be interdecadal, the structural changes in subtropical circulation are significant decadally and in some areas interannually (see, e.g., Jenkins, 1998). Subduction represents the annual average of the seasonal cycle of buoyancy forcing and wind forcing of Ekman layers and gyre circulations. The nature of that seasonality is such that waters imprinted with winter conditions are selected for sequestering. The subduction mechanism was recognized early (e.g., by Iselin, 1939). The existence of some sort of selection process has also been suspected for some time, because the temperature and salinity characteristics observed at depth in subtropical gyres most closely resemble winter mixed-layer properties, rather than the variety of other waters that exist throughout the seasonal cycle at the sea surface. This selection bias was noted by Stommel (1979); it has been called his "mixed-layer demon."

Third, the ocean's active role in dec-cen variability is also manifested through the process of re-entrainment of subsurface waters into the mixed layer, after a period of sequestering, so that the stored properties like heat content are re-exposed through the mixed layer to air-sea exchange. The period of sequestering may be very long—on the time scale expected for water moving along with the THC, or the time required for a parcel subducted into the subtropical gyre to circle the gyre and find its way to a place where accumulated buoyancy forcing re-exposes it to the atmosphere. In the case of "obduction" (Qiu and Huang, 1994), the sequestering can be seasonal: A mixed layer becomes capped by a seasonal thermocline, and is carried downstream during the warm season. During the cold season, the cap erodes, re-exposing the sequestered mixed layer. When the mixed layer is cooled further by its re-exposure, additional water is entrained into it from below.

The most easily recognized example in which subduction and obduction occur is the formation of "mode waters," which involves deep winter convection. The subduction process to a large degree preserves the extraordinary vertical homogeneity found in winter, to the extent that a tapered wedge extends equatorward from the winter outcrop, following the subtropical-gyre circulations.

Internal Ocean Variability Mechanisms

There are a number of interactions internal to the ocean that imply dec-cen variability without the participation of the atmosphere at all; they have been simulated in ocean models where the atmospheric forcing is kept constant. While these are, of course, subject to the criticism that the atmospheric response could change the variability, they illustrate the role of purely oceanic interactions in inducing variability.

Most of the ocean-only mechanisms that have been reported involve anomalies that reach water-formation regions and change the density, thereby affecting the rate of deep-water formation. Weaver and Sarachik (1991) and Chen and Ghil (1996) found purely advective mechanisms for this sort of decadal variability. The weakening (or strengthening) of deep-water formation can induce changes in circulation that may counteract an anomaly or reverse it. Longer-time-scale (centennial) oscillations (see Weaver et al., 1993, or Winton and Sarachik, 1993) were found to originate from the shutoff of deep water, and the entire THC, by strong freshwater fluxes at the surface. The THC returns when heat diffusion warms the deep water enough to produce an unstable high-latitude column, and convection is re-initiated. The time scales on which these processes occur in models can range from centennial to millennial, depending on the parameters involved (especially diffusion of heat).

A number of climate variability mechanisms involve coupling of the ocean and atmosphere in such a way that the ocean plays a dominant role. For example, the alteration of heat content and density structure at depth, because of the variability of the waters subducted from mixed layers in winter, has been suggested by Gu and Philander (1997) to be a mechanism for modifying the equatorial thermocline. These subsurface changes cause alterations in the coupled atmosphere-ocean ENSO oscillation; these lead to atmospheric changes at mid-latitudes, which then complete a feedback loop by affecting conditions in the subduction sites. A simple box model constructed by Gu and Philander (1997) found that this feedback loop caused interdecadal oscillations.

A thermodynamic mechanism for interdecadal dipole SST oscillations in the tropical Atlantic has been proposed by Chang et al. (1997). The cross-equatorial dipole of warm and cold SST anomalies can experience heat-flux anomalies that arise from and are sustained by wind-induced latent-heating anomalies. This positive feedback is countered by a negative feedback that is provided by the advection of heat by the south-to-north flow of the North Brazil Current, the principal cross-equatorial warm current of the THC. This current provides oppositely signed SST (heat-content) anomalies that reverse the atmospheric-circulation anomalies, and then remove the SST anomalies by reversing the evaporative heat-flux anomalies. An alternate mechanism to this cross-equatorial negative feedback has been proposed by Zhou (1996). He suggests that the negative feedback is provided by the action of altered winds, which produce SST anomalies that have opposite signs in lower and higher tropical latitudes. When these anomalies are advected by the surface circulation; they eventually restructure the atmosphere to yield anomalies with the opposite sign, again producing an interdecadal oscillation. Tourre et al. (in press) propose a third mechanism, involving flexure of the thermocline.

Predictability

Hopes for predictability of climate variations intrinsically involving the ocean are based on the slow responsiveness of the ocean relative to the atmosphere, as discussed in Chapter 4. Initialized prediction is based on the concept that once the ocean is set in motion, it will follow a predictable path. Loss of predictability will then be due to two basic factors: imperfect specification of the initial state, so that the initial errors inevitably grow, and random (unpredictable) eddy "noise" in the atmosphere and ocean that progressively contaminates the forecast as it evolves. At this writing, not much progress has been made on characterizing the relative effects of these error-growth mechanisms on decadal time scales. While predictability has been indicated (see Chapter 4), we know little about the ultimate limits of initialized predictability.

For the long-time-scale prediction of ocean evolution, we need improved understanding and modeling of the processes that transport heat in the ocean, the coupled processes that influence SST, and the divergence (net flux) of heat, freshwater, and CO_2. We must improve our understanding of, and ability to model, how these processes vary in time and on local, regional, and global scales, and what their primary dependencies and interdependencies are. To do so will require improved treatment of the interactions between thermodynamic processes (e.g., those associated with boundary-layer fluxes) and dynamic processes. For example, a detailed representation of the mixed layer is required in order to model correctly the responses of upper-ocean SST to a given atmospheric heat flux; the model must be able to take into account vertical property fluxes, lateral property fluxes associated with lateral or isopycnal mixing (including the effects of eddies) and advection, and the proper response of the pycnocline (the band where there is a strong density gradient) to the surface wind-stress curl, which further affects surface water volume and stratification of the upper ocean.

One of the difficulties inherent in dec-cen predictions of ocean-circulation variations is that while we must include more and more of the ocean as we consider longer and longer time scales, we must still treat the small-scale processes as well. A great variety of disparate processes, operating over a broad range of spatial and temporal scales, must therefore be incorporated into predictive models. Performing long integrations over global space scales, while including detailed representation of local-scale processes (and in the case of the carbon system, multi-disciplinary processes), poses a considerable computational challenge.

For example, the thermohaline circulation's rate, volume, properties, and flow paths are sensitive to local-source region processes that control the location, depth, and rate of convection, and the extent of mixing during convection and subsequent flow. THC flow is also sensitive to the treatment of the boundary layer, which will influence the flow path and the mixing along that path, which in turn influence the THC's properties, equilibrium flow depth, and geographic displacement. Many of these local processes are sensitive to small variations in the vertical stratification and the evolution of the mixed-layer. These variations in turn are sensitive to regional-scale dynamic forcing, which influences vertical velocities and pycnocline characteristics such as depth and vertical spread; to surface buoyancy fluxes driven by turbulent and radiative forcings or by freshwater exchanges, which are often associated with the melt, drift, and decay of sea ice; and to the large-scale circulation systems and water-mass distributions that determine the property contrasts across the pycnocline, and the strength of the stratification. The regional-scale processes influence the duration of convection and the preconditioning for convection. If any of these details is incorrect, a model can introduce a small (or large) systematic error (i.e., drift) that can bias the results, or yield an evolution or set of balances that differ from those present in the real world.

Intermediate waters, which to a large degree constrain the upper ocean's volume, are influenced by detailed local processes: convection, confluence of surface waters with outcropping deeper isopycnals, and sea-margin processes. These marginal processes, exemplified by the production of North Pacific intermediate water in the Sea of Okhotsk (Talley, 1991), consist of interactions involving the dynamics of small enclosed basins, ocean-sea-ice interactions, and exchanges through narrow and shallow sills, all of which must be properly represented in models in order to capture the characteristics and sensitivities of the North Pacific intermediate water.

Likewise, the proper treatment of eddies and of isopycnal and diapycnal mixing will require improved understanding and model representation—though some progress has recently been made in both the modeling (e.g., Gent and McWilliams, 1990) and the relevant observations (e.g., Polzin et al., 1997)—of boundaries, open ocean and narrow passages (for instance, the temperature and salinity characteristics of the Indonesian throughflow are altered by strong mixing during the passage from the Pacific to Indian Oceans (Ffield and Gordon, 1996). Boundary-layer treatments at all boundaries—surface, wall, and bottom—including topographic interactions and influences on interior mixing, must still be refined.

In addition to improving our understanding of processes and their model representation, we need to simulate accurately the largest-scale patterns of variability and co-variability. We must also properly capture the general statistical characteristics and evolution of ocean variability (for instance, the variabilities associated with the formation of mesoscale eddies in response to several different mechanisms). Such variability often represents a natural regulator of change or acts as a natural stabilizer. Consequently, if we expect to simulate the ocean under changed conditions, we must improve our ability to represent these fundamental characteristics of turbulent fluids.

Remaining Issues and Questions

The dec-cen ocean issues involve defining and understanding the patterns and mechanisms of the participation of the ocean in climate variability: (1) formation and circulation of water masses that link surface forcings to the subsurface ocean; (2) the variabilities of those water masses and the forcings that alter subsurface ocean properties and circulation; (3) those subsurface changes that can cause heat flux and SST changes, locally or remotely, through the action of advection; and (4) those changes that eventually feed back to alter the atmosphere. Related to these oceanic processes and characteristics are the following questions that are central to advancing our understanding of dec-cen climate variability.

• *What are the dec-cen patterns of near-surface ocean variability, and what dynamical mechanisms, both in the atmosphere and in the ocean, govern them at dec-cen time scales?* To understand this question, we have to answer an associated question: What are the feedbacks and coupling mechanisms in the ocean that change SST, heat, freshwater, sea-ice, and chemical anomalies on dec-cen time scales? The rich literature on patterns of atmospheric variability has no parallel for the ocean. Correlations of SST and its associated forcing fields with these atmospheric patterns have been only partly explored. Much remains to be done on documenting the ocean anomalies that co-vary with the atmospheric anomalies, and their relationship to the subsurface property, circulation, and dynamic changes that underlie the patterns and redistribution of SST anomalies. Differences between basins need to be explored; for example, the participation of the oceans beneath the NAO and the PNA action centers may differ because of the presence of deep overturning in the North Atlantic, and its absence in the North Pacific. Determining the predictability of coherent dec-cen variations and the ocean's role in this predictability depends on untangling their exact mechanisms, including the ocean's role in their maintenance and evolution. We must also identify those processes and ocean scales that dominate pattern evolutions on different time scales. For instance, the ENSO pattern and ENSO-like pattern of Figures 3-6 and 3-8 look similar, yet operate on different time scales; it is not clear whether they represent fundamentally different processes or interactions.

In addition to merely expanding our catalogue of ocean patterns, which may be particularly important in the Southern Hemisphere where the dearth of data has precluded as thorough a search as has been conducted in the Northern Hemisphere, we must also begin to examine and understand the interactions between the patterns, and determine whether the numerous patterns discovered to date are not simply regional subsets of larger patterns. For example, the variability of the tropical Atlantic SST dipole appears to be related to the northern Atlantic SST distributions associated with the NAO, and the PDO has been found to be correlated with the SOI. What are the mechanisms that account for the communication between these co-varying patterns, and what is the ocean's role in these mechanisms? How do the slowly evolving patterns influence the more rapidly evolving ones? The most obvious question here is, how does the interdecadal ENSO-like pattern influence the similarly shaped ENSO pattern? Also, are all these patterns, or documented variations in a number of characteristics in some regions, simply different manifestations of the same phenomenon, or are they indeed different phenomena? Finally, what societally relevant indicators co-vary with these patterns? As was shown earlier, fluctuations in the PDO and in salmon catch resemble each other.

• *What are the dynamic and thermodynamic mechanisms and interactions that control SST on dec-cen time scales?* How are anomalies preserved, relocated, expanded, or dissipated in the oceans? How do the processes of formation and sequestration of water masses vary on dec-cen time scales, and what governs the masses' subsequent modification and eventual return to the surface? How are anomalies of heat, freshwater, and chemical constituents translated into mixed-layer anomalies? How do the mixed-layer anomalies get into the ocean interior? How are they modified as they circulate through the interior, and how are water masses re-entrained back into the mixed layer? How do freshwater fluxes (which are influenced by evaporation-minus-precipitation, sea ice, and runoff) modulate these processes through the creation of salinity anomalies? How do dynamic effects, such as gyre spin-up or thermocline pumping and adjustments, modify the vertical stratification, and thus influence and interact with surface fluxes? More generally, how do the THC, wind-driven circulation systems, and surface fluxes interact to influence SST anomalies and their distribution and evolution? How do the boundary currents influence gyre heat transport and SST, and what is their role in shallow oceanic overturning?

On the largest scales, what is the sensitivity of heat (and freshwater) transport to changes in circulation, eddy activity, and other factors? What effect does the interaction between surface salinity and temperature have on the development and evolution of SST anomalies, especially since temperature and salinity operate on different time scales? Also, what mechanisms are responsible for tropical-extratropical and polar-extrapolar SST relationships? For example, how do changes in intermediate-water formation in subpolar regions influence subtropical thermocline strength and depth, and thus subtropical SST? Or, conversely, how do the latter affect the former? How are SST anomalies, or other anomalies that influence SST, communicated to different basins by the Antarctic Circumpolar Current?

• *How do the ocean circulation and water-mass pathways (and thus the effectiveness of heat, freshwater, and CO_2 sequestration and transport) vary on dec-cen time scales; how are they affected by surface forcing; and what are the governing mechanisms?* What are the relative roles of wind, thermal forcing, and haline forcing, and how do they interact? What are the expressions of these variable forcings in the surface ocean? Processes thought to modulate the intensity of the meridional heat transport effected by gyre circulations and western-boundary currents are eddy-driven sub-basin- and basin-scale recirculations, and the remote influence of wind-stress and buoyancy anomalies via Rossby and coastal waves. What are the relative roles of these mechanisms in dec-cen variability of heat transport and SST? How does the advection of salinity anomalies feed back onto surface freshwater anomalies, heat transport, and SST anomalies? How do anomalies survive the seasonal cycle to reappear in subsequent winters, as the persistently recurrent winter SST anomalies have been observed to do? What is the role of mode waters in maintaining these anomalies and in heat storage? What are the roles of recirculation and mixing in determining the effective propagation rates of SST anomalies, which are usually an order of magnitude less than the mean flow? How do water masses evolve at higher latitudes, and what are the dependencies and sensitivities of deep- and intermediate-water formation? In these regions the sequestering is strongly seasonal; there is recurrent exposure in winter of advected heat-content anomalies, which are manifested as SST anomalies propagating downstream through the warm-to-cold water-transformation pathways. What mechanisms control the strength, heave, and wobble of gyres on dec-cen time scales? What processes control the formation and persistence of large-scale salinity anomalies, and how do these anomalies interact with the atmosphere?

• *What are the processes that determine the uptake of carbon dioxide by the ocean?* The uptake effectiveness of the large oceanic carbon reservoir is sensitive to the ocean's large-scale structure and circulation, and to physical-biogeochemical interactions, all of which have the potential for change under altered climate conditions. This potential, in combination with the ocean's large carbon-storage capacity, suggests that even relatively small changes in uptake or storage effectiveness may have a large influence on atmospheric CO_2 concentrations. It is thus critically important that we gain a better understanding of what changes are possible, and what their impact on the ocean's carbon storage and exchange may be, so that we can ultimately predict future atmospheric CO_2 concentrations as a function of particular emission scenarios. Among the questions that need to be resolved are what the spatial distributions of the boundary reservoirs are, and how they depend on geophysical, biological, and chemical processes and on interactions that occur over local, regional, and global scales. These processes interact over a broad range of scales, and our ability to represent them on these scales is still in its infancy.

Processes and Parameterizations

The behavior of the ocean is critical to our understanding of climate variability and prediction primarily because of its involvement in the air-sea exchanges of heat, water, and carbon dioxide; in a sense, it is a boundary condition for the atmosphere, though a fairly elaborate one, with the potential for internal oscillations and long-term feedbacks. For monthly or even seasonal prediction it may be adequate to use a locally forced mixed-layer model—even one-dimensional—as a representation of the ocean's role. For the longer time scales, interannual through dec-cen, there is an increasing need to include oceanic advection in the predictive models. We need to explore computationally efficient methods to extend to decadal time scales the initialized coupled models that are currently being developed for seasonal-to-interannual forecasting, thereby capitalizing on the decadal-scale predictability that has been suggested for the North Pacific and North Atlantic. A hierarchy of models to improve the treatment of oceanic processes at variety of scales, though, ultimately will require global models capable of resolving (or at least properly representing) the important local-scale processes, including boundary-layer treatments (at all boundaries, surface, wall, and bottom).

The processes and parameterizations that most warrant improved understanding include: ocean mixing (including diapycnal and eddy); the deep-water formation processes; and pycnocline response to surface forcing (strength, wobble, and heave) given different stratification and pycnocline characteristics, and surface forcings. Implicit in improvement in understanding these are better parameterizations of the relationship between internal wave fields, the mixing effects of sea-floor topography and tides, entrainment processes, processes such as salt fingering/double diffusion in a variety of regimes, mixing in high-latitude regions where convection often leads to layered regimes, and a number of processes related to the nonlinear equation of state of sea-water that may prove particularly important for deep-water formation.

There is a distinct need to improve our representation of surface boundary-layer processes, including dynamic and thermodynamic responses and interactions. Good representation of these processes is fundamental to our coupling of the ocean to the atmosphere, yet there are still considerable weaknesses in our three-dimensional treatment of the surface layer. Local process models have sometimes shown that small changes in surface stratification can lead to large changes in the evolution of surface mixed-layer properties over time. Also, few models have fully treated the complex vertical processes, their interaction with the ocean's horizontal structure, and the related advection and diffusion. Similarly, we need to improve our understanding and treatment of how the pycnocline interacts with the surface layer. This interaction takes place across all space and time scales, since the response will be dependent on the overall stratification, which is set by the thermohaline circulation and surface forcings, gyre-scale dynamics, how the pycnocline adjusts, and by local forcing and mixing that may introduce local perturbations and variability that can influence the longer-term evolution of the upper ocean. Representation of these processes also requires knowledge of the sensitivity of the pycnocline to ventilation processes, and to mode- and intermediate-water formation (which sets the characteristics at the base of the pycnocline).

The TOGA Coupled Ocean-Atmosphere Response Experiment (COARE) was a successful series of observations and modeling studies of a large-scale meteorological and oceanographic process. That same philosophy is behind the proposals for basin-wide studies of ENSO and decadal variability in the Pacific, and of the NAO and tropical variability in the Atlantic. Studies of this type may provide the climate information best suited to addressing such phenomena as those mentioned in the paragraph above.

Observations

A dec-cen ocean program should include several observational elements that are directed towards elucidating the physics of key phenomena and processes, both to guide their representation or parameterization (if they cannot be fully resolvable in models) in model simulations, and to provide a framework for interpreting the decadal signals we see in the data and the models. Because ocean measurements are never likely to be dense enough to totally define the state of the ocean, models must be integrated at the very beginning of any program in order to produce a dynamically consistent model-assimilated data set. Furthermore, as discussed above, for dec-cen time scales it is imperative that measurements be made in the surface layer, upper ocean, and sub-pycnocline; these features directly affect the surface conditions, stratification, and lateral displacement of anomalies and heat transport. A number of studies have addressed the specific details required to attain an optimal ocean observing system (e.g., NRC, 1997). These take into account viable new instrumentation and climatic needs, and are highly relevant to a dec-cen program as well.

The description and understanding of the oceanic phenomena that may affect dec-cen-scale variability need comparable degrees of concentrated inquiry. How does subducted water (and the associated anomalies) mix and evolve in flows around the subtropical gyre, and how does it define the vertical density and circulation structure of that gyre? The subduction of subtropical water and its subsequent influence on the eastern equatorial Pacific may be a possible modulator of ENSO on decadal time scales; a similar argument can be made for the relevance of such processes to the tropical Atlantic's dec-cen variability of SST, perhaps through the formation of saline subtropical "underwater" (Saunders and Harris, 1997). Can we define the transport pathways and quantify the mixing that occurs between the times when water is subducted into the subtropical gyre and when it is re-exposed to the atmosphere at the equator? Conversely, a focus on the North Brazil Current and current retroflection would address the potential role of trans-equatorial heat advection as a negative feedback to the Atlantic's dipole SST oscillation.

Occasional concentrated efforts like the IGY (International Geophysical Year) and WOCE have provided high data densities; high-frequency time-series station measurements are invaluable, and occasional measurements in between such surveys allow some continuity of mapping fields. These periodic (decadal) concentrated surveys, repeated measurement sections, and higher-frequency time-series station measurements will continue to be needed in regions of demonstrated dec-cen variability associated with the primary known patterns of atmospheric climate variability. Such measurements will make it possible to continue to extend the quantitative description of the ocean's participation in dec-cen variability. This will be particularly valuable given the observations that we have of slowly propagating SST and subsurface anomalies, indicating that the ocean's dec-cen variability is more complex than mere fluctuations in stationary patterns.

Unfortunately, the international hydrographic/tracer/CO_2 data-collection effort has been fading with time. While various new technologies now provide alternatives to ship-based measurements, these are mainly supplements rather than substitutes, because of sampling and sensor limitations. The global historical datasets, in particular the WOCE sampling completed in 1997, suggest Southern Hemisphere regions that are promising candidates for exploratory revisitation and repeated occupation now that their dec-cen variability has begun to be revealed. Of proven value for defining variability at all frequencies are stations making time-series measurements, yet few of these remain. Based on their contributions to dec-cen variability, an evaluation needs to be made as to which existing stations must be continued, which discontinued stations should be reinitiated, and which new ones started. In a limited way, sparser happenstance measurements can fill in the gaps left by interrupted stations, and can be used to derive a background history for new sites. New time-series stations can make use of the moored profiling conductivity-temperature-depth (CTD) instruments now coming online, reducing the need for ship-based support measurements to mooring-maintenance cruises. Expendable-bathythermograph coverage needs to be enhanced; the coverage area should be expanded, and more high-resolution tracks should be included. In some regions, sampling depths need to be increased, so that in addition to upper-ocean heat-content change, the depth of the thermocline beneath the local winter mixed layer will be sampled. These improvements will provide the knowledge of advection and anomaly propagation that is critical for understanding dec-cen climate variability.

Continued satellite data are needed for global coverage of sea-surface height, SST, winds, and ocean color, but if they are to be useful, corresponding ground truth must be established through ocean observations. Improved XCTD measurements are needed to enhance the present sampling of upper-ocean salinity, so that buoyancy can be better quantified. Subsurface floats with CTDs (PALACE floats, gliders), parked at some depth, and profiling from that depth or deeper to the surface on roughly a weekly schedule for multi-year lifetimes, would provide both subsurface-pathway information and a Lagrangian time series of hydrographic stations. Better shipboard measurements, in support of moored and drifting CTDs are needed, including those of chemical tracers and CO_2 parameters. Improved shipboard measurements could be used to refine the calibration of salinity sensors, thereby allowing CTD measurements to be drift-corrected, and increasing their utility for dec-cen climate studies. Furthermore, enhanced tracer and CO_2 observations would be very useful in documenting and constraining changes in the rates and fluxes of a number of ocean variables, and would be invaluable for model verification. Subsurface floats with trajectory recording (RAFOS and MAVOR) are particularly useful in defining circulation pathways. In the dec-cen context, such observations can be employed to elucidate aspects of the heat budget. A concerted effort needs to be made to improve estimation of heat-flux divergence and heat storage (and their variabilities) from subsurface ocean measurements, and to reduce disparities between those estimates and estimates of air-sea heat exchange.

In order to validate predictions of sea-level rise, better monitoring of global sea-level change and its components will be needed. The prospects for sea-level monitoring are good. A global network of sea-level measuring stations (the Global Sea Level Observing System, GLOSS) is being implemented, and needs to be maintained in the future. At some of these stations, land movements will be measured with satellite geodesy and gravimetric techniques. Satellite altimetry is another important tool that should be fully exploited to measure global sea-level rise.

The estimates of thermal expansion and ice melt over the last century generally add up to less than the measured sea-level rise. An important task for the future is to refine these estimates and close this gap between models and observations. Future global ocean programs, especially CLIVAR, will make it possible to measure the thermal-expansion component directly. Regarding future sea level, the IPCC's (1996a) best estimate is that by the year 2100 it will be 38-55 centimeters higher than it is today. Better observations of SST, salinity, and wind stress will help reduce the uncertainty of the ocean-modeling component of such forecasts. Further development of these global models and their assimilation schemes is necessary to take full advantage of the available and desired observations.

CRYOSPHERE

The part of the Earth's surface that remains perennially frozen, as well as the portion that is near or below the freezing point, constitutes the cryosphere. Our working definition of the cryosphere here is all the forms of frozen water on the land or sea surface, whether admixed (as in permafrost) or pure (as snow or ice). Thus, this section addresses not only glaciers and sea ice (perennial and seasonal), but also vast areas of frozen ground and permafrost, as well as seasonal snow fields that lie beyond the limits of glaciers. At present, glacial ice covers about 3 percent of the Earth's surface, while containing nearly 75 percent of its non-ocean water; sea ice covers another 7 percent; and perennially frozen ground covers 20-25 percent of the exposed land surface.

Influence on Attributes

The cryosphere influences the climate attributes listed in Chapter 2 both directly and indirectly. Its most obvious direct influence is on sea level. Because glacial ice contains a tremendous fraction of the Earth's freshwater supply, it represents a significant source of stored water that could be released to the oceans. If the Antarctic ice sheet, which contains approximately 90 percent of the world's glacial ice, were to melt, global sea level would rise approximately 70 m (IPCC, 1996a). If only the West Antarctic ice sheet were to melt (a more likely scenario), it would be sufficient to raise global sea level about 6 m. Consequently, the mass balance of the glaciers represents a direct influence on sea level.

Currently, the most direct contribution of glacial ice to sea-level change occurs through the melting of alpine glaciers around the world (only a small minority are growing), and the melting of the underside of ice shelves—the portions of continental ice sheets that have extended beyond the continental margins and are currently floating on the seawater. It has recently been estimated that about half of the sea-level rise realized over this last century is the result of the melting of glacial ice (IPCC, 1996a), though this number is highly uncertain. The most uncertain sources of sea-level change are the Greenland and Antarctic ice sheets, since their current and likely future responses to climate variations are poorly known.

In several countries glacier-fed streams contribute directly to water availability by supplying much of the water used for industry, agriculture, energy, and domestic purposes. For these countries, glaciers serve as a natural reservoir that stores water during winter and releases it in summer, giving streams a distinctive pattern of runoff. (Especially large quantities are released in warm summers, just when water from other sources is in short supply.) Mass-balance measurements may be used to estimate how much water can be stored and released in this way, and the amount of variation that can be expected from year to year (Paterson, 1994), allowing water-use planners to allocate water prudently.

The vast expanses of highly reflective surface area in the cryosphere directly affect the global radiation balance by enhancing the equator-to-pole temperature contrast, which represents the heat engine driving the Earth's climate system. Thus, changes in the ice- and snow-covered areas of the polar regions may be expected to influence the large-scale climate system and, through it, temperature, precipitation and evaporation, and possibly storms.

The distribution and extent of permafrost influence the effectiveness of land-surface storage of greenhouse gases, such as methane, and their exchange with the atmosphere. The partitioning of greenhouse gases between climatically benign storage reservoirs and the atmosphere can be significantly altered by a thermally induced change in permafrost extent.

Likewise, several of the major climate patterns discussed in Chapter 3 (e.g., the PNA and NAO) extend well into the polar regions, which are the principal centers of the cryosphere. Variations in the indices describing these patterns seem to be dominated by changes in the low-pressure systems in the northern extremes of these patterns. (Changes in the Aleutian low influence the PNA index, and changes in the Icelandic low influence the NAO index.) This dominance reflects the fact that the subpolar low-pressure systems appear to be less stable than the rather static mid-latitude high-pressure ridges that control the other limits of the indices. These low-pressure cells vary considerably in strength and extent. Since they extend well into the polar regions, and are to some degree dependent on the ice distribution at their lower surface, they might be expected to show some degree of dependence on regional ice and snow distributions as well. This dependence has not been clearly documented, and is not well understood, but it has the potential to extend polar influences to climate attributes elsewhere through their influence on teleconnection patterns.

Evidence of Decade-to-Century-Scale Variability and Change

Ice caps and ice sheets continuously record the chemical and physical nature of the Earth's atmosphere. Ice cores drilled from carefully selected sites often provide paleoenvironmental records with seasonal, annual, decadal, and centennial resolutions. Such cores, some of which date back through the last major glaciation (more than 20,000 years ago), have been recovered from locations as diverse as Antarctica, Greenland, the Tibetan Plateau, and the Andes of Peru and Bolivia. Data from the upper parts of these cores have been integrated with other proxy indicators to yield a high-resolution global perspective on the Earth's climate over the last 1,000 years (see Figure 5-27). These records uniquely capture a history of most of the major climate zones. They indicate considerable natural variability during the present climate epoch, including a number of climate extremes such as the Little Ice Age. Some of the variations

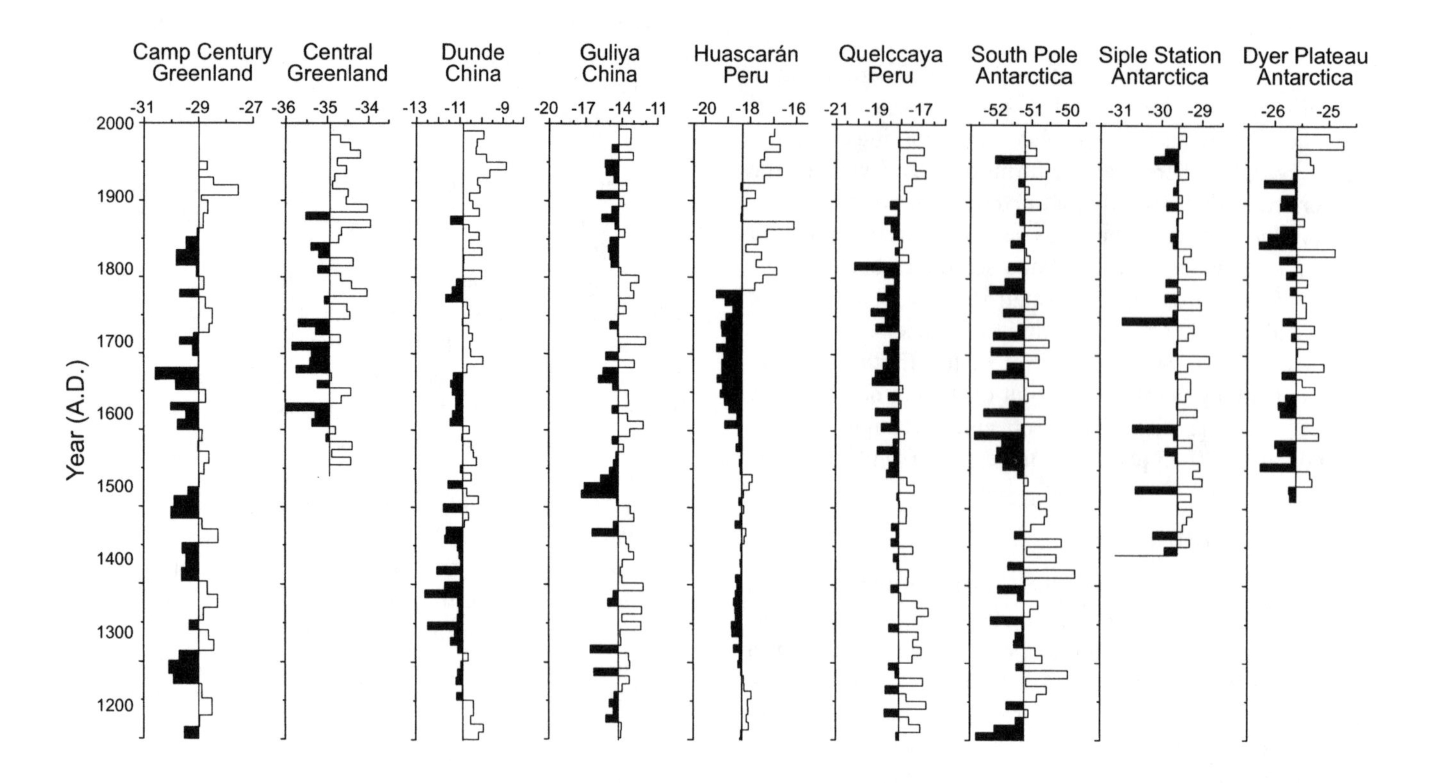

FIGURE 5-27 Decadal averages of the oxygen isotope records are shown for a north-south transect from Camp Century, Greenland, to the South Pole, Antarctica. The shaded areas represent more negative, or cooler, periods (as deduced from $\delta^{18}O$ isotopic evidence) and the unshaded areas represent more positive, or warmer, periods relative to the respective means of the individual records. (After Thompson et al., 1993b; reprinted with permission of Elsevier Science.)

appear to be regional in extent, while others are clearly global or near-global (see, e.g., Bender et al., 1994; Stager and Mayewski, 1997).

Additional evidence of variability has been obtained from boreholes in permafrost terrain, and to a lesser extent (because of the shortness of the record) from modern sea-ice and snow-field observations. Oxygen-isotope records from deep-sea sediment cores also indicate long-term variability in the global volume of landlocked ice (Shackleton and Opdyke, 1973), though most of these records are not of a high enough resolution to address dec-cen-scale variability. A similar statement can be made regarding the geomorphological evidence of glacial changes.

Cryospheric evidence of dec-cen-scale climate variability is presented below by region.

Polar Evidence

The waxing and waning of major continental ice sheets during glacial and interglacial stages are recorded in ice cores from both polar regions, as well as those from China, Peru, and Bolivia. However, climate variations during the Holocene apparently were not globally synchronous; they may appear prominently in some records but be wholly absent from others. For example, a period of cooler conditions, correlative with the so-called Little Ice Age from about AD 1500 to 1880, is prominent in cores from a number of sites in East Antarctica. (The ratio of oxygen isotopes, $\delta^{18}O$, can serve as an atmospheric-temperature proxy (Yao et al., 1996)). An oscillatory temperature relationship observed over the long term between the ice-core records of East and West Antarctica is also present in the temperature records of this century (Mosley-Thompson et al., 1991). The ice-core $\delta^{18}O$ histories from the Antarctic Peninsula, however, reveal a strong, persistent warming trend since the 1930s that is absent in East Antarctica. Ice-core records from the Greenland ice sheet also do not reveal a marked warming in this century. In fact, with the exception of the Antarctic Peninsula, there is little evidence for recent warming in the polar regions. The rates of snow accumulation on the polar ice caps show variability over decades.

The sea-ice record, which is available from satellite observations, is approximately 25 years old. In the Southern Hemisphere, it shows some interdecadal variability, for which a physical explanation is currently lacking. The Antarctic sea-ice fields also show evidence of abrupt change during a several-year period in the 1970s—the appearance of the large Weddell polynya, or open-water space that has not been observed since. The extent of Antarctic sea ice does not show any significant trend over this period (Johannessen et al., 1995). However, Bjorgo et al. (1997) explain that Arctic ice extent appears to have decreased over the last 20 years or so, even when the calibration mismatch associated with the change in microwave sensors in 1987 (from the SMMR to the SSM/I) is taken into account. On longer time scales, more discernible trends and variations in the sea ice appear in historical records and in the extensive logs from fishery, whaling, and sealing fleets. For example, using whaling records, de la Mare (1997) found that the Antarctic sea-ice fields retreated a dramatic 2.8 degrees of latitude between the mid-1950s and early 1970s. Zakharov (1997) and Ogilvie (1984) note striking examples of decadal-scale variations of sea ice in the North Atlantic during the past hundred years, and century-scale sea-ice variations near Iceland. Several studies have found decadal-scale relationships between Arctic sea ice and atmospheric anomalies (e.g., Dickson et al., 1988; Deser and Blackmon, 1993; and Slonosky et al., 1997).

When so-called inverse modeling methods are used, analyses of precisely measured temperature profiles in permafrost boreholes allow reconstructions of past changes in temperature at the permafrost surface, and also of changes in the surface heat balance. Borehole temperatures thus serve as proxies for mean annual air temperatures. Borehole data indicate that temperatures at the permafrost surface in polar and subpolar regions have indeed undergone cyclic changes of as much as 4°C over the last decade or so (Lachenbruch and Marshall, 1986; Osterkamp et al., 1994).

Overpeck et al. (1997) found that over the last 400 years, most of the peaks in atmospheric volcanic-sulfate loading, as reconstructed from Greenland ice cores, correlate with episodes of mean circum-Arctic cooling. For example, the coolest phase of the Arctic Little Ice Age appears to have been precipitated by a period of large and frequent sulfur-producing volcanic eruptions in the early nineteenth century. These correlations suggest that a positive feedback may exist between atmospheric sulfate loading and Arctic temperature (Zielinski et al., 1994; Overpeck et al., 1997).

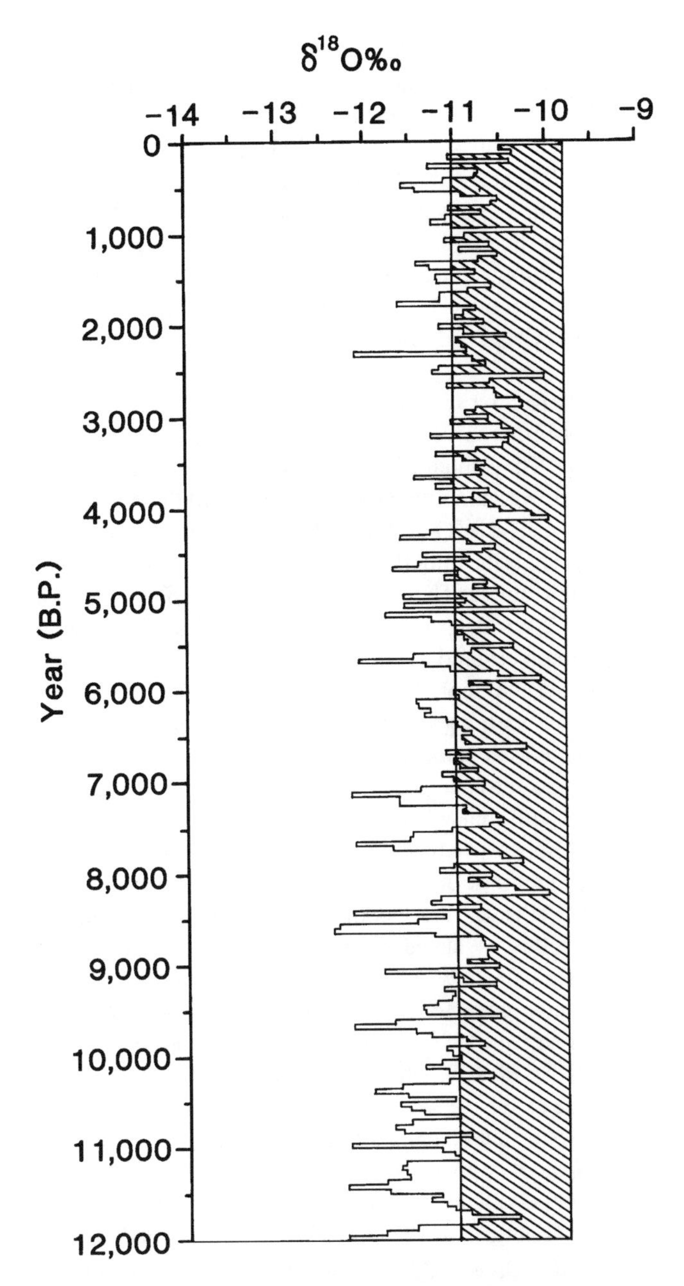

FIGURE 5-28 Fifty-year averages of oxygen isotopes for the last 12,000 years from Cores 1 and 3 on the Dunde Ice Cap, China. The line at -11 represents the long-term average of the records; shaded projections indicate warmer-than-average periods. Note that the most recent 50-year period (1937-1987) is the warmest since the end of the last glacial stage. (After Thompson et al., 1993b; reprinted with permission of Elsevier Science.)

Subtropical and Mid-latitude Evidence

In contrast to the abundance of records from the polar regions, ice-core histories from the mid-latitudes to the tropics are more limited. Recently, ice coring and glaciological investigations have been conducted on the Tibetan Plateau and in Kirghizia. These data, along with meteorological observations, strongly and consistently indicate a twentieth-century warming (Thompson et al., 1993b).

In 1987 two cores were recovered from the Dunde Ice Cap (38°N, 96°E, 5325 m above sea level (masl)) in the Qilian Mountains on the northeastern margin of the Qinghai-Tibetan Plateau (Thompson et al., 1989). These cores' history, presented in Figure 5-28, show that the latest 50 years of the record were the warmest in the last 12 centuries for the northeastern area of the Tibetan Plateau. On a longer time scale, an ice-core record from the Guliya Ice Cap (also on the Qinghai-Tibetan Plateau) provides evidence of regional climatic conditions over the last glacial cycle. ^{36}Cl data suggest that the deepest 20 m of this 308.6 m core may be more than 500,000 years old. The $\delta^{18}O$ change for the most recent deglaciation is ~5.4 per mil, similar to changes shown in cores from Huascarán (Peru) and the poles (Thompson et al., 1997). The oxygen isotopes vary in a pattern similar to that of the CH_4 records from the polar ice cores indicating that the global CH_4 levels and the tropical hydrologic cycle are linked.

In 1990 a joint U.S.-U.S.S.R. team visited the Gregoriev Ice Cap (42°N, 78°E, 4660 masl) in the Tian Shan region of Kirghizia. They obtained two 20 m cores containing records of climatic variation extending back to 1940 (Thompson et al., 1993b). The $\delta^{18}O$ profiles in both cores indicate a warming trend since the mid-1970s. The team also measured borehole temperatures. They found a temperature of -2.0°C in a borehole 20 m deep at 4660 masl, whereas a 1962 Soviet expedition measured a temperature of -4.2°C at 20 m even though their borehole location was at only 4400 masl. This indicates a warming of several degrees in the near-surface mean annual air temperatures since the early 1960s. (Because some refreezing of meltwater occurs on Gregoriev, this 2.2°C difference is considered an upper limit.)

There appears to be evidence of significant variability in the distribution of the snow fields in Canada between 1915 and 1992 (Brown and Goodison, 1996). Remotely sensed data from the Advanced Very High Resolution Radiometer (AVHRR) also show decadal-scale changes in the areal extent of snow cover over both Eurasia and North America (Robinson et al., 1993; Walland and Simmonds, 1997). The AVHRR-based data, which begin in the early 1970s, indicate more extensive snow cover in the 1970s to mid-1980s relative to the later part of the time series; the decline occurs over a span of approximately five years, from 1985-1986 to 1989-1990. Walland and Simmonds (1997) found significant co-variability between Eurasia and North America, with the Eurasian signal lagging behind the North American signal by over a year. The approximate 10 percent reduction in Northern Hemisphere snow cover that occurred in the latter part of the 1980s has likely increased the radiative balance and surface temperature (up to 1.5°C) over the northern extratropical land area, particularly in the spring (Groisman et al., 1994a,b). The Eurasian winter snow fields also appear to co-vary with sea-ice distribution north of Siberia (Maslanik et al., 1996).

Tropical Evidence

There is mounting evidence for a recent, strong warming in the tropics, which is signaled by the rapid retreat and even disappearance of ice caps and glaciers at high elevations. These ice masses are particularly sensitive to small changes in ambient temperatures, since they already exist very close to the melting point. One of the best-studied tropical ice caps is Quelccaya (13°S, 70°W, 5670 masl) in southern Peru. In 1983 two ice cores that went down to bedrock were recovered from there, providing the first conclusive tropical evidence of the Little Ice Age (Thompson et al., 1986). Since 1976 Quelccaya has been visited repeatedly for extensive monitoring. In 1991 and again in 1995 shallow cores were drilled at the summit, near the sites of the 1983 deep drilling and a 15 m coring in 1976. Comparison of the $\delta^{18}O$ records from these four cores (1976, 1983, 1991, and 1995) reveals that the seasonally resolved paleoclimatic record, formerly preserved as $\delta^{18}O$ variations, is no longer being retained within the currently accumulating snowfall on the ice cap. The percolation of meltwater throughout the accumulating snowpack is vertically homogenizing the $\delta^{18}O$.

The extent and volume of Quelccaya's largest outlet glacier, Qori Kalis, was measured six times between 1963 and 1995. These observations documented a drastic retreat that has accelerated over time. Brecher and Thompson (1993) reported that the rate of retreat from 1983 to 1991 was three times that from 1963 to 1983, and in the most recent period (1993 to 1995) the retreat was five times faster. Associated with this retreat was a sevenfold increase in the rate of volume loss, determined by comparing the 1963-to-1978 volume-loss rate to that of 1993 to 1995. Observations made in 1995 confirmed Qori Kalis' accelerating retreat (see the upper portion of Color plate 4), as well as further retreat of the margins of the Quelccaya ice cap, and the development of three adjacent lakes since 1983.

In 1993 two cores were drilled from the col of Huascarán (9°S, 77°W, 6048 masl), a mountain in the north-central Andes of Peru (Thompson et al., 1995). The $\delta^{18}O$ data from these cores (see the lower portion of Color plate 4) indicate that the nineteenth and twentieth centuries were the warmest in the last 5,000 years. Their $\delta^{18}O$ record and meteorological observations made in the region reveal an accelerated rate of warming since 1970, concurrent with the rapid retreat of ice masses throughout the Cordillera Blanca and of the Qori Kalis glacier.

Additional evidence exists for recent warming in the tropics. Hastenrath and Kruss (1992) reported that the total ice cover on Mount Kenya has decreased by 40 percent between 1963 and 1987. Kaser and Noggler (1991) reported that the Speke Glacier in the Ruwenzori Range of Uganda has retreated substantially since it was first observed in 1958. The shrinking of these ice masses in the high mountains of Africa is consistent with similar observations at high elevations along the South American Andes, and indeed throughout most of the world (see Figure 5-29). This general retreat of tropical glaciers is concurrent with an increase in the water-vapor content of the tropical middle troposphere, which may have led to warming in the tropical troposphere (Flohn and Kapala, 1989; Diaz and Graham, 1996).

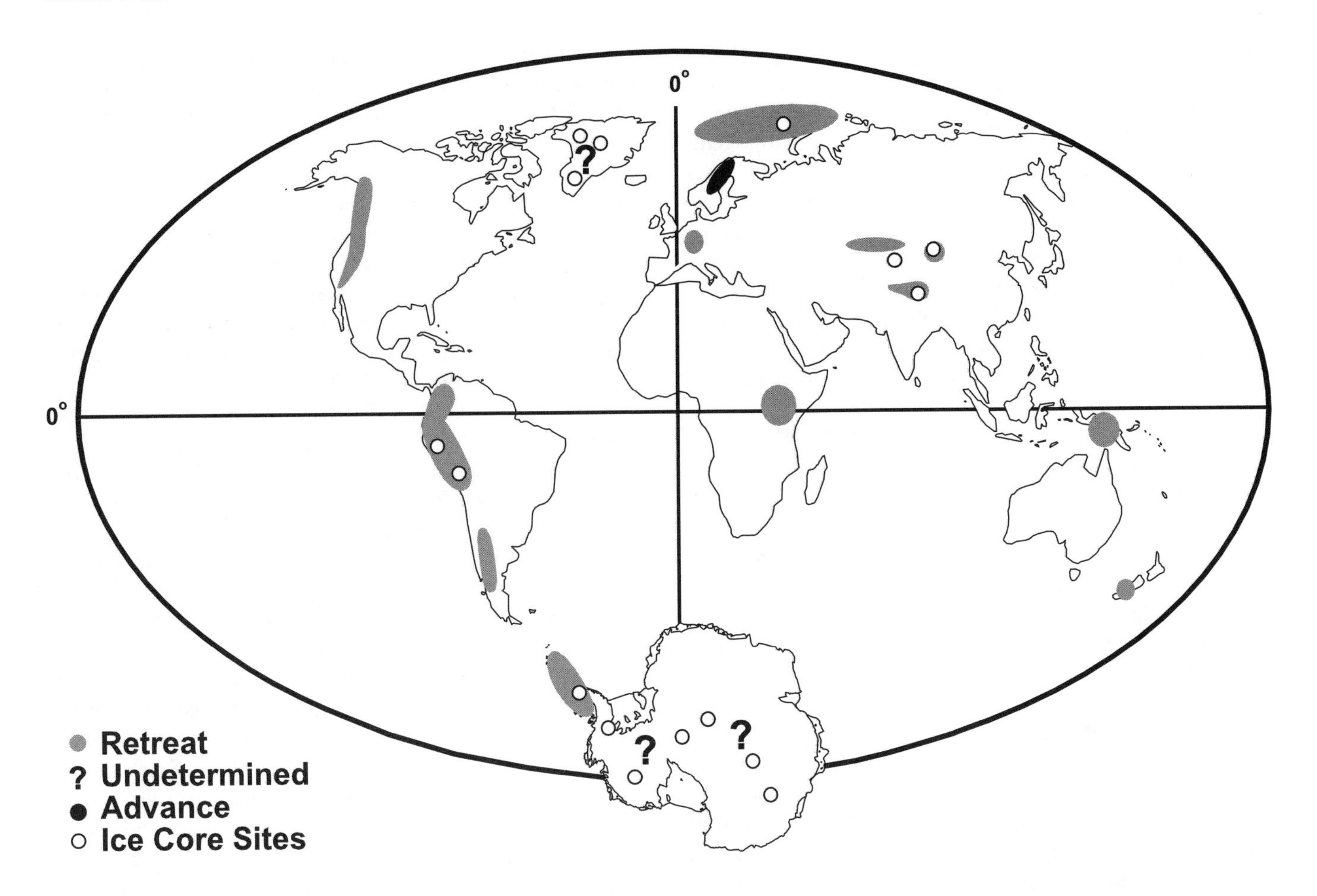

FIGURE 5-29 Changes in global ice cover during the twentieth century. (Figure courtesy of Lonnie G. Thompson, Byrd Polar Research Center.)

Mechanisms

The cryosphere is thought to be one of the most sensitive components of the climate system. Its variability is driven by external forcing, as well as by internal and coupled modes. Its response to external forcing is most dramatically demonstrated by the waxing and waning of the glacial ice sheets, which is paced by changes in solar irradiation. These irradiance changes, which are caused by changes in the Earth's orbital parameters (see, e.g., Hays et al., 1976), are small in magnitude, but significant in their seasonal distribution, specifically in the high Northern Hemisphere latitudes. These changes in total solar irradiance appear to be responsible for the largest variations in climate experienced over the last several million years: the ice-age cycles. While the details by which this small change in forcing is amplified are still unknown, Imbrie et al. (1992) suggest that the Arctic sea-ice fields respond directly to the radiative changes. The resulting change in freshwater exported to the thermohaline source regions of the North Atlantic influences the effectiveness of the thermohaline circulation system, thereby communicating the regional changes elsewhere in the globe. (For instance, large changes may be induced in the Antarctic sea-ice fields). While this particular scenario is controversial, the fact that the extent of the continental ice sheets shows strong correlation with the level of external radiative forcing (see, e.g., Imbrie et al., 1984) suggests a high sensitivity to that external forcing.

Volcanic activity can also have a strong influence on the cryosphere (Overpeck et al., 1997). Volcanic-ash deposition can change ice and snow cover from being one of the most highly reflective media in the climate system (and thus greatly resistant to direct solar radiative melting) to one of the most absorptive (and thus highly susceptible to radiative melting at the surface). This change in albedo can have a large influence on climate when ash is deposited on the relatively thin sea-ice cover; when the ice is melted, the low-albedo ocean surface is exposed, and further surface warming will result. Sea-ice cover greatly reduces the ocean-atmosphere exchange of heat, moisture, and radiatively active gases, so its removal would alter all of those properties. The reduction in the heat flux associated with sea-ice removal is typically between a factor of 20 (thin Antarctic ice) and 100 (thick Arctic ice), levels that can significantly influence regional warming and gas exchange.

Finally, the cryosphere's sensitivity to anthropogenic forcing is not yet known. GCM simulations often show a high sensitivity to changes in sea-ice fields in global-warming simulations. For example, recent numerical simulations of global climate, under conditions of doubled atmospheric CO_2, show that an impressive 38 percent of the annual average global warming could be attributed to the response of sea ice (Rind et al., 1995). Note, however, that while sea-ice changes seem to constitute a considerable positive feedback in the warming, it is highly uncertain what the response of the sea ice to global warming will actually be. This uncertainty is most clearly revealed by similar model experiments regarding the response of the Southern Ocean sea-ice fields to a doubling of atmospheric CO_2: smaller sea-ice changes occur in simulations using the coupled atmosphere-ocean GCM of Manabe and Stouffer (1994) than in the coupled model of Washington and Meehl (1995). Currently, the details of the anthropogenic forcing mechanism and its net influence are still very much in question.

Clearly the most important mechanisms influencing cryospheric variability are its couplings to the atmosphere, ocean, and land surface. They lead to a set of geographically unique polar feedbacks such as the ice/snow-albedo feedback, ice-cloud feedback, ice-ocean feedback (the effects of which apply to a variety of scales, from those influencing the sea-ice distribution to those influencing the vigor of the global thermohaline circulation), and ice-sheet-ocean feedback, including associated instabilities. Each of these feedbacks is discussed below.

The ice-albedo feedback encompasses an entire suite of processes that influence the concentration, spatio-temporal distribution, and surface characteristics of ice. All these characteristics influence surface albedo, which in turn alters the surface radiation balance, which feeds back onto the process itself. For example, sea-ice concentration—which reflects a balance between the advective divergence/convergence of ice and the thermodynamic decay/growth of ice (both processes open/close leads, areas of open water)—influences air-sea heat flux and albedo. In turn, both the air-sea heat flux and albedo influence the thermodynamic growth/decay rate, thereby further altering the ice concentration.

The surface characteristics of ice and snow, which can change albedos from a high of near 0.6 to a low of 0.3, are influenced by moisture content, seawater flooding, ice ridging, age, thickness, and crystalline properties. These are a function of dynamic conditions forced by the winds and ocean currents, ice thickness controlled by the atmosphere and ocean heat fluxes, atmospheric temperatures, snow loads, atmospheric surface history, and more. Many of these processes and characteristics are themselves a function of the albedo.

A comparable set of feedbacks involving land, snow, and atmosphere can influence the distribution of vast snowfields on high-latitude land masses. Because of snowfields' high albedo and great areal extent, even small changes in their distribution can have a considerable influence on regional and hemispheric conditions. The snow insulates the ground, controlling its temperature profile and heat content. These in turn influence the productivity of the local ecosystems, as well as the snow thickness and albedo, by moderating the snow's basal heat flux. The interaction between permafrost, vegetation, and the storage and release of methane also can lead to climatic feedbacks on longer time scales.

Snow-cover variations on the high Tibetan Plateau may be important because of the effect of surface albedo on the strength of the Asian monsoon (Sirocko et al., 1993). The intensities of El Niño events may also be influenced by plateau snow cover. On longer time scales, model simulations indicate that increases in snow and ice cover during the last glacial maximum on the Tibetan Plateau, along with the resulting increases in albedo, may have caused weakening of the monsoonal circulation (Kutzbach et al., 1998).

In addition to the surface albedo of ice and snow, the surface and basal boundary conditions and internal thermodynamics of ice and snow control the surface temperature, which in turn controls the surface longwave back-radiation, and thus atmospheric conditions. The details of this coupled mechanism and the sensitivity of climate to it are not yet known, though model experiments suggest that subtle errors in surface temperature can introduce a systematic bias leading to unrestrained temperature growth in the models.

The ice-cloud feedback includes a variety of processes that determine the local cloud formation and distribution, primarily as a function of the atmospheric column structure, moisture content, and radiative balance. These processes and properties are intimately tied to the surface conditions. For example, increased ice concentration reduces the area of exposed ocean surface, decreasing the surface heat and moisture sources. Surface heat plays a fundamental role in determining planetary boundary-layer characteristics, and surface moisture determines the local availability of water for cloud formation. Visual images of the Antarctic obtained from satellites show that large areas of open ocean (associated with polynyas) in the frigid ice fields lead to considerable convection and local cumulus generation, which are more typical of tropical regions. Even the presence of surface-melt ponds throughout the Arctic pack-ice fields in summer leads to extensive ground fogs, which have considerable influence on the surface heat balance. Limited studies in the polar regions suggest that many of the surface-cloud feedbacks that are observed in lower-latitude regions behave differently in ice-covered regions (Curry et al., 1996).

The ice-ocean feedback influences the formation and ventilation of global deep and bottom waters, while significantly constraining the ice's thickness and spatial/temporal distribution. This feedback reflects the important role that salinity plays in determining the density of water in cold regions, although it is moderated to some extent by the associated latent-heat effects. Sea ice is the predominant source of mobile freshwater, while the growth of sea ice drives surface salinity fluxes. As sea ice grows, it rejects seawater salt, which works its way back into the surface ocean as a brine solution flowing through drain channels (whose density and effectiveness may be a function of temperature). Ice growth thus serves as the chief source of surface salt, which is the dominant determinant of the surface buoyancy flux in winter. Salt rejection is one of the mechanisms that effect the density changes that lead to ocean convection, both shallow and deep. Convection ventilates the ocean and controls the formation of the intermediate, deep, and bottom waters that drive the global thermohaline circulation. Even shallow convection can mix a considerable amount of heat into the surface of the ocean, and heat content strongly influences ice thickness and concentration and moderates the extent and seasonality of lead formation.

The huge margins of the ice sheets, which float on the sea, contribute considerable freshwater to the ocean, through both basal melting and "calving" of icebergs. When this meltwater is injected at depths significantly below sea level (i.e., from the bottom of a marginal ice shelf) in the Antarctic, the high compressibility of the cold water leads to the formation of dense plumes of deep water. When the meltwater is injected at the surface, from icebergs, the freshwater can stabilize the surface water and inhibit deep-water formation. The rates at which all these processes occur will change with ocean temperature, which can reflect local, regional, or global influences.

Both sea ice and its snow cover, which represent the predominant source of freshwater in the polar regions, stabilize the locations of polar ocean convection. Ice growth in one location—along a continental shelf, say—will tend to enhance convection there through salinity rejection. As that ice drifts offshore due to local winds and later melts, the salinity decrease will tend to suppress convection. In either case, convection and melt zones are stabilized by this freshwater input. If the freshwater budget is altered through variations in ice flux, then the vertical stability of the ocean may change, leading to a reduction or enhancement of deep-water formation and ventilation. The significance of this surface freshwater flux is determined by the underlying ocean structure, which is governed by the regional wind-stress forcing and local stratification.

The ice-sheet-ocean feedback is a function of the sensitivity of the huge ice sheets (formidable reservoirs of freshwater), of sea level, and of ocean temperatures. The rate at which an ice sheet flows into the sea is controlled in part by internal pressure gradients reflecting the thickness and height of ice near its center of accumulation relative to that at the margins, and in part by the effectiveness of the frictional coupling at its lower boundary. As glacial ice flows out onto the ocean, the friction is removed, and the ice floats and thins as it spreads away from the continental "grounding line." If sea level is raised, this grounding line may be shifted inland, causing the ice to decouple from the bedrock and move more rapidly out into the ocean. The rate of this acceleration will depend on the nature of the coupling with the bedrock: If the ice is frozen to the Earth, the friction is considerable, and its elimination will have a large relative effect; if the coupling was moderate to begin with due to an unconsolidated Earth base, or due to basal melting from the pressure, the elimination of the friction leads to a smaller relative effect. The slope of the continental landmass inland will also determine the effect of a sea-level rise. A shallow slope can lead to a dramatic shift in grounding line, but the shallow slope will lead to a weaker driving pressure gradient. A steep topography may result in a minimal shift of grounding line, but the considerable pressure gradient may result in great destabilization and rapid drainage of the ice sheet. In either case, the change in sea level can influence the rate of ice drainage and thus the rate of additional sea-level change.

Internal mechanisms of cryospheric variability are usually significant only for the large ice sheets. Internal ice dynamics may be responsible for altering the basal friction coupling and internal flow dynamics of large ice sheets; however, the details of these processes are not well known. Dynamic instabilities can influence the rate at which ice sheets drain into the ocean, and thus the rates of both sea-level rise and freshwater supply to the surface oceans. Ice-sheet surges induced by these instabilities may account for the vast armadas of icebergs that roamed the North Atlantic during the last glacial period, leading to the episodic, brief Heinrich events in that region (MacAyeal, 1994). Some suggest that the influence of Heinrich events reaches well beyond the North Atlantic (see, e.g., Bond and Lotti, 1995). It has been speculated that the West Antarctic ice sheet has collapsed in the past, possibly because of the aforementioned instabilities; if such a collapse were to happen again, its impact on global sea level over centennial time scales would be tremendous.

Yet another set of polar feedbacks is associated with sea-ice rheology. Internal ice deformation and flow influence lead formation (affecting ice concentration), ridging events (affecting ice-surface conditions, seawater flooding, and ice thickness and concentration), and ice-flow directions (affecting freshwater distribution and ice concentration). Each of these will influence the albedo, ice-cloud feedback, and ice-ocean feedback. The relative importance of these internal feedbacks for cryospheric variability is not fully known at this time.

Remaining Issues and Questions

- *How have the sea-ice, snow, and permafrost fields changed on dec-cen time scales, and what is the relationship of these changes to dec-cen patterns of atmosphere, ocean, and land-surface variability?* As discussed earlier, the NAO and PNA extend into the Arctic, and their indices often reflect changes originating in the polar low-pressure cells. It is therefore important to determine the degree of co-variability between changes in ice and snow fields and patterns such as these. Sea-ice changes have been implicated in changes in the thermohaline circulation on a variety of time scales; how do they co-vary? What spatial and temporal patterns of contemporary polar climate change are manifested in changes in permafrost temperature profiles?

• *Through what mechanisms do sea-ice fields, atmosphere, and ocean interact on dec-cen time scales?* Do regional or even local changes in ice divergence alter the albedo, surface heat and moisture fluxes, upper-ocean conditions, and cloud formation enough to influence the Icelandic or Aleutian lows, and thus the NAO and PNA? How do similar near-surface changes influence atmospheric circulation in the Southern Hemisphere? Do changes in such patterns, in large-scale planetary waves, or in ocean circulation alter polar conditions enough to drive other polar changes that may result in changes to other parts of the climate system? For example, would a change in NAO influence the volume of freshwater exported from the Arctic in the form of sea ice enough to significantly alter the thermohaline circulation? Observational evidence says it might (Dickson et al., 1997), as do model experiments (Tremblay, 1997). Also, the thermohaline circulation is sensitive to surface buoyancy fluxes in source regions. These sources lie predominantly in the polar regions, where the growth, decay, and spatial redistribution of ice play dominant roles in the buoyancy flux, and thus may exert a strong influence on the process of mid- and deep-water formation. The ice in turn is highly dependent on the stability of the underlying water column, setting the stage for considerable feedbacks and interactions among the system components.

• *What are the mechanisms of interaction among the snow fields, permafrost, atmosphere, and land systems on dec-cen time scales?* Do snow-related changes in surface albedo, surface heat and moisture fluxes, soil moisture, vegetation cover, and cloud formation significantly influence atmospheric patterns or large-scale planetary waves, and thus drive long-term feedbacks in the climate system? For example, do changes in the seasonal or spatial distribution of extensive winter snowfields alter the surface vegetation or soil moisture enough to drive longer-term influences elsewhere in the climate system? What are the physical relationships between permafrost surface temperature, surface air temperature, and other climatic parameters, and what are the mechanisms controlling these relationships?

• *What are the mechanisms by which changes in the cryosphere of the polar regions are linked or teleconnected to mid-latitude and tropical regions?* Model results suggest that changes in the sea-ice fields alter the nature of the Hadley cell through their influence on the equator-to-pole meridional temperature gradient. Observations suggest that the Antarctic Circumpolar Wave co-varies with ENSO and Indian Ocean monsoons through mechanisms not yet understood. Changes in the thermohaline circulation may be related to changes in the surface freshwater balance associated with the growth and transport of sea ice. The ocean's interaction with ice shelves can alter the surface volume (and thus the gyre characteristics) in the subtropical regions, which alter SST without any change in surface forcing.

• *What are the historical and current global budgets of glacial ice and snow, and what are the primary mechanisms controlling those budgets?* Because glacial ice and snow budgets directly affect sea level, we need to better quantify the mass balance of the continental ice sheets, alpine glaciers, and permanent snowfields. In particular, the ice mass balance at the base of the floating ice shelves is in considerable question, and whether the Greenland and Antarctic ice sheets are gaining or losing mass is still uncertain. Establishing how this ice and snow budget has varied through time will give some indication of the range, rate, and rapidity of change experienced through natural variability. The IPCC (1996a) lists four major gaps that need to be filled to obtain better estimates of glacier contribution to sea-level rise:

1) development of models that link meteorology to glacier mass balance and dynamic response;

2) extension of models to those glaciers expected to have the largest influence on sea level (the valley and piedmont glaciers of Alaska, Patagonian ice caps, and monsoon-fed Asian glaciers);

3) quantification of the refreezing of meltwater inside glaciers; and

4) better understanding of iceberg calving and its interaction with glacier flow dynamics.

Other areas of uncertainty concerning ice budgets are the controls on the melt/growth rate at the base of floating ice sheets, including the rate of ice-sheet drainage as a function of sea level (which alters friction), as well as the precipitation response to cold-region climatic changes.

Processes and Parameterizations

The internal dynamics and thermodynamics of sea ice and ice sheets are generally fairly well understood, and can be readily parameterized despite their complexity. The largest uncertainties in predicting the extent and thickness of sea ice and glacial ice lie in the treatment of the boundaries. For example, the surface albedos of ice and snow under a variety of conditions, and how those conditions arise, are still poorly resolved and understood. Heat fluxes across the boundaries can be reasonably parameterized, but for sea ice, the partitioning of lateral versus vertical heat fluxes at the edge and base of sea ice with the ocean is still not understood. This particular distribution dictates the partitioning between lateral growth/decay of the ice (controlling its geographic distribution) and vertical growth/decay (controlling its thickness). This partitioning in turn controls the lead area and thus the extent of the ice cover, which then affects the albedo, ice-cloud feedback, and the effectiveness of the insulating ice cover and air-sea heat flux. Even if this partitioning were understood, ice modeling would remain problematic because the prediction of lateral wall area for a given concentration of ice, or surface forcing, is conceptually difficult.

Considerable uncertainties are associated with the prognostic treatment of polar clouds. Prognosis is further complicated by the possibility that polar clouds respond differently from clouds in other regions to changes in surface and forcing conditions. Better-polar cloud modeling is needed to more realistically depict the important ice-cloud feedback discussed above, and an extensive, fundamental observational data base is required as well. The observations will improve our theoretical understanding of polar processes as well as our ability to predict polar-cloud behavior.

Other aspects of the surface energy balance need to be better understood to improve long-term cryosphere and climate predictions. These include the details controlling the spatial heterogeneity of ice surface conditions and their net influence on surface fluxes, and the extent to which seawater flooding of thin, seasonal ice cover affects their melting and albedo, particularly in the Antarctic polar oceans. Seawater flooding not only may alter the thermodynamics of the system, but may be corrupting interpretations of satellite images of ice concentration, confounding our ability to monitor the ice and evaluate the models.

Models have treated the basal boundary condition of ice sheets as a function of underlying surface composition and temperature, as well as of the ocean-ice-sheet interaction along ice shelves. The observations needed to test these formulations are lacking, however. Correct model treatment of ice streams, which are small-scale features with high flow rates, will also require further study. These streams represent a major path through which ice sheets are drained, and their distribution may greatly alter estimates of average ice drainage and ice-sheet stability.

Some of the larger-scale polar feedbacks—for instance, the export of ice from the Arctic, its role in the formation of North Atlantic Deep Water, and long-time-scale feedbacks into the polar regions from any such process—are still a long way from being fully understood. Likewise, neither the long-term feedbacks between the climate system and the polar regions, nor the various local and regional feedbacks already discussed, are well understood—in particular, how the small-scale feedbacks affect the larger-scale climate processes. Finally, the prognostic treatment of snow, like that of precipitation in general, is still a difficult prospect. Some progress has recently been made, however, although considerable effort will be required to obtain even a statistically correct representation of the spatial and temporal distribution of the snow fields and their dependencies.

Observations

Several types of observations are critical to the issues articulated above. Long-term monitoring of sea-surface salinity along with SST is important, since salinity represents the dominant control over the water density of high-latitude regions. The sea-ice distribution, motion fields, and thickness need to be known in order to determine the associated freshwater transports and buoyancy fluxes. Permafrost temperature profiles provide unique indications of integrated deccen climate change over vast geographic regions; more such profiles should be collected in order to better define the spatial and temporal distribution of change. Consistent monitoring of iceberg calving and an observational system for determining the basal melt or growth of sea ice (e.g., an array of moored buoys measuring temperature and salinity across the floating ice shelves) must be established before the sea-ice budget can be closed. Finally, both field and satellite studies are needed to refine the mass budgets of the Greenland and Antarctic ice sheets. On-site studies focused on changes in ice flow, melting, and calving should be continued and extended. Observations of water-vapor net flux (divergence) will help to pin down the source of the ice sheets' mass. A laser altimeter on a polar-orbiting satellite is needed to augment the existing radar altimetry. These instruments will provide accurate estimates of ice-sheet volume and give early warning of possible ice-sheet collapse.

Model parameterizations must be improved to better represent the ice-albedo feedback, snow-climate feedbacks, ice-cloud feedback, ice-ocean feedback, ice-sheet-ocean feedback, and ice-sheet instabilities. Also, simulation of sea-ice and snow distribution and related impacts must be improved. Randall et al. (1998) have described some of the observational requirements necessary to improve our ability to model these processes on large scales, and some of the existing research programs that have been designed to fulfill these requirements.

The feedbacks among the hydrologic cycle (including river runoff into the Arctic), the atmospheric circulation, and the thermohaline circulation must be better understood on a variety of scales, because such larger-scale feedbacks may play a fundamental role in polar climate. The potential for extracting high-resolution records of past climate change from polar sediments along the Antarctic continental shelves and slopes and in polar fjords and Arctic lakes and estuaries (the latter being the primary focus of the Paleoclimates of Arctic Lakes and Estuaries (PALE) program of the Arctic System Science initiative) should be evaluated, and pursued if proven feasible.

LAND AND VEGETATION

Influence on Attributes

The state of the land and its vegetation affect the climate in a number of ways. The fraction of solar irradiance absorbed by different landscapes depends on vegetation. For example, deserts reflect a greater portion of the incoming solar radiation than vegetated regions do. Of the vegetated regions, grasslands reflect more radiation than surfaces covered by forests. The influence of vegetation on albedo is amplified when the angle of incident solar radiation is low, as it is during high-latitude winters, and when snow covers the ground; forests present a light-absorbing layer above the snow, whereas bare ground and grasses do not. To assess the sensitivity of high latitude regions to the albedo effect of boreal vegetation, Bonan et al. (1992) computed the climate response of a GCM in which the forests north of 45°N were replaced by bare ground. The zonally-averaged temperatures in this deforestation simulation were generally between 3 and 10°C colder in the mid- to high-latitudes, relative to a simulation conducted with the same GCM in which forests north of 45°N were present. Whitlock and Bartlein (1997) suggest that changes in vegetation may have played a significant role in climate changes in northwest America over the last 125,000 years. In a set of GCM simulations of Cretaceous-era climate, Otto-Bliesner and Upchurch (1997) found that the globally averaged temperature was over 2°C warmer in a model run that included a best-guess estimate of global vegetation cover than in one in which bare soil covered the land surface. The vegetation decreased the surface albedo, causing high-latitude areas to warm and delaying sea-ice formation, which in turn further decreased albedo and increased temperatures.

Processes in the soil and plants both absorb and produce long-lived greenhouse gases (CO_2, CH_4, N_2O), thereby influencing the atmosphere's infrared-radiation budget. Vegetation emits chemically reactive organic gases (terpenes, isoprene, methanol, etc.) that are involved in atmospheric reactions that lead to production of ozone in the troposphere. Photochemical processes in the lower atmosphere also cause small particles to be created from hydrocarbons emitted by plants. These particles scatter light, causing a visible bluish haze that decreases the transmission of solar radiation to the ground. Soil and mineral dust, whose atmospheric entrainment is influenced by vegetation cover, also affects the scattering of light. Both surface temperature and turbulent mixing of air in the planetary boundary layer are functions of wind friction at the Earth's surface. Surface roughness is greatly influenced by both the stature and density of vegetation, in addition to the effects of topography.

Plants also partly control the hydrologic cycle through evapotranspiration, as noted earlier in this chapter. Leaves open their stomata during photosynthesis, causing them to lose water vapor into the atmosphere while taking up CO_2. Roughly two-thirds of precipitation over land is recycled water vapor from plants. Model simulations of the Amazon confirm that the vegetation plays an active role in maintaining the regional hydrologic regime; simulated deforestation resulted in dramatically decreased precipitation and increased temperature and evaporation (Shukla et al., 1990). Soils, which are a very slowly created mixture of rock-weathering products and organic material derived from plants, function as a reservoir of water. They thus influence the timing of evaporation from the land surface. Therefore, plants indirectly influence surface temperature through their effect on soil moisture, which has a large heat capacity, and thus influences latent heating. Evapotranspiration also changes the balance between the fluxes of sensible and latent heat at the surface, causing local surface cooling. When plants are water-stressed, their stomata may close to reduce transpiration and conserve water, thereby warming the surrounding air. The expected physiological response of plants to a high-CO_2 world would be to close their stomata somewhat, reducing their evaporative loss, but furthering warming over the continents (Sellers et al., 1996).

The urban landscape has a marked influence on climate; a recognized problem in studies of long-term temperature change is that many meteorological measurement sites have gradually become absorbed into expanding metropolitan areas, known as "urban heat islands." The artificial heat output of the greater New York metropolitan area is about one-eighth of the solar energy absorbed there on the ground. Furthermore, wind speed has diminished, particle loadings have increased, anthropogenic emissions of many trace gases have increased, and precipitation and other weather features have changed markedly for tens of miles downwind from many urban areas (Barry and Chorley, 1992).

Evidence of Decade-to-Century-Scale Variability and Change

The major ecosystem zones (biomes) of the Earth, such as tundra, temperate grassland, and wet tropical forest, are determined in part by the range and variability of a region's temperature and precipitation. The type of vegetation prevalent in the past at a given location is sometimes recorded in pollen buried in ancient soils and sediments. Such data show that large vegetation changes have occurred in many areas in response to climate change. For example, pollen data tell us that large parts of the Sahara, although currently completely barren, supported vegetation (savanna woodland and desert grassland) from about 9500 to 4500 BP (see, e.g., Ritchie et al., 1985). Evidence has been found of increased lake levels in the area during the same time period; both conditions have been linked to a strengthened monsoon circulation in that period (Kutzbach and Street-Perrott, 1985). In western Europe, many tree species such as pine, elm, and oak migrated northward and westward with surprising rapidity (typically 150 to 500 m per year) after the close of the last ice age, replacing a shrub-dominated vegetation (Huntley, 1988). Species adapted to Arctic and alpine tundra suffered a crisis in western Europe during the warm period in the mid-Holocene, around 6000 BP, when their habitat was at a minimum. More recently, the succession in which the dominant tree species changed from beech to oak to pine during the Little Ice Age has been recorded in pollen in southern Ontario (Campbell and McAndrews, 1993).

Between one-third and one-half of the Earth's surface has been transformed by human actions (Vitousek et al., 1997). The evidence of ecosystem variations is especially pronounced during the last 150 years. Vast tracts of temperate forest were cut down during the nineteenth and early twentieth centuries. The location of greatest deforestation has shifted to the tropics in the most recent decades. One-fifth of the tropical forest area was lost between 1960 and 1990, and it is estimated that the remaining area is being lost at a rate of 7 percent per decade (WRI, 1996). Today, almost 40 percent of the Earth's land area (excluding Antarctica) is devoted to cropland and permanent pasture (WRI, 1996). Most of this agricultural expansion has occurred at the expense of forests and grasslands; only a few small patches of original prairie remain on the North American continent. The majority of wetlands in the United States have been drained (Kusler et al., 1994) during the last half-century. Aerial photography shows clearly how dominant society's influence over the land is—human settlements and structures, roads, a checkerboard of croplands, artificial lakes, coastal modifications, and so on. Currently about 8 percent of the land in Western Europe is (sub)urbanized or covered by roads, and 45 percent is devoted to cropland and pasture; in the United States the corresponding figures are 4 percent and 45 percent, respectively (WRI, 1996). It is possible, of course, that the relatively large portion of the land surface that is managed in some way by humans may permit us to exert a modest amount of deliberate climate control, since we can control the reflective and absorptive properties of man-made structures.

Evidence for changes in the amount of carbon stored in vegetation and soils derives principally from our knowledge of changes in land use. Deforestation results in the loss of carbon in standing wood, and causes oxidation of part of the organic carbon stored in forest soils. Agricultural practices and reforestation also affect the carbon balance. Combining the recorded global history of land use with time-dependent models of carbon dynamics, Houghton et al. (1987) estimated the loss of carbon to the atmosphere that can be attributed to direct human intervention to be 1.0-2.6 Gt of carbon per year in 1980; this flux has varied through time since at least 1850 (Houghton, 1993; IPCC, 1996a). While land-use changes are important, recent climate variability has probably also led to substantial changes in vegetation-related carbon fluxes. Using historical temperature and precipitation data in conjunction with a carbon-cycle model, Dai and Fung (1993) found that climate may have caused significant interdecadal variations in regional and global terrestrial carbon storage since 1940. Observations of increasing amplitude of intra-seasonal atmospheric CO_2 variations (Keeling et al.,1996a) and remotely-sensed, large-scale increases in terrestrial photosynthetic activity (Myneni et al., 1997) suggest that plant growth has increased in recent years.

Measurements of CO_2 concentrations in ice cores provide a very clear record of changes in carbon storage between glacial and interglacial times, but these measurements do not directly distinguish the respective roles played by the oceans and the terrestrial systems in causing the atmospheric concentration changes. Obviously changes in carbon storage can be expected when the geography of vegetation is significantly altered (Prentice and Sykes, 1995; Friedlingstein et al., 1995). It has also been inferred from ice-core records that the emissions of CH_4 varied between glacial and interglacial periods (Chappellaz et al., 1993b; Thompson et al., 1993a), and that they have strongly increased in recent years. Changes in ecosystem types and land use (wetlands, rice paddies, cattle grazing, etc.) clearly have had a major impact on these emissions. Global N_2O emissions have also increased during recent decades, but there is still considerable uncertainty as to the cause.

Mechanisms

Past vegetation changes have been driven by natural climate variations. Regional and global models of vegetation dynamics are based on the sensitivity of species and ecosystems to variables such as the mean coldest-month temperature, the annual accumulated temperature over 5°C, precipitation, and soil moisture capacity (see, e.g., Prentice et al., 1992; VEMAP, 1995). These variables reflect vegetation characteristics, such as: most woody tropical plants are killed when the temperature drops below 0°C, and for a species to sustain growth, the air temperature must exceed a species-specific minimum value for a species-specific minimum length of time (expressed as growing degree days). The climate warming that has been projected for the coming centuries could induce changes to natural vegetation as great as those at the end of the last ice age; species distributions in North America could be shifted by as much as 500 or 1,000 km (Overpeck et al., 1991). The variability and types of disturbance are another significant factor in determining ecosystem composition and distribution. For instance, the frequency and severity of wind storms and fires affect the migration and establishment of species and ecosystems, and need to be taken into account in predicting the geography of future ecosystems (Overpeck et al., 1990).

At present, the dominant reason for changes in vegetation is direct human intervention, both purposeful and inadvertent. More than half of the ice-free surface of the continents has been altered substantially by human uses (Kates et al.,

1990). Our need for food and resources drives land-use patterns that result, either rapidly or gradually, in land-cover changes. Vitousek et al. (1986) estimate that 31 percent of all net primary production on land directly serves humans as fiber, food, or fuel; 2.3 percent is actually consumed by us or by animals that we use for food. As human populations grow and place even more demands on our natural environment, this already pervasive influence of humans on the Earth's biota is likely to concomitantly increase. Anthropogenic change in ecosystem functioning can also result from the removal of predators or the introduction of invasive species.

Both direct and indirect effects of CO_2 have been recognized as mechanisms of change in the interaction of, and competition between, species composing the vegetation on undisturbed land. Fertilization of plant growth by higher atmospheric CO_2, and by moderate amounts of wet and dry deposition of nitric acid, has a demonstrable influence on vegetation. To test the response of intact ecosystems to CO_2 changes more realistically than in laboratory settings, so-called free-air CO_2 enrichment (FACE) experiments are being carried out, in which CO_2 is pumped over ecosystems in their natural environment. One such experiment, carried out in Chesapeake Bay wetlands, is reported by Drake (1992). Long-term CO_2-enrichment studies with manipulated microclimate have also been carried out in enclosures, an example of which is the assessment by Tissue and Oechel (1987) of the effects of temperature and CO_2 change on Arctic tundra. The responses to elevated CO_2 in the Arctic and Chesapeake Bay cases were quite different, which suggests that nutrient (especially nitrogen) availability may play an important role in regulating response to increased CO_2 and temperature (Rastetter et al., 1992). Another effect of enhanced levels of CO_2 is that plant root-to-shoot ratio tends to increase (Rogers et al., 1994). Schindler and Bailey (1993) estimated that the amount of carbon storage stimulated by anthropogenic nitrogen deposition may be between 1.0 and 2.3 Gt of carbon per year. Others, such as Asner et al. (1997), consider the potential of this effect to be lower.

Not only does enhanced CO_2 fertilization tend to increase the amount of carbon stored in live vegetation, but it can alter the species balance of ecosystems. For instance, under greater ambient CO_2 concentrations, C_3 plants (the majority of crops) tend to be favored over C_4 plants (some essential warm-weather crops, including corn and sugar cane) (Poorter, 1993). Shifts in species composition may also arise from changes in the availability of nutrients (Wedin and Tilman, 1996). By affecting climate, elevated CO_2 levels may also indirectly influence the number and balance of species in ecosystems (Davis and Zabinski, 1992; Barry et al., 1995).

There is increasing awareness that future ecosystems may not represent a simple, climatically driven redistribution of ecosystems as they are currently composed. Rather, mechanisms such as those mentioned above must be taken into account in predictions about future ecosystems. Furthermore, factors such as changes in land use and management, which are quite difficult to predict, are likely to be at least as important as climate-related changes. Direct human intervention and climate change do not act as independent agents of vegetation change. Both the U.S. Dust Bowl of the 1930s and the desertification in the Sahel are examples of how unfavorable climatic conditions and societal demands may synergistically lead to environmental degradation.

Variations in fire frequency and intensity can often be related to variations in climatic conditions. For instance, the widespread fires in southeast Asia in 1997 have been attributed to the extreme El Niño-related drought at that time. In addition to having a direct economic impact when fires affect forestry and personal property, such changes can also influence ecosystems in a number of ways. Active suppression of forest fires, while benefiting humans in many ways, can be detrimental to certain species that depend on fire for various reasons (e.g., facilitating germination). Moreover, fire suppression can lead to age homogenization, in which forests tend to become dominated by single-age stands. Although uniform forests are often high in timber productivity, the decrease in diversity leaves them generally more vulnerable to fire, windstorms, disease, and other naturally occurring events (Noss and Cooperrider, 1994). Increased fire frequency can also cause local extinctions of species, even in mature forest stands (Gill, 1994).

Acid deposition has led to widespread dieback of trees, especially at higher elevations. Elevated surface ozone can reduce photosynthesis, increase respiration, and lead to leaf senescence earlier in the season (Chameides et al., 1994), all of which reduce productivity. Increased UV-B radiation has been shown to reduce photosynthesis and growth in many species in greenhouses, although the effects are less marked under field conditions where light levels are high (Allen and Amthor, 1995).

By causing changes in the vegetation and the soils, the above-described processes will have an impact on the biogeochemical cycles. Because climate depends in part on the chemistry of the atmosphere, large-scale atmospheric chemistry-vegetation-climate feedbacks may exist. For instance, higher atmospheric CO_2 concentrations may directly (via the fertilization effect) and indirectly (via climate-induced changes) increase carbon sequestration in vegetation (see, e.g., Woodwell and Mackenzie, 1995), yielding a negative feedback to the level of atmospheric CO_2. Several other chemistry-vegetation-climate feedbacks have been proposed, many of which are discussed in Woodwell and MacKenzie.

Predictability

Future changes in the composition and distribution of ecosystems, and the accompanying biogeochemical cycles of carbon and nitrogen, are hard to predict. Not only are a large number of factors simultaneously undergoing change, but we cannot be certain of future human actions. Pollution, fertilization, climate change, land use, succession, the use of pesticides, species extinction and the introduction of new species, and the fragmentation of once-widespread ecosystems are all occurring at once, often making it difficult to decipher cause and effect. For example, atmospheric measurements have established the existence of a carbon sink of appreciable magnitude at temperate latitudes on the continents of the Northern Hemisphere (Tans and White, 1998), but it has proven difficult to choose among the competing explanatory hypotheses (Houghton et al., 1998). Candidates are CO_2 fertilization, nitrogen fertilization, afforestation, and a climate-driven increase in carbon storage (see IPCC, 1995, for an overview).

Changes in land use do not follow a predictable progression. The progression will depend on local factors, most of them economic, social, and technological. For example, population growth has in many cases contributed to the conversion of forested land to farmland, but in the eastern United States and western Europe the process has been reversed during the last 50 years (McKibben, 1995). Accurate modeling of climate and atmospheric chemistry require the accurate specification of land-surface parameters that are intimately tied to the fluxes of heat, water vapor, and trace gases. Although attempts have been made to predict land-use changes (Zuidema et al., 1994), our skill in this regard is still low, largely because we lack sufficient insight into what has been called the human dimension of global change. The International Geosphere-Biosphere Programme's Human Dimensions Project has outlined a science/research plan designed to increase our skill in predicting the progression from human needs to land-cover change (Turner et al., 1993). The plan proposes to classify the world's land area by similar social and environmental circumstances into a manageable number of categories, and probe the causal connections in each category in more detail according to a common protocol or framework.

Remaining Issues and Questions

• *What are the effects of human activity and climate change on ecosystem structure and function?* From paleoclimatic records, we know that both vegetation and animal species respond to climate variations according to their individual tolerances. Competitive and trophic interactions among species are thereby altered, redefining where organisms can survive and reproduce and changing ecosystem compositions. The ability of organisms to respond to future climate variations or change will be greatly influenced by human land-use patterns and other anthropogenic influences. Associated with structural changes in ecosystems are changes in the biogeochemical cycling of carbon and nutrients, in ways that remain difficult to anticipate. Finally, the distribution of disease-carrying organisms will change with ecosystem restructuring and redistribution (IPCC, 1996b).

• *What are the relative contributions of the different processes by which vegetation and soils store or lose carbon?* Vegetation and soils store three times as much carbon as the atmosphere or upper ocean, yet large uncertainties remain regarding the quantitative contributions of various processes. The carbon sink in the Northern Hemisphere has increased over recent decades; forest regrowth resulting from changing land-use patterns, or perhaps increased fertilization by CO_2 and nitrogen, or simply climate change may have been factors in this increase.

• *At what rates will vegetation and soils emit CH_4, N_2O, and volatile organic carbon (VOC) compounds in the future?* CH_4 production in soils depends strongly on moisture conditions (including the extent of the permafrost, which is slowly melting). N_2O production is a result of denitrification processes that occur in soils. The rates at which VOC compounds (ozone precursors) are emitted depend heavily on the species involved. Changes in these emissions will depend on a combination of factors involving both ecosystems and climate.

• *How do dec-cen-scale changes in land use and land cover affect the energy balance of the land surface on dec-cen time scales?* The nature of land cover, which determines its reflectivity, is expected to change with changing climate and human activities. For example, a warmer high-latitude climate will favor the expansion of boreal forest into tundra-dominated regions, with a concomitant lowering of the albedo. Desertification, which may result from human or natural activity or both, increases surface albedo. The thermal structure, moisture content, and dynamics of the atmosphere are influenced by the proportions of sensible and latent heat transferred from the surface, which is a function of the type and extent of land cover.

• *How does vegetation influence the transfer of freshwater through the land surface on dec-cen time scales?* The extent of stomatal opening influences the rate of evapotranspiration from the land surface. Higher atmospheric CO_2 concentrations will cause CO_2 to more readily enter plants; plants will then be able to keep their stomata somewhat more closed, which will decrease their transpiration losses and increase their water-use efficiency. An increase in vegetation density tends to decrease runoff and increase evaporative fluxes, resulting in greater atmospheric water-vapor content and precipitation over land.

• *How does changing vegetation cover influence the loading and composition of atmospheric aerosols on dec-cen time scales?* Vegetation naturally emits aerosol precursors (e.g., non-methane hydrocarbons), and the nature and amount of these compounds depends on the species. The distribution of aerosol precursors will therefore change as ecosystems and species respond to climate variations and human perturbations. Biomass burning generates aerosols (particularly soot) that influence the regional radiation balance. Desertification produces mineral dust that is transported into the troposphere and exerts a regional radiative forcing. The distribution of all these aerosols can be expected to vary on dec-cen time scales in response to climatic and human influences.

Processes, Parameterizations, and Observations

Changes in land-surface characteristics—including surface vegetation, topsoil extent, and soil moisture—must be monitored on a long-term basis. Not only do these changes alter the distribution of surface reservoirs of radiatively active gases and the surface-atmosphere exchange of those gases, they also influence albedo and, through stress effects on plant evapotranspiration efficiency, the hydrologic cycle.

Long-term monitoring of near-surface aerosol distributions will be required to assess whether perturbations of stable gradients of these aerosols could induce stationary changes in the surface radiation balance, which could lead to large-scale alteration of circulation.

In order to improve models' abilities to predict dec-cen-scale variability, we need to more realistically parameterize many land-surface processes, such as: interactions between soil and vegetation under various conditions (including frozen soils); surface-atmosphere gas exchange and net uptake (including biogeochemical and physical feedbacks); and the effect of land-surface processes on atmospheric conditions, (including evaporation and precipitation). Clearly our understanding of most of these processes must be improved first.

Land-surface characteristics and radiatively active atmospheric constituents are vital sets of climate-model parameters, and are generally not prognostic variables that can be used interactively by models. At present, because changes in these factors cannot yet be adequately predicted, they are considered to be an external forcing in most models, and their characteristics must be specified in advance. Even in the absence of any significant skill in predicting land-cover change, however, we can usefully run different vegetation scenarios in physical global-change models. This approach would at least yield some insight into likely climatic and environmental consequences of those scenarios, and provide some guidance for setting environmental-policy goals pertaining to land cover. In addition, as with greenhouse gases, the transient evolution of land cover (including wetlands) under a slowly changing climate and rapidly exploding population must be monitored to provide the boundary conditions needed for model simulations and assessment of plausible future trends.

6

Crosscutting Issues

Certain issues cut across all the disciplines involved in climate research. Among them are the nature of our available climate information, modeling efforts, prediction efforts, detection and attribution issues, and linkages across time scales. Progress in these areas will go far to advance our understanding of climate change and variability over decade-to-century time scales.

CLIMATE INFORMATION

Fundamental to our understanding of climate change is the information we use to investigate and study such change. For dec-cen time scales, the information is available through a variety of forms, and each of these must be exploited if we are to make significant headway. We have: (1) instrumental data for the past 100 years or so, which will be improved and extended into the future; (2) proxy indicators of climate change, both historical records and paleoclimate evidence; (3) analysis products, the results of applying some form of analysis to observations to yield a more focused picture or consistent interpolations; and (4) the output of models, our only tool for estimating future climate states. Each of these information products is briefly discussed here, with suggestions for future additions to our climate database.

Instrumental Data

To build the requisite understanding of climate change and variability, we must first recognize our current state of understanding and its limitations. The inadequacies of the current instrumental data available for exploring climate change and variability on dec-cen time scales are readily exposed when we try to use them to answer some of our most fundamental questions. For example: Is the planet getting warmer? Is the hydrologic cycle changing? Is the atmosphere-ocean circulation changing? Are the weather and climate becoming more extreme or variable? Is the radiative forcing of the climate changing? These questions cannot be answered definitively, because there is no global climate observing system that gathers all the information needed. Each of these apparently simple questions is actually quite complex, both because of its multivariate aspects, and because the spatial and temporal sampling required to adequately address it must be considered on a global scale.

A brief review of our ability to answer these questions reveals many areas of success, but also some glaring inadequacies that must be addressed if we are to understand and predict climate change, and to refine this database for future generations. The basic problems are the shortness and inaccuracy of the instrumental record and its lack of spatial coverage, together with the difficulty of interpreting the paleoclimatic proxy record, which is discussed in the next section. What we can do to improve our ability to detect and monitor climate change is outlined at the end of this section, following the discussion of the fundamental questions.

Is the Planet Getting Warmer?

Measurements show that near-surface air temperatures are increasing. Best estimates suggest that the overall warming has been around 0.5°C since the late nineteenth century (IPCC, 1996a). Nonetheless, many questions have arisen regarding the adequacy of these estimates (IPCC, 1996a), and the relative scarcity of global measurements throughout the century is only the first. Changes in the methods of measuring land and marine surface-air temperatures from ships, buoys, and land-surface stations; changes in instrumentation, instrument exposures, and sampling times; urbanization effects—these are but a few of the time-varying biases that have plagued the interpretation of the surface air-temperature records for the twentieth century. Only by also considering other temperature-sensitive variables—e.g., snow cover, glaciers, sea level, and even some proxy non-real-time measurements such as ground temperatures from boreholes—can we be confident that the planet has indeed warmed. The measurements we rely on to calculate global changes of temperature were never collected for that purpose; they were made primarily to aid in navigation, agriculture, commerce, and, in recent decades, weather forecasting. Thus, many uncertainties remain about important details of the temperature increase. The IPCC (1996a) has summarized known changes in the temperature record; this summary is presented in graphic form in the upper panel of Figure 6-1.

Recent global-scale measurements of layer-averaged atmospheric temperatures and sea surface temperatures from instruments aboard satellites have greatly aided our ability to monitor global temperature change (Spencer and Christy, 1992a,b; Reynolds, 1988), but the situation is far from satisfactory (Hurrell and Trenberth, 1996). Changes in satellite temporal sampling (e.g., orbital drift), changes in atmospheric composition (e.g., volcanic emissions), and techni-

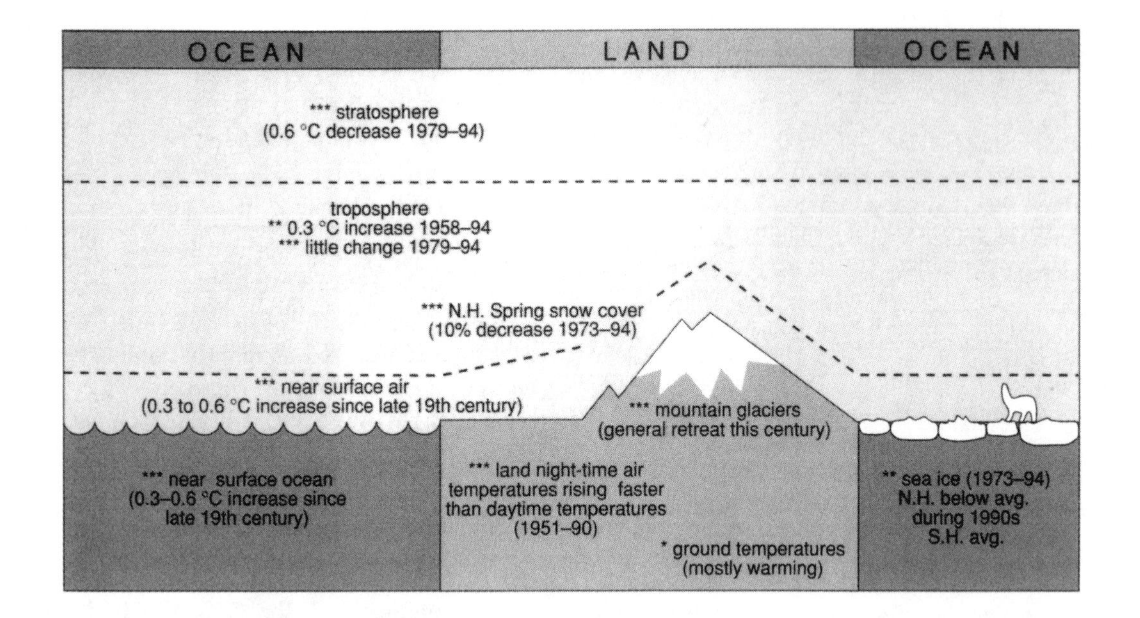

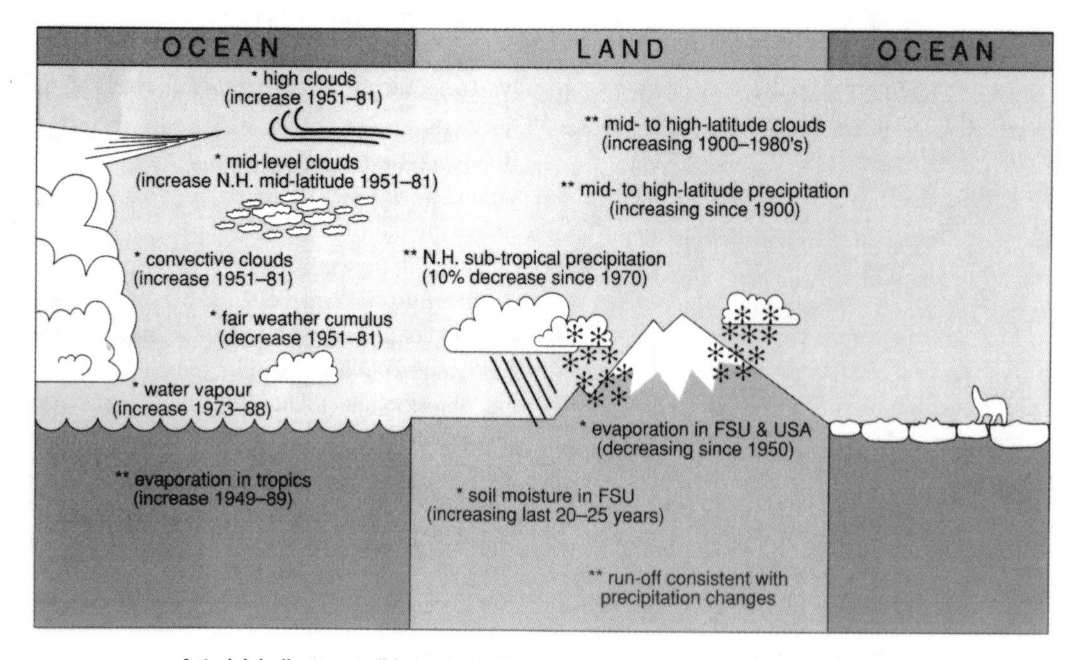

Asterisk indicates confidence level (i.e., assessment): * high, ** medium, * low**

FIGURE 6-1 Schematic of observed variations of selected climate indicators. Upper panel, temperature indicators; lower panel, hydrologic indicators. (From IPCC, 1996a; reprinted with permission of the Intergovernmental Panel on Climate Change.)

cal difficulties related to overcoming surface-emissivity variability limit our ability to produce highly reliable products of near-surface global temperature change. Nonetheless, space-based measurements have shown that stratospheric temperatures have decreased over the past two decades, though perhaps not as much as suggested by measurements from weather balloons. (It is now known that the data from these balloons high in the atmosphere have an inadvertent temporal bias, because of improvements in shielding from direct and reflected solar radiation (Leurs and Eskridge, 1995).)

Even if the instrumental records of ground, atmosphere, and sea surface temperatures were accurate enough, the length of the records would still be an issue. To be certain that the system is getting warmer, we need a record long enough to enable us to distinguish a steady trend of warming from long-period variations, which we may be seeing only partially. Identifying and understanding dec-cen variability is crucial to determining whether the planet is getting warmer.

Is the Hydrologic Cycle Changing?

The source term for the hydrologic water balance, precipitation, has been measured for over two centuries in some locations. Even today, however, it is acknowledged that in many parts of the world we still cannot reliably measure true precipitation (Sevruk, 1982; IPCC, 1996a). For example, annual biases of more than 50 percent are not uncommon in cold climates (Karl et al., 1995), and even for more moderate climates precipitation is believed to be underestimated by 10 to 15 percent (IPCC, 1992). Improvements in instrumentation have also introduced time-varying biases (Karl et al., 1995). Satellite-derived measurements of precipitation are the only ones that provide large-scale ocean coverage. Although comprehensive estimates have been made of large-scale spatial precipitation variability over the oceans, where few measurements exist, problems inherent in developing such estimates limit our confidence in using them to identify global-scale decadal changes. For example, even the recent work of Spencer (1993) in estimating worldwide ocean precipitation using a microwave sounding unit aboard the NOAA polar orbiting satellites has several limitations: The observations are limited to ocean coverage (and hindered by the requirement of an unfrozen ocean), do not adequately measure solid precipitation, have low spatial resolution, and are affected by the diurnal sampling inadequacies associated with polar orbiters (e.g., limited overflight capability).

Documentation of past changes in land-surface precipitation has been compared with other hydrologic data, such as changes in streamflow, to ascertain its robustness. The lower panel of Figure 6-1 summarizes some of the more important known changes in precipitation, such as the increase in the middle to high latitudes and the decrease in the subtropics. There is also evidence to suggest that much of the increase in mid- to high-latitude precipitation arises from increased autumn and early-winter precipitation in much of North America and Europe. Color plate 5 depicts the spatial aspects of the changes in precipitation during this century; rather large-scale coherent patterns are apparent.

Other changes related to the hydrologic cycle are also summarized in Figure 6-1. Confidence is low for many of the changes noted, which is particularly distressing given the important role of clouds and water vapor in climate-feedback effects. Records of cloud amount are the result of surface-based (human) observations (now being replaced by automated measurements in the United States) and satellite measurements. Neither surface-based nor space-based datasets have proven to be entirely satisfactory for detecting changes in clouds. Polar-orbiting satellites have enormous difficulties related to sampling aliasing and satellite drift (Rossow and Cairns, 1995). For human observations, changes in observer schedules, observing biases, and incomplete sampling have created serious problems in data interpretations, now compounded by the change to automated measurements at many stations. Nonetheless, there is still some indication (but low confidence) that global cloud amounts have tended to increase. This finding is supported by reductions in evaporation (as measured by pan evaporimeters) over the past several decades in Russia and the United States, and by a worldwide reduction in land-surface diurnal temperature range. Moreover, an increase in water vapor has been documented over much of North America and in the tropics (IPCC, 1996a).

Water vapor is the most important greenhouse gas in the atmosphere, so changes in water vapor, especially at upper levels of the troposphere, are very important for understanding climate change. The measurement of changes in atmospheric water vapor is hampered by data-processing and instrumental difficulties for both weather-balloon and satellite retrievals. Satellite data also suffer from discontinuities among successive satellites and from errors introduced by changes in orbits and calibrations. Upper-tropospheric water vapor is believed to be a particularly important climate-feedback quantity, but little can be said as yet about how it has varied over the course of the past few decades, how it responds to climate change, or how successful models are in simulating these changes.

Is the Atmosphere-Ocean Circulation Changing?

A number of the responses presented above suggest that the atmosphere and ocean circulation systems have been changing. This evidence is surprisingly meager, however. Daily analyses of circulation are performed routinely, but the analysis schemes have changed over time, so they are of limited use for monitoring climate change. Moreover, even the recent re-analysis efforts by the world's major numerical weather-prediction centers, for which the analysis scheme is fixed over the historical record, contains time-varying biases; some of the data carry embedded biases, and the data mix changes over the course of the re-analysis (Trenberth and Guillemot, 1998). Even less information is available on measured changes and variations in ocean circulation. Only recently has a single, coarse snapshot of the ocean been taken in the World Ocean Circulation Experiment (WOCE).

Are the Weather and Climate Becoming More Extreme or Variable?

Perhaps one of society's greatest concerns about weather and climate is their extremes. Only a limited quantity of reliable information is available about large-scale changes in extreme weather or climate variability, for two reasons. First, there is inadequate monitoring of the necessary quantities. Second, access to weather and climate data held by the world's national weather and environmental agencies is prohibitively expensive. The time-varying biases that affect climate means are even more difficult to effectively eliminate from the extremes of the distributions of various weather and climate elements. There are a few areas, however, where regional and global changes in weather and climate extremes have been reasonably well documented.

Interannual temperature variability has not changed significantly over the past century. On shorter time scales and higher frequencies, however (e.g., days to a week), there is some evidence for a decrease in high-frequency temperature variability across much of the Northern Hemisphere (Karl et al., 1996). Related to this decrease has been a tendency for fewer low-temperature extremes to occur, but widespread changes in extreme high temperatures have not been noted.

Trends in intense rainfalls have been examined for a variety of countries. There is some evidence for an increase in intense rainfalls (in the United States, tropical Australia, Japan, and Mexico), but analyses are far from complete and the record contains many discontinuities. The strongest increases in extreme precipitation are documented in the United States and Australia.

There are grounds for believing that intense tropical-cyclone activity has decreased in the North Atlantic, the one basin for which we have reasonably consistent tropical-cyclone data throughout the twentieth century. Even here, though, it is difficult to be certain of the tropical-cyclone strengths reflected in data obtained prior to World War II. Elsewhere, tropical-cyclone data do not reveal any long-term trends, or if they do, the trends are most likely to be the result of inconsistent analyses. Changes in meteorological assimilation schemes have introduced very difficult problems in interpreting changes in extratropical cyclone frequency. In some regions, however, such as the North Atlantic, a clear trend toward increased storm activity has been noted. This tendency has also been apparent in significant increases in wave heights in the northern half of the North Atlantic. In contrast, decreases in storm frequency and wave heights have been noted in the southern half of the North Atlantic over the past few decades. These changes are also reflected in the prolonged positive excursions of the NAO since the 1970s.

Is the Radiative Forcing of the Planet Changing?

Without an adequate time history of the important agents of climate change—that is, those factors that affect the radiative-heat balance of the planet—it is impossible to understand global change. The atmospheric concentration of CO_2, an important greenhouse gas because of its long atmospheric residence time and relatively high atmospheric concentration, has increased substantially over the past few decades. This rise is quite certain; it is revealed by precise measurements made at Mauna Loa Observatory since the late 1950s (Keeling et al., 1976, 1989), at the South Pole since the mid-1960s (Keeling and Whorf, 1994), and at a number of other stations around the world that began operating in subsequent decades (WMO, 1984). Since CO_2 is a long-lived atmospheric constituent and it is well mixed through the atmosphere, a moderate number of well-placed stations can provide a very robust estimate of global changes in carbon dioxide as long as they operate for the primary purpose of monitoring seasonal-to-decadal changes.

To understand the causes of the increase in atmospheric carbon dioxide, however, we must understand how the carbon cycle operates and how the anthropogenic carbon budget is balanced. Understanding the carbon cycle, which is discussed in greater detail in the first part of Chapter 5, requires estimates of anthropogenic sources of carbon: the emissions resulting from fossil-fuel burning and production and cement production, as well as the net emission from changes in land use, such as deforestation. These estimates are derived from a combination of modeling, sample measurements, high-resolution satellite imagery, and sophisticated analysis of different types of CO_2 and climate records (Keeling et al., 1996a; Dettinger and Ghil, 1998). Understanding the carbon budget also requires measuring carbon storage in the atmosphere, the ocean uptake, and uptake by forest regrowth; assessing the effect on vegetation of CO_2 and nitrogen fertilization; and taking into account climate-feedback effects, such as the increase in vegetation resulting from increased temperatures. At present many of these factors are still uncertain, because so few sustained ecosystem measurements have been made. It is clear, however, that anthropogenic emissions are the primary cause of the atmospheric increase of CO_2. Indeed, a major unresolved issue is why the atmospheric concentration of CO_2 is not even higher than observed.

Several other radiatively important anthropogenic atmospheric trace constituents have been measured over the past few decades. These measurements have confirmed significant increases in atmospheric concentrations of methane

(CH_4), nitrous oxide (N_2O), and the halocarbons (including the stratospheric-ozone-destroying agents, the chlorofluorocarbons and the bromocarbons). Because of their long lifetimes, implying spatial homogeneity in the atmosphere, a few well-placed, high-quality in situ stations have been able to provide good estimates of global changes in the concentration of these constituents. Stratospheric ozone depletion has been monitored by satellite and by in situ ozonesondes. Both observing systems have been crucial in ascertaining changes in stratospheric ozone. (Note that ozone was originally of interest not because of its role as a radiative-forcing agent, but because of its ability to absorb UV radiation before it reached the Earth's surface.) The combination of the surface- and space-based observing systems has yielded much more precise measurements than either system could have provided alone. Over the past few years it has been possible to improve the ozonesonde and satellite data by using information about past calibration methods, in part because differences in trends between the two observing systems could be identified.

Color plate 6 depicts the IPCC (1996a) best estimate of the radiative forcing associated with various atmospheric constituents. Unfortunately, measurements of most of the forcings other than those already discussed have low or very low confidence, both because of our uncertainty about their role in the physical climate system and because we have not adequately monitored their change. For example, our current estimates of changes in sulfate-aerosol concentrations are derived from model estimates of source emissions, not from measured atmospheric concentrations. Monitoring sulfate aerosol is complicated because its short atmospheric lifetime causes its concentration to vary spatially. Another example of low confidence is measurements of solar irradiance. These measurements have been taken by balloons and rockets for several decades, but continuous measurements of top-of-the-atmosphere solar irradiance did not begin until the late 1970s with the Nimbus 7 and the Solar Maximum Mission satellites.

Significant absolute differences in total irradiance are found between different satellites' measurements, emphasizing the critical need for overlap between satellites and for absolute calibration of the irradiance measurements to determine decadal changes (NRC, 1994). Spectrally resolved measurements will be a key element in our ability to model the effects of solar variability, but at present no long-term commitment has been made to take such measurements. Another important forcing currently estimated through measured, modeled, and estimated changes in optical depth is that related to the aerosols sporadically injected high into the atmosphere by major volcanic eruptions. However, aerosols of volcanic origin usually persist in the atmosphere for at most a few years. Improved measurements of the size distribution, composition, and radiative properties of volcanic aerosols will help us better understand this agent of climate change.

What Can We Do to Improve Our Ability to Detect Climate and Global Changes?

Even after extensive re-working of past data, in many instances we are incapable of resolving important aspects of climate and global change. Better quality and continuity, and fewer time-varying biases, will be required of virtually every monitoring system and dataset if we expect to conclusively answer questions about how the planet has changed. Our inability to do so now is often a result of having to rely on observations that were never intended to be used to monitor the physical characteristics of the planet over the course of decades. Long-term monitoring capable of resolving deccen changes requires different strategies of operation.

In situ measurement systems are in a state of decay or decline, or are undergoing poorly documented change. Surface-based automated measurement is being introduced without adequate precautions to explore and record the differences between the old and new observing systems. Satellite-based systems alone cannot provide all the measurements necessary for detecting changes. Much wiser implementation and monitoring practices must be adhered to for both space-based and surface-based observing systems if we are to adequately understand global changes. A number of steps can be taken to improve our ability to monitor climate and global change:

- When changes are made to existing environmental monitoring systems, or new observing systems are introduced, standard practices should include an assessment of the impact of these changes on our ability to monitor environmental variations and alteration.
- For critical environmental variables, it should be standard practice to overlap measurements in time and space when a new observing system replaces an old one.
- Information on instrument calibration and validation, as well as the history of an observing station or platform, are essential for data interpretation and use. Changes in instrument sampling time, local environmental conditions, and any other factors pertinent to the interpretation of the observations and measurements should be recorded as a mandatory part of the observing routine and be archived with the original data. The algorithms used to process observations need to be well documented, and accessible to the scientific community. Documentation of changes and improvements in the algorithms should be carried along with the data throughout the archiving process.
- Timely, regular assessments of the quality and homogeneity of the instrumental databases are needed, and should include all data used to monitor past and present environmental variations and change. Special attention should be given to the long-term, high-resolution instrumental data required to identify change or variations in the occurrence of extreme environmental events.
- Societal and policymaking requirements for knowledge or observations of environmental variations and change

should be taken into account in laying out a strategy for a global, comprehensive system for observing climate.

- Stations, platforms, and observation systems with long, uninterrupted records should be maintained. Every effort should be made to protect the datasets that document long-term homogeneous observations, particularly those encompassing a century or more. Priorities for sites or observation systems should be assigned on the basis of their contribution to long-term monitoring of each element of the climate system.
- In the design and implementation of new environmental observing systems, highest priority should be given to data-poor regions, regions sensitive to change, and key measurements that currently have inadequate temporal resolution. In addition, the appropriateness of the variables to be measured should be verified.
- Network designers, operators, and instrument engineers must be provided with long-term environmental monitoring requirements when they begin to design an observing system. Most observing systems now in place were designed for purposes other than long-term monitoring, and many of them do not acquire information in a suitable form. Instruments must have adequate precision, and their biases must be small enough, to resolve the environmental variations and changes that are of primary interest to climate research.
- Much of the development of new observational capabilities, and much of the evidence supporting the value of these observations, stemmed originally from research needs or research-oriented programs. Stable, long-term commitments to these observation systems, and a clear plan for their transition from research to operations, are two requirements for the development of adequate long-term environmental monitoring capabilities.
- Data-management systems that facilitate the use and interpretation of observational data are essential. Freedom of access, low cost, mechanisms that encourage use (directories, catalogs, browsing capabilities, and availability of metadata, including station histories, algorithm accessibility, documentation, and so on), and quality control should guide data management. International cooperation is critical for effective management and exchange of data used to monitor long-term environmental variability and change.

An ongoing study by the NRC Panel on Climate Observing Systems' Status is in the process of identifying and characterizing existing and emerging factors that could lead to a deterioration in the quality of climate data or in its availability from operational and research observational networks. The study will also make recommendations for maintaining the quality and availability of climate data.

How Can Other Observations Address Climate Models?

While detection of climate change is, as discussed above, a critical problem for instrumental observations, it is not the only one. Ultimately, the problem of assessing the importance of climate variability that occurs over decades and longer, and particularly the task of predicting it, will rest on using physically based models. It is unlikely that these models can be adequately validated through comparison of predicted and observed climate change alone. Such validation would require observing a large number of examples to separate random unmodeled noise from simulation skill, and each example would require decades to observe. Confidence in any understanding of climate variability on dec-cen time scales must rest on the verified realism of the components of our models. This confidence cannot be gained from theoretical arguments alone—each model component must be tested against appropriate observations. Much of this report addresses a large number of processes that affect climate and must be included in models and, therefore, must be verified before the models can be trusted to simulate climate variability. It would be exhausting to reiterate these processes and to point out what observations are needed to verify their model representations. Rather, it will suffice to emphasize that quantitative verification is needed, and to review some of the kinds of observations and studies that will yield the needed climate information:

- Accurate measurements for developing and verifying model components. A model that quantitatively simulates many of the important climate processes deserves greater confidence than one that does not. Thus accurate measurements of such quantities as air-sea fluxes of heat, water, and momentum at specific sites; the directional and wavelength distribution of atmospheric radiation under various meteorological and cloud conditions; the flow of water from the Pacific to Indian Oceans through the Indonesian Seas; or the flux of ice from Antarctica into the Southern Ocean would provide valuable tests of climate models. Measurements of this type are most likely to come from research projects aimed at specific processes. This approach was first applied to relatively small-scale processes that could be studied by a single discipline in a few weeks of extensive observation. TOGA COARE, however, provided an example of applying the same philosophy to a large-scale meteorological and oceanographic process. Studies of this type may provide the climate information best tuned to addressing specific climate phenomena.
- Surveys of climate indicators for assessing overall model performance. The large-scale spatial and seasonal variations of properties distributed by the climate system provide critical tests of the transport processes simulated by climate models. For example, the distribution and rates of accumulation of anthropogenic or chemically reactive gases and aerosols in the atmosphere, or the distribution of inorganic carbon and freshwater in the ocean, are fields whose simulation critically tests both the source/sink and transport processes in models. Diagnostic quantities such as the ratios of certain oceanic properties can in some cases be an even more useful

tool for assessments of model simulations than the use of directly observed quantities. Global surveys of these property fields are large undertakings, so it will be useful to identify a few key property fields that will provide the most stringent and useful tests, and to develop ways of measuring them.

Proxy Data

If we had only the recent instrumental record as our guide, we might consider the climate system to be relatively stable on decade-to-century time scales. Instrumental records of atmospheric and oceanic conditions are too short to adequately define dec-cen variability: Atmospheric records rarely predate the current century, and high-quality ocean records are limited to recent decades. Moreover, the instrumental record is spatially biased; multidecadal temperature records from the oceans are concentrated in shipping lanes, and long atmospheric records originate mostly in western Europe and North America. The existing instrumental record of climate is thus insufficient to reveal most natural modes of climate variability on multidecadal to multicentury time scales. Models can simulate the climate systems over these time scales, but they must rely on the questionable assumption that the processes they incorporate behave and interact similarly over a range of time scales that far exceeds the instrumental baseline.

Fortunately, long-term records of climate exist in paleoclimatic archives worldwide. They offer the opportunity to extend our observational baseline into the dec-cen range of the spectrum of climate variability. Such records provide new information on the natural variability and sensitivity of climate, and they constitute an observational basis for evaluating the behavior of the numerical models used for climate prediction.

Paleoclimatic Contributions

Paleoclimatic records contribute in important ways to a better understanding of dec-cen climate variability. Simply extending the record of climate at a particular location makes possible evaluation of natural variability over time scales not represented by instrumental data. For example, proxy records of moisture balance from California and the Great Plains reveal that the droughts of this century pale by comparison with droughts over the rest of the millennium as regards both amplitude and duration (Muhs and Maat, 1993; Madole, 1994; Stine, 1994; Laird et al., 1996). Such long-term records allow us to place this century's warming in perspective—in particular, to evaluate whether it is unprecedented (Bradley and Jones, 1993; Briffa et al., 1995; Jacoby et al., 1996).

We can use information on past climate variability to assess whether the short-term (seasonal-to-interannual) modes and patterns observed today (see Chapter 3) have the same spatial patterns and global teleconnections as long-term (dec-cen) modes. For example, seasonal-to-interannual modes such as ENSO are apparently modulated on dec-cen time scales (Cole et al., 1993; Dunbar et al., 1994). The African and Asian monsoons probably respond in phase to large-scale forcing over multicentury time scales (Overpeck et al., 1996), even though over shorter periods the Asian monsoon weakens and East African rainfall increases during ENSO warm extremes. In addition, proxy records of snow accumulation (Alley et al., 1993) and temperature from ice cores (Dansgaard et al., 1993) can provide some insight into how rapidly climate regime changes can occur.

Paleoclimate reconstructions also allow us to observe the response of climate to changes in forcings or boundary conditions—for example, the relationship between drought and solar variability, or the behavior of ENSO during periods of warmer background SST or lower sea level. By improving our understanding of the natural variability and sensitivity of climate over dec-cen time scales, paleoclimatic reconstructions enable us to evaluate model behavior over these time scales (see, e.g., Knutson et al., 1997). Finally, paleoclimatic archives preserve records of climate forcings as well as responses (Zielinski et al., 1994; Lean et al., 1995). Reconstructions of volcanic aerosol loading or solar variability (for instance) can be incorporated into model simulations to test the system's responses (Rind and Overpeck, 1993). Fields of SSTs can also be interpolated from point reconstructions for use in initializing transient simulations with atmospheric GCMs.

Sources of Proxy Data

Many geologic and biologic archives preserve useful information on previous climate conditions. The primary sources of information useful on dec-cen time scales are described below. Certain of these archives (ice cores, coral reefs, and old-growth trees) are under threat of destruction from human and environmental influences, which will spell the loss of valuable climate information.

Ice cores have provided important records of quantities of atmospheric constituents from both high-latitude and high-altitude regions. Temperature can often be derived from the isotopic content of the ice (e.g., Thompson et al., 1995) on the basis of the relationship between the stable isotopic content of precipitation and the temperature of condensation, although large-scale climate features also control significant isotopic variance over dec-cen time scales (White et al., 1997a). Borehole temperature measurements permit verification of the temperature-isotope relationship for periods when precipitation, season, or other factors alter this dependence (Cuffey et al., 1995). Accumulation rates allow reconstruction of the hydrologic balance (Meese et al., 1994). Certain ice cores preserve pristine air bubbles throughout, making possible unique observations of past changes in atmospheric greenhouse-gas concentrations (Raynaud et al.,

1993; Etheridge et al., 1996) and yielding chemical ratios that reflect biogeochemical processes (Whung et al., 1994). Ice cores preserve records of aerosol loading and chemistry (Zielinski et al., 1994; O'Brien et al., 1995) that reflect both source strengths and transport patterns, and ice-core pollen records track regional vegetation changes (Thompson et al., 1995). Record lengths range from millennia in many tropical sites to hundreds of millennia in the coldest regions; resolution is decadal or less at the base of the oldest cores, because the more deeply buried ice thins under pressure, but the resolution can be seasonal in records of recent centuries.

Corals represent a multivariate record of tropical surface-ocean variability. Oxygen isotopic variations in the geochemistry of coral skeletons track SST (Fairbanks and Dodge, 1979; Dunbar et al., 1994; Gagan et al., 1994; Wellington et al., 1996) and salinity where such variations are strong (Cole and Fairbanks, 1990; Linsley et al., 1994). Certain metals that substitute for calcium in the aragonite lattice also reflect temperature (Beck et al., 1992; Shen and Dunbar, 1996; Shen et al., 1996; Mitsuguchi et al., 1996), while others are incorporated in proportion to their concentration in surface water, which may be governed by processes such as upwelling, runoff, or wind mixing (Shen et al., 1987, 1992; Lea et al., 1989). In some locations, coral growth rates reflect SST (Lough et al., 1996), and fluorescent bands in their skeletons may track river discharges (Isdale, 1984). Radiocarbon concentrations in corals reflect surface water ^{14}C variations, which make possible the reconstruction of aspects of ocean circulation over recent centuries (Druffel, 1987, 1997; Druffel and Griffin, 1993). Coral record lengths range from 100 to 800 years, with weekly to quarter-year resolution depending on growth rate; shorter records at similar resolution can be found in fossil sequences over the past 10^3 to 10^5 years (Beck et al., 1997; Gagan et al., 1998).

Tree-ring climate reconstructions are generally derived from suites of many (10-50) cores from a single site, cross-dated to provide absolute age control and pooled to eliminate noise associated with individual tree responses (Fritts, 1976; Cook and Kairiukstis, 1990). Statistical calibration with local climate, including independent validation intervals, yields quantitative records of tree sensitivities to aspects of climate like summertime drought (Meko et al., 1993; Hughes and Graumlich, 1996), summer temperature (Briffa et al., 1990, 1995; Cook et al., 1991), or annual temperature (Jacoby and D'Arrigo, 1989; Jacoby et al., 1996). Different aspects of tree growth, including ring widths and the density of various aspects of the wood architecture, can be measured and calibrated against climate. Tree-ring reconstructions, which are already widely available over northern mid-latitudes, can be pooled and gridded to generate reconstructed fields of such parameters as drought (Cook et al., in press).

Sediments in oceans and freshwater environments preserve a wealth of information on global, regional, and local climate variations; when sedimentation rates are sufficiently high, dec-cen climate variability is recorded. Climate parameters interpreted from sedimentary evidence include a wide range of physical, ecological, hydrologic, and chemical aspects particular to a given site. In anoxic environments, where the uppermost layers of sediment are not disturbed by burrowing organisms, the sediments are laid down in annual couplets (varves) that make possible extremely high-resolution chronology and interpretations (see, e.g., Overpeck, 1996; Hughen et al., 1996; Behl and Kennett, 1996). Where sedimentation rates are high but varves do not form, decadal resolution is achievable (Hodell et al., 1995; Laird et al., 1996). Sediment records with this resolution can span on the order of 10^4 years. Signs of lake-level changes, including geomorphologic features and submerged forest remains, provide another type of sedimentary evidence for dec-cen variability of hydrologic controls on lake levels (Stine, 1994). These are not continuous records, but they offer a snapshot of hydrologic conditions useful in addressing century-scale climatic change.

Historical information about climate offers another source of information on past variability in attributes deemed important to a given society. Often these proxies are related to practical considerations, such as droughts, harvest records, fish catch, death records, the freezing of waterways, or the level of a river (Quinn, 1993; Frenzel et al., 1994). Although such records require a transfer function to produce strictly climatological quantities, their immediate relevance to societally important attributes can also be considered an advantage.

Other proxy sources of climatic information may prove useful for understanding dec-cen variability in specific locations, or as new archives are explored and calibrated. For example, relict tree stumps rooted in modern Sierra Nevada wetlands (Stine, 1994) and recently active dune fields in Colorado grasslands (Muhs and Maat, 1993; Madole, 1995) provide snapshots of considerably dryer weather in North America during the Holocene. Dating of tropical and temperate mollusc remains from archaeological sites in Peru has led Sandweiss et al. (1996) to conclude that ENSO variability was not present until about 5000 BP. A combination of reconstructions from ice cores and findings at archaeological sites in Greenland support the theory that Norse settlements were abandoned because of colder summertime temperatures associated with the early stages of the transition into the Little Ice Age which gripped the North Atlantic region for several centuries (Pringle, 1997).

Calibration of Proxy Records

The extraction of climatological quantities from proxy data requires an understanding of how the signal is incorporated into the proxy, and of any competing influences on the record. Calibration procedures vary among the different archives, and include both process-based and statistical approaches. For tree rings, precise chronologies permit the compositing of many individual records to reduce noise;

standard practices include a model-development process that incorporates both calibration procedures and an independent period of validation against instrumental data. Although the processes that control tree growth are understood to the extent that certain relationships can be expected, statistical calibration provides the primary basis for most tree-ring reconstructions of climate. For most other types of archives, poorer replication and chronological uncertainties mean that compositing dozens of records is not feasible. Thus a more process-based approach is usually taken, involving such steps as automatic weather stations set up at ice-coring sites, automated temperature monitors and water-collection programs at coral sites, and sediment traps in regions of sediment coring. The resulting measurements indicate which processes are relevant to the incorporation of the paleoclimate signal, and statistical methods are used to evaluate the degree to which the chosen interpretation is correct.

Complications specific to calibration of paleoclimatic records include geochronology, biology, and the seasonality of response. All paleoclimatic data must contend with the issue of assigning accurate ages; annual precision is achievable in many cases where annual layers are deposited (particularly tree rings, but also many corals and ice cores). Age uncertainties for all methods need to be quantifiable in order to determine the limits of useful interpretations. Many paleoclimate archives are living organisms, whose biology can affect how a climate signal is recorded. Organisms may stop growing during a time of climate-induced stress, or long-term non-climatic growth trends may exist and must be removed. Biological processes can cause a consistent offset of a carbonate skeleton's geochemistry from the thermodynamically expected concentrations of isotopes and metals. Finally, many proxies reflect a seasonally specific or weighted response, based on nonconstant growth or deposition rates throughout the year or increased sensitivity during certain seasons. These potential complications need to be recognized if optimal reconstructions are to be developed.

Data Products

Once calibration has been established, paleoclimate interpretations produce a variety of types of data products. The simplest records are histories of past variations at a single site. These histories may be a snapshot or window on the past that reveals a scenario different from the modern climate, like the tree stumps in a lake bed or dune fields under modern grasslands mentioned earlier. A more typical reconstruction from ice and sediment cores, corals, and tree rings is a continuous, well-dated time series from which information about seasonal to centennial modes of temporal variability can be extracted. Just like a single instrumental temperature record, a single paleoclimatic record is of limited value in understanding large-scale climate variability. However, if the site is sensitive to large-scale variability, or if many sites are developed, indices of large-scale modes such as the NAO or ENSO can be reconstructed (Cook et al., 1997).

Taking this approach a step further, the synthesis of results from many sites can provide a basis for interpolating spatial fields of climate reconstructions. The development of paleoclimatic data into fields of climatic quantities improves the interface between these data and GCM simulations, which produce, and are usually forced by, such spatial fields. For example, CLIMAP (1981) produced SST and ice-extent maps for the last glacial maximum that have been used extensively as boundary conditions and even validation fields for GCM simulations of that period; similar approaches can be used for studies of dec-cen variability to assess the climate response to large forcing changes. Cook et al. (in press) describe how 388 drought-sensitive tree-ring chronologies from the continental United States can be combined into a reconstruction of the annual Palmer Drought Severity Index (PDSI) for the past 300 years. Overpeck et al. (1997) present a multi-proxy 400-year temperature reconstruction from the Arctic that confirms recent unprecedented warming associated with dramatic environmental changes. EOF-based interpolation techniques can be used to fill gaps between single-site reconstructions in order to generate fields from sparser datasets (Kaplan et al., in press; Mann et al., 1998), although this approach does assume that dominant EOF patterns are consistent through time. A global gridded annual-mean time series of, for example, surface temperature and precipitation for the past 1,000-2,000 years from reconstructed paleodata would allow the application of standard statistical techniques for comparison with modern data.

Summary

Proxy data provide a unique contribution to the objectives of a research program aimed at understanding climate change on dec-cen time scales. Indeed, there are few other continuous sources of observations on climate variability that extend beyond the middle of the nineteenth century. Paleoclimatic reconstructions provide a test bed for numerical climate models, and suggest new conceptual models for long-term climate variations. To take full advantage of this information resource will require broadening support for climate research in the following ways:

- Cross-disciplinary interaction among those climate scientists using proxy and instrumental data, and those developing and using numerical models, needs to be fostered. The combination of instrumental, paleoclimatic, and modeling approaches can provide answers to questions about calibration, limitations, and interpretation that cannot otherwise be resolved for more than the last hundred years or so.
- Climate-monitoring programs can provide important process and calibration baselines for paleoclimatic interpre-

tations, particularly if such programs are designed from the start to include aspects of climate that proxy data can record.

• In some parts of the world, potential sources of paleoclimatic information are disappearing as tropical glaciers melt and old-growth trees and coral-reef environments are exploited. Quick efforts to sample these endangered resources will pay off in new understanding.

• Searches for new proxy records and refinements of existing ones are needed to expand spatial coverage, provide measures of additional climate parameters, and clarify the limitations of the various types of proxies.

• Open access to paleoclimate data can be assured through continued support for database activities, both to encourage data submission and to integrate a variety of proxy information into a single database for climate-variability studies. The World Data Center-A for Paleoclimatology archives paleodata and makes it available to the climate-science community. As can be seen from its website at <http://www.ngdc.noaa.gov/paleo/paleo.html>, the WDC-A is continually developing this resource.

• The collection, processing, and interpretation of proxy data to yield a final climatological product require considerable effort. These efforts should be vigorously pursued in order to turn our currently sparse, though tantalizing, picture of past climate changes into a more complete set of time-varying spatial fields. This is the only currently available means by which we can advance our understanding of dec-cen variability from an observational perspective, and it should be fully exploited.

Analysis Products and Model Output

A special hybrid climate-information product involves the interpolation of climatic datasets into spatially consistent fields, through the use of climate modeling and assimilation techniques. In essence, the inaccurate, irregularly spaced, and often sparse observations are blended with a model simulation to produce a globally consistent sequence of climatological fields. In particular, major numerical and simulation centers, such as the European Centre for Medium-Range Weather Forecasting, NOAA's National Centers for Environmental Prediction, and NASA's Data Assimilation Office, have been carrying out retrospective analysis ("re-analysis") projects over the past few decades, using a single model and data-assimilation scheme (Kalnay et al., 1996; Todling et al., 1998) to reconstruct climate evolution since World War II. These re-analysis products can be used as spatially and temporally interpolated data. Alternatively, statistical techniques such as optimal interpolation can be used, in which the spatial covariance structure is created from the observed data and then used to fill in spatial gaps that preserve this structure (assuming that the structure has not changed).

These methods allow us to expand the spatial and temporal coverage of the data in a way that is completely consistent with physical laws and observed data relationships. In this manner, the data are more accessible for incorporation into models or comparison with model output, facilitating the integration of models and observations that will be critical to the success of dec-cen-scale research. To the same end, it is important that observational data and model output be subjected to the same kind of processing (e.g., smoothing or gridding) so that they can be compared on an equal footing. Note too that if the instrumental record is to grow in a regular and systematic way, a single re-analysis will not suffice. A new re-analysis of the entire record must be performed every 5 to 10 years, using the best available model and any additional data that have become available since the previous re-analysis.

COUPLED-MODEL DEVELOPMENT AND INFRASTRUCTURE

All scientific understanding is crystallized, and all predictions are made, using models. They may be descriptive and intuitive, or more formal mathematical ones. Climate dynamics currently employs models that embody equations reflecting our admittedly incomplete knowledge of the physical, chemical, and biological laws that govern each of the climate system's subsystems, and also their interactions. These models vary in their degree of completeness and detail; some are very simple, mechanistic ones, whereas others are highly elaborate and require huge resources in people and computing power. Models are used to assimilate incomplete, imperfect, and irregularly distributed observations; to simulate known climate phenomena; to discover or help understand new, as yet unknown phenomena; and to predict the climate system's evolution over time. This section briefly reviews existing models, and then describes a model configuration that should permit achievement of the major goals of a research program in decade-to-century-scale climate variability.

Existing Models

Models for each of the climate system components—atmosphere, biosphere, cryosphere, and hydrosphere—and their predictive capabilities have been presented in the relevant sections of Chapter 4. This information is merely integrated and summarized here as a basis for building the model configuration described in the next subsection.

The modeling of each one of the climate system components is at a slightly different point in its evolution, both from simple to more detailed models and from the study of one model or class of models to an organic integration of all relevant models. Only a combined analysis of all these models with each other and with the observations can provide all the different types of information needed about the (sub)-system. The models in use range from scalar evolution equations to general-circulation models. A scalar equation can describe, for instance, the evolution of global surface air temperature due to various heat sources and sinks and climatic feedbacks, both positive and negative. The GCMs use numerically solvable forms of the most complete systems of evolution equations that describe local variations in the temperature, pressure, velocity, and composition of the atmosphere, ocean, and other subsystems. They have been successfully utilized for data assimilation, simulation of various atmospheric and oceanic phenomena, and weather prediction.

No single class of model can satisfy all needs. The GCMs' spatially detailed, local description of phenomena and their short-term variability limits these models' ability to provide such a description for sufficiently long time intervals and sufficiently broad ranges of the many unknown parameters that enter the governing equations. Simple models are flexible enough to explore broad swaths of parameter space and study variability on the long time scales of interest for dec-cen climate variability, but they can be evaluated against observations only through the use of intermediate models with greater spatial and physical detail. Such intermediate models describe variations in latitude or height only, in latitude and height, or in latitude and longitude, while averaging or integrating over the directions not explicitly included.

For the atmosphere, experience with GCMs as well as with the systematic use of the full hierarchy of models mentioned above is extensive, and can provide a pattern for similar usage in the other subsystems. Experience with ocean, cryosphere, biosphere, and continental hydrosphere models in climate studies is considerably less, but is rapidly reaching a similar level of sophistication.

Models that consider the tropical ocean coupled with the global atmosphere have played a key role in the advancement of coupled models, and have also played an important role in the TOGA program. TOGA has clearly had a great impact on the rapid and successful development of such models (NRC, 1996). A full hierarchy of TOGA-type models, from simple through intermediate to coupled GCMs (CGCMs), is now being used in data assimilation, in simulation of air-sea interactions, and in experimental but routine seasonal-to-interannual prediction. CGCMs and simple and intermediate coupled models for the global or mid- to -high-latitude oceans are also evolving rapidly. Models for snow, surface hydrology, and near-surface vegetation have already evolved, at least in part, as modules or appendices of atmospheric GCMs.

A Target Modeling Structure

As noted earlier, the enterprise of monitoring, understanding, and eventually predicting climate variability on dec-cen time scales is inseparable from that of modeling. Thus, the organized structure (or hierarchy) of coupled models described below is part of the process of building our understanding, and should evolve with it. A substantial part of the structure needs to be set up early in this process, and most of it should be in place by the time a concerted scientific program is ramped up.

At the top of this organized structure there should be at least one modular CGCM, with a very flexible interface between the subsystem models, or "modules," of which it is composed, and an architecture that permits an easy interchange of the modules. The complexity inherent in the climate system's behavior on the dec-cen scale will require the participation of the entire climate-dynamics community, at universities and at government laboratories, in the modeling enterprise. The interchangeable-module architecture is certainly one that would permit all members of this community, from graduate students to senior researchers, to distill their evolving knowledge of the subsystems into improved modules, and to use the full power of the CGCM in a configuration well adapted to their specific area of application. While a modular approach would have significant advantages, it is recognized that to obtain maximum computational efficiency, the module's computer-program codes may have to be optimized for specific machines.

This modular CGCM (or CGCMs) needs to be supported by a full hierarchy of simple and intermediate models of each subsystem and of subsets of the entire system. For example, an intermediate model of the global ocean or of a particular ocean basin may be coupled to a simple model of the global atmosphere, as one rung in such a hierarchy. This complementarity is required by the learning process involved in developing better modules for CGCMs, as well as by the need to evaluate CGCM performance and build knowledge of climate variability and its predictability on dec-cen time scales.

This target structure, in turn, requires adequate resources in data, people, and computing. The observations described in the preceding section are necessary for model evaluation and eventual climate prediction. Observations have to be combined, by the process of data assimilation using the models, to provide physically consistent descriptions of past and present climate change, as well as to permit the prediction of the future from an optimally described present.

A new generation of scientists needs to be educated to deal with the enormous complexities of modeling the entire climate system and predicting its behavior. These scientists will have to master not only the traditionally relevant disciplines of dynamic meteorology and physical oceanography, but also important aspects of analytic chemistry and biochemistry, ecology, and other biological disciplines. They must be at ease with small and large models—that is, have both theoretical and numerical skills, as well as a high degree of sophistication in the use of the most modern computing, communications, and observing technology. To achieve this, their education will have to bridge not only disciplines but also institutions.

The resources available for climate research will have to

include computing, communications, and multimedia support at the highest levels of performance provided by current technology. A minimal network should incorporate at least one fully dedicated device capable of the highest throughput available at a single site at any given time. It should include communication links capable of sustaining information flow at a level consistent with this device. These resources will enable the various members of the community to operate as if housed on a single campus. Powerful workstations will be needed at many locations to support work on: simple and intermediate models; module development, testing, and applications; and analysis of simulations and experimental predictions with the modular CGCM (or CGCMs). Both central and distributed memory will have to be sufficient for the observations gathered, as well as for model simulations, model-assimilated datasets, and experimental predictions. Finally, hypertext information exchange—text, datasets, still and moving images—should be supported to facilitate the integration by many scientists of the rich information gathered and generated about the climate system, its components, and its decade-to-century-scale variability.

DETECTION, ATTRIBUTION, AND SIMULATION

The global-mean surface air temperature has increased by about 0.5°C since the mid-nineteenth century, as was indicated in Figure 2-6. This increase is not monotonic; there is a fairly rapid rise from 1910 to 1940 and again from the mid-1970s to the present, and some decrease in the 1940s and 1960s-1970s. (The increase was more uniform in the Southern Hemisphere than in the Northern Hemisphere, where the cooling between 1960 and 1980 was more pronounced.) It has been suggested that the overall warming trend can be attributed to the observed increase in greenhouse gases, including CO_2; the lack of monotonicity in the increase must then be interpreted as the result of natural climate variability or other forcings. Recent modeling studies indicate that it may be necessary to consider in particular the cooling effect of anthropogenic sulfate aerosols, which reflect solar radiation, in order to explain quantitatively the magnitude of the observed warming (IPCC, 1996a). However, the inclusion of sulfate aerosols is not sufficient to account for all of the variations in globally averaged surface temperature over the last century, especially the rapid warming during the 1920s and early 1930s.

One of the most important goals of dec-cen research is the reliable projection of future climate change (e.g., global warming) through the use of coupled ocean-atmosphere-land models. The best way of lending credence to future model predictions of climate is to successfully simulate past climate change. In order to assess how realistic the sensitivity of a climate model is to external forcing, it is necessary to simulate observed long-term change of climate by driving the model with time series of actual thermal forcings, such as increased concentrations of greenhouse gases and aerosols in the atmosphere. The performance of the model can then be evaluated by comparing the simulated and the observed long-term climate changes. The large uncertainties about the believability of model projections of future climate change must be reduced by assessing the models through the hindcast of climate, as described above. To carry out such a hindcast, long-term observational data for driving the model are essential. The highest priority should be given to:

(1) Obtaining reliable, long-term observations of factors that change thermal forcing, which drives the long-term variations of a climate model. These factors include not only the increase in greenhouse gases, but also changes in solar irradiance and in aerosol loading in the atmosphere.

(2) Making reliable, long-term observations of a carefully chosen set of basic climatic variables. Important variables include spectra of radiative fluxes at the top of the atmosphere, as well as satellite-observable radiances that are indicators of levels of cloudiness, snow cover, sea ice, vegetation, and possibly soil wetness. Other excellent candidates for long-term observation are sea level, surface salinity, and the water-mass structure of the oceans. The monitoring of variables such as total carbon content, alkalinity, and the partial pressure of CO_2 will also be valuable for reliably projecting the future increase of atmospheric CO_2.

(3) Reconstructing past changes of the variables listed above. This task will require comprehensive compilations of historical and other proxy data, with careful attention to their quality.

In order to evaluate a climate model through the hindcast strategy described above, it is also necessary to distinguish the anthropogenic change of climate from its natural, internally generated fluctuations. For this purpose, we need to improve our knowledge of internally generated (rather than externally forced) climate variability, with a view to separating the anthropogenic changes from the natural variations. The latter seem to contain quasi-periodic, and hence predictable, components with interannual and interdecadal periods (Plaut et al., 1995; Mann et al., 1995b).

Several groups have recently achieved some success in exploring this topic by analyzing the climate variability from very-long-term (thousand-year) integrations of coupled ocean-atmosphere-land models. For example, Stouffer et al. (1994) and Manabe and Stouffer (1996) have found that (with the notable exception of the tropical eastern Pacific, where SST anomalies are underestimated) their model approximates the standard deviation of annual mean surface air temperature and its geographical distribution. The model also simulates the broad-band spectrum of observed global-mean surface air temperature, from interannual to interdecadal time scales. However, it fails to reproduce the warming trend of centennial time scale (i.e., ~0.5°C per century) that has been observed since the end of the last century. If the model is assumed to be realistic—in spite of its failure to reproduce the quasi-periodic components of the natural variability—this result suggests that the observed centennial-scale warming trend is not generated within the climate system by nonlinear interaction among the atmosphere, ocean, and continental surface. Instead, the trend must be caused by a sustained trend in natural and/or anthropogenic thermal forcing, such as changes in solar irradiance, greenhouse gases, and aerosol loading in the atmosphere. Essentially similar results have also been obtained from the long-term integration of a coupled model at the U.K. Meteorological Office (see, e.g., Mitchell et al., 1995).

Identifying predominant patterns associated with the natural, internally generated climate variability would aid in the detection of patterns of anthropogenic change (see, for example, Barnett and Schlesinger, 1987; Hasselmann, 1993; and Santer et al., 1995). If the effect of sulfate aerosols is considered together with the effect of greenhouse gases in GCMs, the spatial distribution of the model-generated change of atmospheric temperature over the decadal time scale appears to become more realistic (Santer et al., 1996). These and other recent results (see, e.g., IPCC, 1996a, for an overview) are leading to a more reliable estimate of the anthropogenically induced climate change, as well as of the natural variability caused by mechanisms internal to the climate system.

In the future, major effort will need to be devoted to observational and modeling studies of internally generated climate variability, so that this variation can be distinguished from anthropogenic climate change. Records from past observations of both ocean and the atmosphere should be compiled and analyzed for variables such as concentration of greenhouse gases in ice cores, sea-level pressure, surface and subsurface temperature and salinity in oceans, air temperature and humidity at the surfaces, and temperature and geopotential height at selected pressure levels in the atmosphere. It is also essential to improve model parameterizations of various feedback processes, in particular those involving cloud, snow, and sea-ice cover, all of which substantially affect incoming solar and/or outgoing terrestrial radiation at the top of the atmosphere. Other factors of critical importance are cumulus convection and land-surface heat and water budgets. Greater use of data from remote sensing and in situ measurements of radiative emissions and river runoff facilitate evaluating and improving the parameterizations of the important processes identified above.

LINKAGE ACROSS TIME SCALES

As was noted in Chapter 4, there are practical reasons for dividing the study of climate variations by the time scales on which they occur. The climate system clearly evolves over a continuum of time scales, however, and no "spectral gap" in nature justifies such a separation. To advance our understanding of overall climate change and variability most efficiently, it is important that we explicitly recognize those processes that cannot easily be categorized by scale, and that we particularly emphasize those mechanisms which affect climate variability and change over a range of time scales.

A few specific examples of climate-variability patterns and their possible causal mechanisms that appear on more than one time scale are (1) the interdecadal variability of ENSO, in amplitude, periodicities, and warm or cold anomaly distribution; (2) the North Atlantic Oscillation (NAO) and its purely atmospheric, purely oceanic, or coupled mechanisms; and (3) changes in the carbon cycle, over land, ocean, and the tropical or mid-latitude or polar regions. The modes of variability that cross two or more time scales can arise either from the intrinsically broad-band behavior in time of a specific spatial mode, or from the nonlinear coupling between narrow-band spatio-temporal modes that share certain regional characteristics. Which one of these overall types of behavior is at the root of a given dec-cen climate phenomenon has important implications for its predictability.

Currently, the national and international organizations devoted to the study of physical climate are structured to address separately the high-frequency variability (GEWEX), seasonal-to-interannual variability (GOALS), decade-to-century-scale variability (CLIVAR DecCen), and millennial and longer-scale variability (e.g., PAGES). Each of these groups has identified a suite of high-priority issues that must be addressed. Many of the detailed processes involved in these issues are common to all four units. For example, improved understanding of air-sea exchanges is of fundamental importance to the study of climate, regardless of time scale. Similarly, the patterns of climate variability and the coupled modes are of equal importance to all groups, because their regional manifestations occur on a broad range of time scales. Issues related to these common processes and patterns warrant particular attention, and a dec-cen program that is highly coordinated with GOALS and GEWEX would enable them to be studied most effectively. Furthermore, the physically based studies of climate must be fully integrated with those investigating the chemical-biological aspects, which are currently being addressed by elements of the International Geosphere-Biosphere Programme.

7

Conclusions and Recommendation

CONCLUSIONS

Climate is constantly changing and will continue to do so; we cannot assume a stable climate system even in the absence of anthropogenic influences. Such variability manifests itself over a continuum of time scales, from seasonal and interannual to decadal and centennial, or longer.

This report documents decadal-to-centennial variations of those climatic attributes—freshwater, temperature, sea level, solar radiation, storms, and ecosystems—that directly affect societies and economies. It also suggests how this variability is governed by the interacting components of the climate system: atmospheric composition and radiative forcing, atmospheric circulation, the hydrologic cycle, ocean circulation, the cryosphere, and land and vegetation.

Because of the duration of dec-cen-scale climate changes, and the potentially large magnitude of their effects, mitigation and/or adaptation measures are likely to involve investments in infrastructure and changes in policy. Unfortunately, the subtlety of slow change over long time scales, relative to the obvious diurnal, seasonal, and interannual variations, can disguise its potential long-term severity. Society's willingness to address problems of climate variability in advance is limited both by the inconspicuousness of the changes and by uncertainty in our ability to forecast them. This underscores the importance of improving our understanding of dec-cen climate change, the rate and range of its variability, the likelihood and distribution of its occurrence, and the sensitivity of climate to changes in the forcings, both natural and anthropogenic.

A firm understanding of these characteristics will constitute the foundation on which future policy decisions and infrastructure management can be rationally based. Development of this understanding will call for the ability to answer questions that are both fundamental and overarching. Specifically, we need to know:

- What are the spatio-temporal patterns of dec-cen variability, and what mechanisms give rise to them?
- What is the relationship between natural dec-cen variability and observed global warming? What do we have to know about the natural variability in order to detect anthropogenic change?
- How does variability in the forcings, both natural and anthropogenic, affect dec-cen variability?
- What is the role of interaction among the climate components in generating and sustaining dec-cen variability?
- To what extent is dec-cen variability predictable?

The resolution of these questions will depend on progress in addressing a number of more specific scientific issues, which are presented at the conclusion of Chapter 3, near the end of each of the climate-component discussions of Chapter 5, and in Chapter 6. The panel has selected those issues they consider most urgent for effectively advancing our understanding of dec-cen variability and change. In general, these issues revolve around basic concerns such as: how the various climate components have changed in space and time, and what mechanisms drive the changes; the interactions and feedbacks among the various atmospheric constituents, radiative forcing, and surface boundary conditions, and how they influence dec-cen climate variability and change; how the sources and sinks of greenhouse trace gases and the partitioning of carbon between reservoirs vary on dec-cen time scales, and what mechanisms drive their variations; and identifying the externally forced, internally forced, and coupled modes of variability and change in the Earth's climate system, and the processes and mechanisms driving them.

RECOMMENDATION

The Dec-Cen panel recommends that the United States initiate a Dec-Cen Program designed to increase understanding of climate variability on decade-to-century time scales, and determine its predictability. The initial design of this program would address those issues currently identified in Chapters 3, 5, and 6. Flexibility and adaptability will have to be maintained, so that new directions and opportunities can be pursued as our understanding is improved and research directions are refined. In order to achieve its goal, the U.S. Dec-Cen Program must represent a balance between the following elements:

1. *A long-term, stable observing system* to constantly monitor, with sufficient accuracy and resolution, a subset of crucial earth-system variables (i.e., key state variables and primary forcings) on dec-cen time scales. *A modeling activity* should be an integral part of this system, to add value to the observations by assimilating them into suitable models. By more clearly delineating the actual nature of decadal variability in the climate system, the capability for detecting anthropogenic climate change will be improved, thereby making an important contribution to the IPCC's goals. The observing system can be built incrementally, but must be initiated immediately, because a long, sustained climate record is required to document the wide range of climate variability and change over dec-cen time scales. (Specific details are outlined in Chapter 6.)

2. *A hierarchical program of modeling studies* that has the ultimate goal of simulating and predicting the entire Earth system on dec-cen time scales. This element requires an infrastructure that makes powerful computational and communications resources readily available to the full modeling community. (Again, specific details are outlined in Chapter 6.)

3. *Process studies* that address those geophysical, chemical, and biological processes which are so poorly understood that they hinder our ability to define, understand, and predict dec-cen variability. These studies should include analytic, model-based, and observational ones, as well as combinations of the three. (Fundamental processes requiring particular attention are listed following the issues in each subsection of Chapter 5.)

4. *Producing and disseminating long-term proxy and instrumental datasets* for use in the study of dec-cen variability. These datasets provide the only means for immediately securing a reliable, albeit preliminary, assessment of past dec-cen variability, and for establishing the pre-anthropogenic-effect baseline of natural variability. Because the sources of some of the most well-established proxy records are at risk of permanent destruction in the very near future (e.g., alpine glaciers and long-lived stable tropical corals), it is imperative that these records be secured while it is still possible to do so.

The recommended U.S. Dec-Cen Program would be part of the U.S. contribution to the DecCen and ACC (Anthropogenic Climate Change) components of the Climate Variability and Predictability (CLIVAR) Programme of the WCRP. It would also contribute to the International Geosphere-Biosphere Programme (IGBP). The scope of the U.S. program would be broader than that of either the WCRP or IGBP alone; the research objectives would deal in an integrated manner with the biological, terrestrial, oceanic, and atmospheric aspects of climate variability on dec-cen time scales. It would thus draw on the contributions already being made by such programs and projects as the IGBP's PAGES, the WCRP's GEWEX, SPARC, ACSYS, and WOCE, and GCOS, GOOS, and GTOS.

The U.S. Dec-Cen Program must also be well coordinated with the U.S. and WCRP seasonal-to-interannual program, the Global Ocean-Atmosphere-Land System (GOALS) Program, because the dec-cen influence on short-term variability and change may aggravate the difficulties of short-term climate prediction. Indeed, several components of the proposed Dec-Cen Program cannot be separated from those of the GOALS Program (e.g., determining the influence of dec-cen variability on the frequency, intensity, and duration of El Niño events).

The ultimate research objective of a U.S. Dec-Cen Program would be to define, understand, and model dec-cen climate variability and change—natural and anthropogenic—so that the extent to which they are predictable can be reliably determined. If it can be shown that they are predictable, the ultimate practical aim of the Dec-Cen Program would be to design and implement a system that predicts the various aspects of climate at the temporal range permitted by their predictability. This system would build on the emerging seasonal-to-interannual prediction systems now being constructed, predict future decadal-to-centennial variations to the extent possible, and learn to use these predictions for the benefit of all.

The dec-cen prediction system would have to investigate the future responses to whatever external-forcing scenarios are imposed, both naturally and anthropogenically, upon the Earth's climate system. It would provide a basis for distinguishing natural variability from anthropogenic change, to the extent theoretically and practically possible. The understanding of this distinction will yield a strategy for the detection and attribution of any anthropogenic climate change that is superimposed on the background of natural variability.

Decadal-to-centennial climate variability is a subtle phenomenon. It proceeds too slowly to be perceptible to our senses, but its cumulative effects ultimately define the life prospects of future generations. Informed stewardship of the Earth's resources for the generations to come must draw on the insights afforded by model predictions and model-aided extrapolation of observations that can give us a glimpse of the future. A U.S. Dec-Cen Program will be the first step toward assuming this responsibility.

References

Aagaard, K., and E.C. Carmack, 1989. The role of sea ice and other fresh waters in the Arctic circulation. *J. Geophys. Res.* **94**, 14485-14498.

Alam, M., 1996. Subsidence of the Ganges-Brahmaputra delta of Bangladesh and associated drainage, sedimentation, and salinity problems. In *Sea-Level Rise and Coastal Subsidence: Causes, Consequences, and Strategies*, J.D. Milliman and B.U. Haq (eds.), Kluwer Academic Publishers, Dordrecht, The Netherlands, pp. 169-192.

Allan, R.J., J.A. Lindesay, and C.J.C. Reason, 1995. Multidecadal variability in the climate system over the Indian Ocean region during the austral summer. *J. Climate* **8**, 1853-1873.

Allen, L.H., and J.S. Amthor, 1995. Plant physiological responses to elevated CO_2, temperature, air pollution, and UV-B radiation. In *Biotic Feedbacks in the Global Climate System*, G.M. Woodwell and F.T. MacKenzie (eds.), Oxford University Press, New York, pp. 51-84.

Alley, R.B., D.A. Meese, C.A. Shuman, A.J. Gow, K.C. Taylor, P.M. Grootes, J.W.C. White, M. Ram, E.D. Waddington, P.A. Mayewski, and G.A. Zielinski, 1993. Abrupt increase in Greenland snow accumulation at the end of the Younger Dryas event. *Nature* **362**, 527-529.

Anderson, L.G., G. Bjork, O. Holby, E.P. Jones, G. Kattner, K.P. Koltermann, B. Liljeblad, R. Lindegren, B. Rudels, and J.H. Swift, 1994. Water masses and circulation in the Eurasian Basin: Results from the Oden 91 expedition. *J. Geophys. Res.* **99**, 3273-3283.

Asner, G.P., T.R. Seastedt, and A.R. Townshend, 1997. The decoupling of terrestrial carbon and nitrogen cycles. *Bioscience* **47**, 226-234.

Bacon, S., and D.J.T. Carter, 1993. A connection between mean wave height and atmospheric pressure gradient in the North Atlantic. *Int. J. Climatol.* **13**, 423-436.

Baldocchi, D., R. Valentini, S. Running, W. Oechels, and R. Dahlman, 1996. Strategies for measuring and modelling carbon dioxide and water vapour fluxes over terrestrial ecosystems. *Global Change Biol.* **2**, 159-168.

Baldwin, C., and U. Lall, in press. Seasonality of Streamflow: The Upper Mississippi River. *Water Resour. Res.*

Barnett, T.P., 1985. Variations in near global sea level pressure. *J. Atmos. Sci.* **42**, 478-501.

Barnett, T.P., 1995. Monte Carlo climate forecasting. *J. Climate* **8**, 1005-1022.

Barnett, T.P., and M.E. Schlesinger, 1987. Detecting changes in global climate induced by greenhouse gases. *J. Geophys. Res.* **92**, 14772-14780.

Barnett, T.P., N. Graham, M.A. Cane, S. Zebiak, S. Dolan, J. O'Brien, and D. Legler, 1988. On the prediction of the El Niño of 1986-1987. *Science* **241**, 192-196.

Barnola, J.M., D. Raynaud, Y.S. Korotkevich, and C. Lorius, 1987. Vostok ice core provides 160,000-year record of atmospheric CO_2. *Nature* **329**, 408-414.

Barnston, A., and R.E. Livezey, 1987. Classification, seasonality and persistence of low-frequency circulation patterns. *Mon. Wea. Rev.* **115**, 1083-1126.

Barnston, A.G., H.M. van Dool, S.E. Zebiak, T.P. Barnett, M. Ji, D.R. Rodenhuis, M.A. Cane, A. Leetmaa, N.E. Graham, C.R. Ropelewski, V.E. Kousky, E.O. O'Lenic, and R.E. Livesey, 1994. Long-lead seasonal forecasts-where do we stand? *Bull. Amer. Meteor. Soc.* **75**, 2097-2114.

Barry, R.G., and R.J. Chorley, 1992. *Atmosphere, Weather and Climate,* 6th ed., Routledge, London, 550 pp.

Barry, J.P., C.H. Baxter, R.D. Sagarin, and S.E. Gilman, 1995. Climate-related, long-term faunal changes in a California rocky intertidal community. *Science* **267**, 672-675.

Barsugli, J.J., and D.S. Battisti, 1998. The basic effects of atmosphere-ocean thermal coupling on midlatitude variability. *J. Atmos. Sci.* **55**, 477-493.

Battisti, D.S., and E.S. Sarachik, 1995. Understanding and predicting ENSO. *Rev. Geophys.* **33**, 1367-1376.

Battle, M., M. Bender, T. Sowers, P. Tans, J. Butler, J. Elkins, J. Ellis, T. Conway, N. Zhang, P. Lang, and A. Clarke, 1996. Atmospheric gas concentrations over the past century measured in air from firn at the South Pole. *Nature* **383**, 231-235.

Beck, J.W., R.L. Edwards, E. Ito, F.W. Taylor, J. Recy, F. Rougerie, P. Joannot, and C. Henin, 1992. Sea-surface temperature from coral skeletal strontium/calcium ratios. *Science* **257**, 644-647.

Beck, J.W., J. Recy, F. Taylor, R.L. Edwards, and G. Cabioch, 1997. Abrupt changes in early Holocene tropical sea-surface temperature derived from coral records. *Nature* **385,** 705-707.

Beer, J., U. Siegenthaler, G. Bonani, R.C. Finkel, H. Oeschger, M. Suter, and W. Wölfli, 1988. Information on past solar activity from ^{10}Be in the Camp Century ice core. *Nature* **331**, 675-679.

Behl, R.J., and J.P. Kennett, 1996. Brief interstadial events in the Santa Barbara basin, NE Pacific, during the past 60 kyr. *Nature* **379**, 243-246.

Bell, G.D., and J.E. Janowiak, 1995. Atmospheric circulation associated with the Midwest floods of 1993. *Bull. Amer. Meteor. Soc.* **76**, 681-695.

Bender, M., T. Sowers, M.-L. Dickson, J. Orchardo, P. Grootes, P.A. Mayewski, and D.A. Meese, 1994. Climate correlations between Greenland and Antarctica during the past 100,000 years. *Nature* **372**, 663-666.

Bender, M., T. Sowers, and E. Brook, 1997. Gases in ice cores. *Proc. Nat. Acad. Sci. USA* **94**, 8343-8349.

Berner, E.K., and R.A. Berner, 1987. *The Global Water Cycle. Geochemistry and Environment.* Prentice-Hall Inc., Englewood Cliffs, New Jersey, 387 pp.

Betts, R.A., P.M. Cox, S.E. Lea, and F.I. Woodward, 1997. Contrasting physiological and structural vegetation feedbacks in the climate change simulations. *Nature* **387**, 796-799.

Bindoff, N.L., and J.A. Church, 1992. Warming of the water column in the southwest Pacific Ocean. *Nature* **357**, 59-62.

Bindoff, N.L., and T.J. McDougall, 1994. Diagnosing climate change and ocean ventilation using hydrographic data. *J. Phys. Oceanogr.* **24**, 1137-1152.

Bindoff, N.L., and T.J. McDougall, in press. Decadal changes along an Indian Ocean section at 32°S and their interpretation. *J. Phys. Oceanogr.*

Bingham, F.M., 1992. Formation and spreading of subtropical model water in the North Pacific. *J. Geophys. Res.* **97**, 11177-11189.

Biondi, F., C.B. Lange, M.K. Hughes, and W.H. Berger, 1997. Interdecadal signals during the last millennium (AD 1117-1992) in the varve record of Santa Barbara basin, California. *Geophys. Res. Lett.* **24**, 193-196.

Bitz, C.C., 1997. A Model Study of Natural Variability in the Arctic Climate, Ph.D. thesis, University of Washington, Seattle, Washington, 200 pp.

Bitz, C.M., D.S. Battisti, R.E. Moritz, and J.A. Beasley, 1996. Low-frequency variability in the Arctic atmosphere, sea ice, and upper-ocean climate system. *J. Climate* **9**, 394-408.

Bjorgo, E., O.M. Johannessen, and M.W. Miles, 1997. Analysis of merged SMMR-SSMI time series of Arctic and Antarctic sea ice parameters 1978-1995. *Geophys. Res. Lett.* **24**, 413-416.

Blunier, T., J. Chapellaz, J. Schwander, J. Barnola, T. Desperts, B. Stauffer, and D. Raynaud, 1993. Atmospheric methane record from a Greenland ice core over the last 1000 years. *Geophys. Res. Lett.* **20**, 2219-2222.

Blunier, T., J. Schwander, B. Stauffer, T. Stocker, A. Dällenbach, A. Indermühle, J. Taschumi, J. Chappellaz, D. Raynaud, and J.-M Barnola, 1997. Timing of temperature variations during the last deglaciation in Antarctica and the atmospheric CO_2 increase with respect to the Younger Dryas event. *Geophys. Res. Lett.* **24**, 2683-2686.

Blunier, T., J. Chappellaz, J. Schwander, A. Dällenbach, B. Stauffer, T.F. Stocker, D. Raynaud, J. Jouzel, H.B. Clausen, C.U. Hammer, and S.J. Johnsen, 1998. Asynchrony of Antarctic and Greenland climate change during the last glacial period. *Nature* **394**, 739-743.

Bonan, G., D. Pollard, and S.L. Thompson, 1992. Effects of boreal forest vegetation on global climate. *Nature* **359,** 716-718.

Bond, G.C., and R. Lotti, 1995. Iceberg discharges into the North Atlantic on millennial time scales during the last glaciation. *Science* **267**, 1005-1010.

Bond, G.C., W. Showers, M. Cheseby, R. Lotti, P. Almasi, P. deMenocal, P. Priore, H. Cullen, I. Hajdas, and G. Bonani, 1997. A pervasive millennial-scale cycle in North Atlantic Holocene and glacial climates. *Science* **278**, 1257-1266.

Booth, C.R., and S. Madronich, 1994. Radiation amplification factors: Improved formulation accounts for large increases in ultraviolet radiation associated with Antarctic ozone depletion. In *Ultraviolet Radiation in Antarctica: Measurements and Biological Research*, C.S. Weiler and P.A. Penhale (eds.), AGU Antarctic Research Series, v. 62, American Geophysical Union, Washington, D.C., pp. 39-42.

Bradley, R.S., 1988. The explosive volcanic eruption signal in Northern Hemisphere continental temperature records. *Clim. Change* **12**, 221-243.

Bradley, R.S., and P.D. Jones, 1992. *Climate Since A.D. 1500*. Routledge, New York, 679 pp.

Bradley, R.S., and P.D. Jones, 1993. "Little Ice Age" summer temperature variations: Their nature and relevance to recent global warming trends. *Holocene* **3**, 367-376.

Brecher, H.H., and L.G. Thompson, 1993. Measurement of the retreat of Qori Kalis in the tropical Andes of Peru by terrestrial photogrammetry. *Photogram. Eng. Remote Sensing* **59**, 1017-1022.

Briffa, K.R., T.S. Bartholin, D. Eckstein, P.D. Jones, W. Karlen, F.H. Schweingruber, and P. Zetterburg, 1990. A 1400-year tree-ring record of summer temperatures in Fennoscandia. *Nature* **346**, 434-439.

Briffa, K.R., P.D. Jones, F.H. Schweingruber, S.G. Shiyatov, and E.R. Cook, 1995. Unusual twentieth-century warmth in a 1000-year temperature record from Siberia. *Nature* **376**, 156-159.

Broccoli, A.J., and S. Manabe, 1990. Can existing climate models be used to study anthropogenic changes in tropical cyclone climate? *Geophys. Res. Lett.* **17**, 1917-1920.

Broccoli, A.J., N.-C. Lau, and M.J. Nath, in press. The cold ocean-warm land pattern: Model simulation and relevance to climate change detection. *J. Climate*.

Broecker, W., 1991. The great ocean conveyor. *Oceanography* **4**, 79-89.

Broecker, W., 1994. Massive iceberg discharges as triggers for global climate change. *Nature* **372**, 421-424.

Broecker, W., and G.H. Denton, 1989. The role of ocean-atmosphere reorganizations in glacial cycles. *Geochim. Cosmochim. Acta* **53**, 2465-2501.

Brown, R.D., and B.E. Goodison, 1996. Interannual variability in reconstructed Canadian snow cover 1915-1992. *J. Climate* **9**, 1299-1318.

Bruun, P., 1962. Sea-level rise as a cause of shore erosion. *J. Waterways Harbors Div., Proc. Soc. Civil Eng.* **88**, 117-130.

Bryden, H.L., D. Roemmich, and J. Church, 1991. Oceanic heat transport across 24 degrees N in the Pacific. *Deep-Sea Res.* **37**, 297-324.

Bryden, H.L., M.J. Griffiths, A.M. Lavin, R.C. Millard, G. Parrilla, and W.M. Smethie, 1996. Decadal changes in water mass characteristics at 24°N in the subtropical North Atlantic Ocean. *J. Climate* **9**, 3162-3186.

Budyko, M.I., 1969. The effect of solar radiation variations on the climate of the Earth. *Tellus* **21**, 611-619.

Bush, A.B.G., and S.G.H. Philander, 1998. The role of ocean-atmosphere interactions in tropical cooling during the Last Glacial Maximum. *Science* **279**, 1341-1344.

Businger, J.A., and S.P. Oncley, 1990. Flux measurement with conditional sampling. *J. Atmos. Ocean. Tech.* **7**, 349-352.

Campbell, I.D., and J.H. McAndrews, 1993. Forest disequilibrium caused by rapid Little Ice Age cooling. *Nature* **366**, 336-338.

Cane, M.A., and S.E. Zebiak, 1987. Predictability of El Niño events using a physical model. In *Atmospheric and Oceanic Variability*, H. Cattle (ed.), Royal Meteorological Society Press, London, pp. 153-182.

Cane, M.A., S.E. Zebiak, and S.C. Dolan, 1986. Experimental forecasts of El Niño. *Nature* **321**, 827-832.

Cane, M.A., A.C. Clement, A. Kaplan, Y. Kushnir, D. Pozdnyakov, R. Seager, S. Zebiak, and R. Murtugudde, 1997. Twentieth-century sea surface temperature trends. *Science* **275**, 957-960.

Carmack, E.C., R.W. Macdonald, R.G. Perkin, F.A. McLaughlin, and R.J. Pearson, 1995. Evidence for warming of Atlantic water in the southern Canadian basin of the Arctic Ocean: Results from the Larsen-93 expedition. *Geophys. Res. Lett.* **22**, 1061-1064.

Carmack, E.C., K. Aagaard, J.H. Swift, R.W. Macdonald, F.A. McLaughlin, E.P. Jones, R.G. Perkin, J.N. Smith, K. Ellis, and L. Kilius, in press. Changes in temperature and contaminant distributions within the Arctic Ocean. *Deep-Sea Res.*

Cavalieri, D.J., P. Gloersen, C.L. Parkinson, J.C. Comiso, and H.J. Zwally, 1997. Observed hemispheric asymmetry in global sea ice changes. *Science* **278**, 1104-1106.

Cayan, D.R., 1996. Interannual climate variability and snowpack in the western United States. *J. Climate.* **9**, 928-948.

Cayan, D.R., and D.H. Peterson, 1989. The influence of the North Pacific atmospheric circulation and streamflow in the West. In *Aspects of Climate Variability in the Western Americas,* D.H. Peterson (ed.), Geophys. Monogr. 55, American Geophysical Union, Washington, D.C., pp. 375-397.

Cessi, P., and G.R. Ierley, 1995. Symmetry-breaking multiple equilibria in quasigeostrophic, wind-driven flows. *J. Phys. Oceanogr.* **25**, 1196-1205.

Chahine, M.T., 1992. The hydrological cycle and its influence on climate. *Nature* **359**, 373-380.

Chameides, W.L., P.S. Kasibhatla, J. Yienger, and H. Levy III, 1994. Growth of continental-scale metro-agro-plexes, regional ozone production, and world food production. *Science* **264**, 74-77.

Chang, P., L. Ji, and H. Li, 1997. A decadal climate variation in the tropical Atlantic Ocean from thermodynamic air-sea interactions. *Nature* **385**, 516-518.

Chang, P., L. Ji, H. Li, C. Penland, and L Matrisova, 1998. Prediction of tropical Atlantic sea surface temperature. *Geophys. Res. Lett.* **25**, 1193-1196.

Changnon, S.A. (ed.), 1996. The Great Flood of 1993: Causes, impacts, and responses. Westview Press, Boulder, Colorado, 300 pp.

Chappellaz, J., T. Blunier, D. Raynaud, J.M. Barnola, J. Schwander, and B. Stauffer, 1993a. Synchronous changes in atmospheric CH_4 and Greenland climate between 40 and 8 kyr BP. *Nature* **366**, 443-445.

Chappellaz, J.A., I.Y. Fung, and A.M. Thompson, 1993b. The atmospheric CH_4 increase since the last glacial maximum: 1. Source estimates. *Tellus* **45B**, 228-241.

Charney, J.G., 1975. Dynamics of deserts and drought in the Sahel. *Quart. J. Roy. Meteor. Soc.* **101**, 193-202.

Chen, F., and M. Ghil, 1995. Interdecadal variability of the thermohaline circulation and high-latitude surface fluxes. *J. Phys. Oceanogr.* **25**, 2547-2568.

Chen, F., and M. Ghil, 1996. Interdecadal variability in a hybrid coupled ocean-atmosphere model. *J. Phys. Oceanogr.* **26**, 1561-1578.

Chen, D., S.E. Zebiak, A.J. Busalacchi, and M.A. Cane, 1995. An improved procedure for El Niño forecasting: Implications for predictability. *Science* **269**, 1699-1702.

Chin, M., and D.D. Davis, 1995. A reanalysis of carbonyl sulfide as a source of stratospheric background aerosol. *J. Geophys. Res.* **100**, 8993-9005.

Ciais, P., P.P. Tans, M. Trolier, J.W.C. White, and R.J. Francey, 1995. A large Northern Hemisphere terrestrial CO_2 sink indicated by the $^{13}C/^{12}C$ ratio of atmospheric CO_2. *Science* **269**, 1098-1102.

CIESIN (Consortium for International Earth Science Information Network), 1995. Thematic Guide to Integrated Assessment Modeling of Climate Change [online]. University Center, Michigan. Available: http://sedac.ciesin.org/mva/iamcc.tg/TGsec4.html [1998, October 21].

Clement, A., R. Seager, M.A. Cane, and S.E. Zebiak, 1996. An ocean thermostat. *J. Climate* **9**, 2190-2196.

CLIMAP Project Members, 1981. Seasonal reconstruction of the Earth's surface at the last glacial maximum. MC-36 Geological Society of America Map and Chart Series.

Cohen, J.E., 1995. *How Many People Can the Earth Support?* Norton, New York, 532 pp.

Cole, J.E., and R.G. Fairbanks, 1990. The Southern Oscillation recorded in the oxygen isotopes of corals from Tarawa Atoll. *Paleoceanography* **5**, 669-683.

Cole, J.E., R.G. Fairbanks, and G.T. Shen, 1993. The spectrum of recent variability in the Southern Oscillation: Results from a Tarawa Atoll coral. *Science* **262**, 1790-1793.

Coles, V.J., M.S. McCartney, D.B. Olson, and W.J. Smethie, Jr., 1996. Changes in Antarctic Bottom Water properties in the western South Atlantic in the late 1980s. *J. Geophys. Res.* **101**, 8957-8970.

Conway, T.J., P.P. Tans, L.S. Waterman, K.W. Thoning, D.R. Kitzis, K.A. Masarie, and N. Zhang, 1994. Evidence for interannual variability of the carbon cycle from the NOAA/CMDL global air sampling network. *J. Geophys. Res.* **99**, 22831-22855.

Coohill, T.P., 1991. Stratospheric ozone depletion as it affects life on Earth—the role of ultraviolet action spectroscopy. In *Impact of Global Climatic Changes on Photosynthesis and Plant Productivity, Proceedings of the Indo-US Workshop*, Oxford and IBH Publishing Co., New Delhi, pp. 3-22.

Cook, E.R., and L.A Kairiukstis, 1990. *Methods of Dendrochronology: Applications in the Environmental Sciences*, Kluwer Academic Publishers, Dordrecht, The Netherlands, 408 pp.

Cook, E.R., T. Bird, M. Peterson, M. Barbetti, B. Buckley, R. D'Arrigo, R. Francey, and P. Tans, 1991. Climatic change in Tasmania inferred from a 1089-year tree-ring chronology of Huon pine. *Science* **253**, 1266-1268.

Cook, E.R., D.M. Meko, and C.W. Stockton, 1997. A new assessment of possible solar and lunar forcing of the bidecadal drought rhythm in the western United States. *J. Climate* **10**, 1343-1356.

Cook, E.R., D.M. Meko, D.W. Stahle, and M.K. Cleaveland, in press. Drought reconstructions for the continental United States, *J. Climate.*

Crutzen, P., 1976. The possible importance of CSO for the sulfate layer of the stratosphere. *Geophys. Res. Lett.* **3**, 73-76.

Cuffey, K.M., G.D. Clow, R.B. Alley, M. Stuiver, E.D. Waddington, and R.W. Saltus, 1995. Large Arctic temperature change at the Wisconsin-Holocene glacial transition. *Science* **270**, 455-458.

Curry, J.A., W.B. Rossow, D. Randall, and J.L. Schramm, 1996. Overview of Arctic cloud and radiation characteristics. *J. Climate* **9**, 1731-1764.

Dai, A., and I. Fung, 1993. Can climate variability contribute to the "missing" CO_2 sink? *Global Biogeochem. Cycles* **7**, 599-609.

Dai, A., A.D. Del Genio, and I.Y. Fung, 1997. Clouds, precipitation, and temperature range. *Nature* **386**, 665-666.

Dansgaard, W., S.J. Johnsen, H.B. Clausen, D. Dahl-Jensen, N.S. Gundestrup, C.U. Hammer, C.S. Hvidberg, J.P. Steffensen, A.E. Sveinbjornsdottir, J. Jouzel, and G. Bond, 1993. Evidence for general instability of past climate from a 250-kyr ice-core record. *Nature* **364**, 218-220.

Davis, M.B., and C. Zabinski, 1992. Changes in geographical range resulting from greenhouse warming: Effects on biodiversity in forests. In *Global Warming and Biological Diversity,* R.L. Peters and T.E. Lovejoy (eds.), Yale University Press, New Haven, pp. 297-308.

de la Mare, W., 1997. Abrupt mid-twentieth-century decline in Antactic sea-ice extent from whaling records. *Nature* **389**, 57-60.

Delworth, T., S. Manabe, and R.J. Stouffer, 1993. Interdecadal variations of the thermohaline circulation in a coupled ocean-atmosphere model. *J. Climate* **6**, 1993-2011.

de Ronde, J. G., 1993. What will happen to The Netherlands if sea level rise accelerates? In *Climate and Sea Level Change: Observations, Projections and Implications*, R.A. Warrick, E.M. Barrow, and T.M.L. Wigley (eds.), Cambridge University Press, Cambridge, U.K., pp. 322-335.

Deser, C., and M.L. Blackmon, 1993. Surface climate variations over the North Atlantic Ocean during winter: 1900-1989. *J. Climate* **6**, 1743-1753.

Deser, C., M.A. Alexander, and M.S. Timlin, 1996. Upper ocean thermal variations in the North Pacific during 1970-1991. *J. Climate* **9**, 1840-1855.

Deser, C., M.A. Alexander, and M.S. Timlin, in press. Evidence for a wind-driven intensification of the Kuroshio Current extension from the 1970s to the 1980s. *J. Climate.*

Dettinger, M.D., and D.R. Cayan, 1995. Large scale atmospheric forcing of recent trends towards early snowmelt runoff in California. *J. Climate* **8**, 606-623.

Dettinger, M.D., and M. Ghil, 1998. Seasonal and interannual variations of atmospheric CO_2 and climate. *Tellus* **50B**, 1-24.

Dettinger, M.D., D.R. Cayan, and G.J. McCabe, Jr., 1993. Decadal trends in runoff over the western United States and links to persistent North Pacific sea-surface-temperature and atmospheric-circulation patterns. In *Proceedings of the 18th Annual Climate Diagnostics Workshop,* U.S. Department of Commerce, Washington, D.C., pp. 240-243.

Dettinger, M.D., M. Ghil, and C.L. Keppenne, 1995. Interannual and interdecadal variability in United States surface-air temperatures, 1910-87. *Clim. Change* **31**, 35-66.

Diaz, H.F., and C.A. Anderson, 1995. Precipitation trends and water consumption related to population in the southwestern United States: A reassessment. *Water Resour. Res.* **31**, 713-720.

Diaz, H.F., and N.E. Graham, 1996. Recent changes in tropical freezing heights and the role of sea surface temperature. *Nature* **383**, 152-155.

Dickson, R.R., and J. Namias, 1976. North American influence on the circulation and climate of the North Atlantic sector. *Mon. Wea. Rev.* **104**, 1255-1265.

Dickson, R.R., J. Meincke, S.A. Malmberg, and A.J. Lee, 1988. The "Great Salinity Anomaly" in the northern North Atlantic 1968-1982. *Prog. Oceanogr.* **20**, 103-151.

Dickson, R., J. Lazier, J. Meincke, and P. Rhines, 1996. Long-term coordinated changes in the convective activity of the North Atlantic. In *Decadal Climate Variability: Dynamics and Predictability*. D.L.T. Anderson and J. Willebrand (eds.). NATO ASI Series, Vol. I44, Springer-Verlag, Heidelberg, pp. 211-261.

Dickson, R., J.R.N. Lazier, J. Meincke, P. Rhines, and J. Swift, 1997. Long-term coordinated changes in the convective activity of the North Atlantic. *Prog. Oceanogr.* **38**, 241-295.

Dlugokencky, E.J., L.P. Steele, P.M. Lang, and K.A. Masarie, 1994. The growth rate and distribution of atmospheric methane. *J. Geophys. Res.* **99**, 17021-17043.

Dobson, G.M.B., 1966. Annual variation of ozone in Antarctica. *Quart. J. Roy. Meteor. Soc.* **92**, 549-552.

Douglas, B.C., 1991. Global sea level rise. *J. Geophys. Res.* **96**, 6981-6992.

Douglas, B.C., 1995. Global sea level change: Determination and interpretation. In *Reviews of Geophysics, Supplement, U.S. National Report to the International Union of Geodesy and Geophysics 1991-1994*, pp. 1425-1432.

Drake, B.G., 1992. A field study of the effects of elevated CO_2 on ecosystem production. *Water Air Soil Pollut.* **64**, 25-44.

Druffel, E.R.M., 1987. Bomb radiocarbon in the Pacific: Annual and seasonal time scale variations. *J. Mar. Res.* **45**, 667-698.

Druffel, E.R.M., 1997. Pulses of rapid ventilation in the North Atlantic surface ocean during the past century. *Science* **275**, 1454-1457.

Druffel, E.R.M., and S. Griffin, 1993. Large variations of surface ocean radiocarbon: Evidence of circulation changes in the southwestern Pacific. *J. Geophys. Res.* **98**, 20249-20259.

Dugam, S.S., S.B. Kakade, and R.K. Verma, 1997. Interannual and long-term variability in the North Atlantic Oscillation and Indian summer monsoon rainfall. *Theor. Appl. Climatol.* **58**, 21-29.

Dunbar, R.B., G.M. Wellington, M.W. Colgan, and P.W. Glynn, 1994. Eastern Pacific sea surface temperature since 1600 A.D.: The $\delta^{18}O$ record of climate variability in Galapagos corals. *Paleoceanography* **9,** 291-316.

Dutton, E.G., and J.R. Christy, 1992. Solar radiation forcing at selected locations and evidence for global cooling following the eruptions of El Chichón and Pinatubo. *Geophys. Res. Lett.* **19**, 2313-2316.

EIA, 1995. Household Energy Consumption and Expenditures 1993, Residential Energy Consumption Survey. Energy Information Administration, Department of Energy, DOE/EIA-0321(93).

Elkins, J.W., T.M. Thompson, T.H. Swanson, J.H. Butler, B.D. Hall, S.O. Cummings, D.A. Fisher, and A.G. Raffo, 1993. Decrease in the growth rates of atmospheric chlorofluorocarbons 11 and 12. *Nature* **364**, 780-783.

Elliot, W.P., J.K. Angell, and K.W. Thoning, 1991. Relationship of atmospheric CO_2 to tropical sea and air temperatures and precipitation. *Tellus* **43B**, 144-155.

Emanuel, K., 1987. The dependence of hurricane intensity on climate. *Nature* **326**, 483-485.

Enfield, D.B., and D.A. Mayer, 1997. Tropical Atlantic sea surface temperature variability and its relation to El Niño-Southern Oscillation. *J. Geophys. Res.* **102**, 929-945.

Epps, D.S., 1997. Weather impacts on energy activities in the U.S. Gulf Coast. In the proceedings of the *Workshop on the Social and Economic Impacts of Weather, Boulder, Colorado, April 2-4, 1997,* R.A. Pielke, Jr. (ed.), NCAR, Boulder, Colorado.

Epstein, P.R., 1993. Algal blooms in the spread and persistence of cholera. *Biosystems* **31**, 209-221.

Esbensen, S., 1984. A comparison of intermonthly and interannual teleconnections in the 700mb geopotential height field during the Northern Hemisphere winter. *Mon. Wea. Rev.* **112**, 2016-2032.

Etheridge, D.M., G.I. Pearman, and P.J. Fraser, 1992. Changes in tropospheric methane between 1841 and 1978 from a high accumulation rate Antarctic ice core. *Tellus* **44B**, 282-294.

Etheridge, D.M., L.P. Steele, R.L. Langenfelds, R.J. Francey, J.M. Barnola, and V.I. Morgan, 1996. Natural and anthropogenic changes in atmospheric CO_2 over the last 1000 years from air in Antarctic ice and firn. *J. Geophys. Res.* **101**, 4115-4128.

Fahrbach, E.R.M., G. Rohardt, M. Schroeder, and R.A. Woodgate, in press. Gradual warming of the Weddell Sea deep and bottom water. *J. Geophys. Res.*

Fairbanks, R.G., and R.E. Dodge, 1979. Annual periodicity of the $^{18}O/^{16}O$ and $^{13}C/^{12}C$ ratios in the coral *Montastrea annularis. Geochim. Cosmochim. Acta* **43**, 1009-1020.

Farmer, G., T.M.L. Wigley, P.D. Jones, and M. Salmon, 1989. Documenting and explaining recent global-mean temperature changes. Final Report to NERC (Contract GR3/6565), Climatic Research Unit, Norwich, U.K., 141 pp.

Ffield, A., and A.L. Gordon, 1996. Tidal mixing signatures in the Indonesian seas. *J. Phys. Oceanogr.* **26**, 1924-1936.

Fine, R.A., W.H. Peterson, and H.G. Ostlund, 1987. Penetration of tritium into the tropical Pacific. *J. Phys. Oceanogr.* **17**, 553-564.

Fiocco, G., D. Fua, and G. Visconti (eds.), 1996. *The Mount Pinatubo Eruption: Effects on the Atmosphere and Climate.* NATO ASI Series, Vol. I42, Springer-Verlag, Heidelberg, 310 pp.

Flohn, H., and A. Kapala, 1989. Changes in tropical sea-air interaction processes over a 30-year period. *Nature* **338**, 244-246.

Flohn, H., A. Kapala, H.R. Knoche, and H. Machel, 1990. Recent changes of the tropical water energy budget and of mid-latitude circulations. *Climate Dyn.* **4**, 237-252.

Folland, C.K., T.N. Palmer, and D.E. Parker, 1986. Sahel rainfall and worldwide sea temperatures 1901-85. *Nature* **320**, 602-607.

Folland, C.K., J. Owen, M.N. Ward, and A. Colman, 1991. Prediction of seasonal rainfall in the Sahel region using empirical and dynamical methods. *J. Forecasting* **10**, 21-56.

Frankignoul, C., and K. Hasselmann, 1977. Stochastic climate models. Part 2. Application to sea-surface temperature anomalies and thermocline variability. *Tellus* **29**, 284-305.

Frenzel, B., C. Pfister, and B. Gläser (eds.), 1994. *Climate Trends and Anomalies in Europe, 1675-1715,* G. Fisher Verlag, Stuttgart, 480 pp.

Friedlingstein, P., I.Y. Fung, E. Holland, J. John, G. Brasseur, D. Erickson, and D. Schimel, 1995. On the contribution of CO_2 fertilization to the missing biospheric sink. *Global Biogeochem. Cycles* **9**, 541-556.

Friis-Christensen, E., and K. Lassen, 1991. Length of the solar cycle: An indicator of solar activity closely associated with climate. *Science* **254**, 698-700.

Fritts, H.C., 1976. *Tree Rings and Climate*. Academic Press, London, 567 pp.

Fung, I., J. John, J. Lerner, E. Matthews, M. Prather, L.P. Steele, and P.J. Fraser, 1991. Three-dimensional model synthesis of the global methane cycle. *J. Geophys. Res.* **96**, 13033-13065.

Gagan, M.K., A.R. Chivas, and P.J. Isdale, 1994. High-resolution isotopic records from corals using ocean temperature and mass spawning chronometers. *Earth Planet. Sci.* **121**, 549-558.

Gagan, M.K., L.K. Ayliffe, D. Hopley, J.A. Cali, G.E. Mortimer, J. Chappell, M.T. McCulloch, and M.J. Head, 1998. Temperature and surface-ocean water balance of the mid-Holocene tropical western Pacific. *Science* **279,** 1014-1018.

Garzoli, S.L., G.J. Goni, A.J. Mariano, and D.B. Olson, 1997. Monitoring the upper southeastern Atlantic transports using altimeter data. *J. Mar. Res.* **55**, 453-481.

Gent, P.R., and J.C. McWilliams, 1990. Isopycnal mixing in ocean circulation models. *J. Phys. Oceanogr.* **20**, 150-155.

Ghil, M., and S. Childress, 1987. *Topics in Geophysical Fluid Dynamics: Atmospheric Dynamics, Dynamo Theory and Climate Dynamics*, Springer-Verlag, New York, 485 pp.

Ghil, M., and R. Vautard, 1991. Interdecadal oscillations and the warming trend in global temperature time series. *Nature* **350**, 324-327.

Gill, A.M., 1994. How fires affect biodiversity. In *Fire and Biodiversity: The Effects and Effectiveness of Fire Management*, Biodiversity Series, Paper No. 8., Dept. of the Environment, Sport, and Territories, Melbourne, Australia.

Giorgi, F., and M.R. Marinucci, 1996. Improvements in the simulation of surface climatology over the European region with a nested modeling system. *Geophys. Res. Lett.* **23**, 273-276.

Glantz, M.H., 1994. The West African Sahel. In *Drought Follows the Plough,* M.H. Glantz (ed.), Cambridge University Press, Cambridge, U.K., 197 pp.

Glantz, M.H., 1996. *Currents of Change: El Niño's Impact on Climate and Society*, Cambridge University Press, Cambridge, U.K., 194 pp.

Gleason, J.F., P.K. Bhartia, J.R. Herman, R. McPeters, P. Newman, R.S. Stolarski, L. Flynn, G. Labow, D. Larko, C. Seftor, C. Wellemeyer, W.D. Komhyr, A.J. Miller, and W. Planet, 1993. Record low global ozone in 1992. *Science* **260**, 523-526.

Gordon, A.L., and R.A. Fine, 1996. Pathways of water between the Pacific and Indian Oceans in the Indonesian Seas. *Nature* **379**, 146-149.

Gornitz, V., 1995. A comparison of differences between recent and late Holocene sea level trends from eastern North America and other selected regions. *J. Coast. Res.* **17**, 287-297.

Grace, J., J. Lloyd, J. McIntyre, A. Miranda, P. Meir, H. Miranda, C. Nobre, J. Moncrieff, J. Massheder, Y. Malhi, I. Wright, and J. Gash, 1995. Carbon dioxide uptake by an undisturbed tropical rainforest in southwest Amazonia, 1992 to 1993. *Science* **270**, 778-780.

Graham, N.E., 1994. Decadal-scale climate variability in the tropical and North Pacific during the 1970s and 1980s: Observations and model results. *Climate Dyn.* **10**, 135-162.

Gray, W.M., 1979. Hurricanes: Their formation, structure and likely role in the tropical circulation. In *Meteorology over the Tropical Oceans*, D.B. Shaw (ed.), Royal Meteorological Society, Bracknell, U.K., pp. 155-218.

Gray, W.M., 1990. Strong association between West African rainfall and U.S. landfall of intense hurricanes. *Science* **249**, 1251-1256.

Gray, W.M., C.W. Landsea, P.W. Mielke, and K.J. Berry, 1992. Predicting Atlantic seasonal hurricane activity 6-11 months in advance. *Wea. Forecasting* **7**, 440-455.

Griffies, S.M., and K. Bryan, 1997. A predictability study of simulated North Atlantic multidecadal variability. *Climate Dyn.* **13**, 459-487.

Groisman, P.Y., and D.R. Easterling, 1994. Variability and trends of total precipitation and snowfall over the United States and Canada. *J. Climate* **7**, 184-205.

Groisman, P.Y., T.R. Karl, and R.W. Knight, 1994a. Observed impact of snow cover on the heat balance and the rise of continental spring temperatures. *Science* **263**, 198-200.

Groisman, P.Y., T.R. Karl, R.W. Knight, and G.L. Stenchikov, 1994b. Changes of snow cover, temperature, and radiative heat balance over the Northern Hemisphere. *J. Climate* **7**, 1633-1656.

Grootes, P.M., 1995. Ice cores as archives of decade-to-century-scale climate variability. In *Natural Climate Variability on Decade-to-Century Time Scales*, D.G. Martinson, K. Bryan, M. Ghil, M.M. Hall, T.R. Karl, E.S. Sarachik, S. Sorooshian, and L.D. Talley (eds.), National Academy Press, Washington, D.C., pp. 544-554.

Gruber, N., 1998. Anthropogenic CO_2 in the Atlantic Ocean. *Global Biogeochem. Cylces* **12**, 165-191.

Gu, D., and S.G.H. Philander, 1997. Interdecadal climate fluctuations that depend on exchanges between the tropics and extratropics. *Science* **275**, 805-807.

Guilderson, T.P., and D.P. Schrag, 1998. Abrupt shift in subsurface temperature in the tropical Pacific associated with changes in El Niño. *Science* **281**, 240-243.

Günther, H., W. Rosenthal, M. Stawarz, J.C. Carretero, M. Gomez, I. Lozano, O.Serano, and M. Reistad, 1998. The wave climate of the Northeast Atlantic over the period 1955-1994: The WASA wave hindcast. *Global Atmos. Ocean Sys.* **6**, 121-163.

Haarsma, R.J., J.F.B. Mitchell, and C.A. Senior, 1993. Tropical disturbances in a GCM. *Climate Dyn.* **8**, 247-257.

Hakkinen, S., 1993. An Arctic source for the Great Salinity Anomaly: A simulation of the Arctic ice-ocean system for 1955-1975. *J. Geophys. Res.* **98**, 16397-16410.

Hall, N.M.J., B.J. Hoskins, P.J. Valdes, and C.A. Senior, 1994. Storm tracks in a high resolution GCM with doubled CO_2. *Quart. J. Roy. Meteor. Soc.* **120**, 1209-1230.

Hall, M.M., M. McCartney, and J.A. Whitehead, 1997. Antarctic bottom water flux in the equatorial western Atlantic. *J. Phys. Oceanogr.* **27**, 1903-1926.

Hansen, D.V., and H.F. Bezdek, 1996. On the nature of decadal anomalies in the North Atlantic sea surface temperature. *J. Geophys. Res.* **101**, 8749-8758.

Hansen, J., and A. Lacis, 1990. Sun and dust versus greenhouse gases: An assessment of their relative roles in global climate change. *Nature* **346**, 713-719.

Hansen, J., and S. Lebedeff, 1988. Global surface temperature through 1987. *Geophys. Res. Lett.* **15**, 323-326.

Hansen, J., A. Lacis, R. Ruedy, and M. Sato, 1992. Potential climate impact of Mount Pinatubo eruption. *Geophys. Res. Lett.* **19**, 215-218.

Hansen, J., R. Ruedy, M. Sato, and R. Reynolds, 1996. Global surface air temperature in 1995: Return to pre-Pinatubo level. *Geophys. Res. Lett.* **23**, 1665-1668.

Hartmann, D.L., and F. Lo, 1998. Wave-driven zonal flow vacillation in the Southern Hemisphere. *J. Atmos. Sci.* **55**, 1303-1315.

Hasselmann, K., 1976. Stochastic climate models: I. Theory. *Tellus* **28**, 473-485.

Hasselmann, K., 1993. Optimal fingerprints for the detection of time dependent climatic change. *J. Climate* **6**, 1957-1971.

Hastenrath, S., 1990. Decadal scale changes of the circulation in the tropical Atlantic sector associated with Sahel drought. *Int. J .Climatol.* **10**, 459-472.

Hastenrath, S., 1995. Recent advances in tropical climate prediction. *J. Climate* **8**, 1519-1532.

Hastenrath, S., and L. Druyan, 1993. Circulation anomaly mechanisms in the tropical Atlantic sector during the northeast Brazil rainy season: Results from the GISS general circulation model. *J. Geophys. Res.* **98**, 14917-14923.

Hastenrath, S., and L. Greischar, 1993. Circulation mechanisms related to northeast Brazil rainfall anomalies. *J. Geophys. Res.* **98**, 5093-5102.

Hastenrath, S., and L. Heller, 1977. Dynamics of climate hazards in northeast Brazil. *Quart. J. Roy. Meteor. Soc.* **103**, 77-92.

Hastenrath, S., and P.D. Kruss, 1992. The dramatic retreat of Mount Kenya's glaciers between 1963 and 1987: Greenhouse forcing. *Ann. Glaciol.* **16**, 127-133.

Hastenrath, S., L.C. Castro, and P. Acietuno, 1987. The Southern Oscillation in the tropical Atlantic sector. *Contrib. Atmos. Phys.* **60**, 447-463.

Hayden, B.P., 1981. Secular variation in Atlantic coast extratropical cyclones. *Mon. Wea. Rev.* **109**, 159-167.

Hays, J.D., J. Imbrie, and N.J. Shackleton, 1976. Variations in the Earth's orbit: Pacemaker of the ice ages. *Science* **194**, 1121-1132.

Hebert, P.J., J.D. Jarrell, and M. Mayfield, 1996. *The Deadliest, Costliest, and Most Intense United States Hurricanes of this Century (and Other Frequently Requested Hurricane Facts).* NOAA Technical Memorandum NWS NHC-31 (February), National Hurricane Center, Coral Gables, Florida.

Herbert, T.D., 1997. A long marine history of carbon cycle modulation by orbital-climatic changes. *Proc. Nat. Acad. Sci. USA* **94**, 8362-8369.

Hewitson, B.C., and R.G. Crane, 1996. Climate downscaling techniques and applications. *Climate Res.* **7**, 85-95.

Hirst, A.C. 1986. Unstable and damped equatorial modes in simple coupled ocean-atmosphere models. *J. Atmos. Sci.* **43**, 606-630.

Hirst, A.C., D.R. Jackett, and T.J. McDougall, 1996. The meridional overturning cells of an ocean model in neutral density co-ordinates. *J. Phys. Oceanogr.* **26**, 775-791.

Hodell, D.A., J.H. Curtis, and M. Brenner, 1995. Possible role of climate in the collapse of the Classic Maya civilization. *Nature* **375**, 391-394.

Hofmann, D.J., 1990. Increase in stratospheric background sulfuric acid aerosol mass in the past 10 years. *Science* **248**, 996-1000.

Hofmann, D.J., S.J. Oltmans, J.A. Lathrop, J.M. Harriss, and H. Vömel,

1994. Record low ozone at the South Pole in the spring of 1993. *Geophys. Res. Lett.* **21**, 421-424.

Hood, L.L,. and J.P. McCormack, 1992. Components of interannual ozone change based on Nimbus 7 TOMS data. *Geophys. Res. Lett.* **19**, 2309-2312.

Horel, J.D., and C.R. Mechoso, 1988. Observed and simulated intraseasonal variability of the wintertime planetary circulation. *J. Climate* **1**, 582-599.

Hoskins, B.J., 1983. Feedback of the transient midlatitude, synoptic time-scale waves onto the longer time-scale flow. In *Workshop on Intercomparison of Large-Scale Models Used for Extended-Range Forecasts*, Proceedings., European Centre for Medium-Range Weather Forecasting, Reading, U.K., pp. 53-61.

Hoskins, B., and D.J. Karoly, 1981. The steady linear response of a spherical atmosphere to thermal and orographic forcing. *J. Atmos. Sci.* **38**, 1179-1196.

Houghton, R.A., 1993. Is carbon accumulating in the northern temperate zone? *Global Biogeochem. Cycles* **7**, 611-617.

Houghton, R.W., and Y. Tourre, 1992. Characteristics of low frequency sea surface temperature fluctuations in the tropical Atlantic. *J. Climate* **5**, 765-771.

Houghton, R.A., R.D. Boone, J.R. Fruci, J.E. Hobbie, J.M. Melillo, C.A. Palm, B.J. Peterson, G.R. Shaver, and G.M. Woodwell, 1987. The flux of carbon from terrestrial ecosystems to the atmosphere in 1980 due to changes in land use: Geographic distribution of the flux. *Tellus* **39B**, 122-139.

Houghton, R.A., E.A. Davidson, and G.M. Woodwell, 1998. Missing sinks, feedbacks, and understanding the role of terrestrial ecosystems in the global carbon balance. *Global Biogeochem. Cycles* **12**, 25-34.

Hoyt, D.V., and K.H. Schatten, 1993. A discussion of plausible solar irradiance variations, 1700-1992. *J. Geophys. Res.* **98**, 18895-18906.

Huang, J., K. Higuchi, and A. Shabbar, 1998. The relationship between the North Atlantic Oscillation and El Niño-Southern Oscillation. *Geophys. Res. Lett.* **25**, 2707-2710.

Hughen, K.A., J.T. Overpeck, L.C. Peterson, and S. Trumbore, 1996. Rapid climate changes in the tropical Atlantic region during the last deglaciation. *Nature* **380**, 51-54.

Hughes, M.K., and H.F. Diaz, 1994. Was there a Medieval Warm Period, and if so, where and when? *Clim. Change* **26**, 109-142.

Hughes, M.K. and L.J. Graumlich, 1996. Multimillennial dendroclimatic studies from the western United States. In *Climatic Variations and Forcing Mechanisms of the Last 2000 Years,* P.D. Jones, R.S. Bradley, and J. Jouzel (eds.), Springer-Verlag, NATO ASI Series, Vol. I41, Heidelberg, pp. 109-123.

Hulme, M., Y. Biot, J. Borton, M. Buchanan-Smith, S. Davies, C. Folland, N. Nicholds, D. Seddon, and N. Ward, 1992. Seasonal rainfall forecasting for Africa. Part I: Current status and future developments. *Int. J. Environ. Stud.* **39**, 245-256.

Huntley, B., 1988. Europe. In *Vegetation History,* B. Huntley and T. Webb III (eds.), Kluwer Academic Publishers, Dordrecht, The Netherlands, pp. 341-383.

Hurrell, J.W., 1995. Decadal trends in the North Atlantic Oscillation: Regional temperature and precipitation. *Science* **269**, 676-679.

Hurrell, J., 1996. Influence of variations in extratropical wintertime teleconnections on Northern Hemisphere temperature. *Geophys. Res. Lett.* **23**, 665-668.

Hurrell, J.W., and K.E. Trenberth, 1996. Satellite vs. surface estimates of air temperature since 1979. *J. Climate* **9**, 2222-2232.

Hurrell, J.W., and H. van Loon, 1996. Decadal variations in climate associated with the North Atlantic Oscillation. *Clim. Change* **36**, 301-326.

Imbrie, J., J.D. Hays, D. Martinson, A. McIntyre, A. Mix, J. Morley, N. Pisias, W. Prell, and N.J. Shackleton, 1984. The orbital theory of Pleistocene climate: Support from a revised chronology of the marine $\delta^{18}O$ record. In *Milankovitch and Climate. Part 1*, A.L. Berger et al. (eds.), D. Reidel, Boston, pp. 269-305.

Imbrie, J., E. Boyle, S. Clemens, A. Duffy, W. Howard, G. Kukla, J. Kutzbach, D. Martinson, A. McIntyre, A. Mix, B. Molfino, J. Morley, L. Peterson, N. Pisias, W. Prell, M. Raymo, N. Shackleton, and J. Toggweiler, 1992. On the structure and origin of major glaciation cycles: 1. Linear responses to Milankovich forcing. *Paleoceanography* **7**, 701-738.

IPCC (Intergovernmental Panel on Climate Change), 1990. *Climate Change: The IPCC Scientific Assessment.* J.T. Houghton, G.J. Jenkins, and J.J. Ephraums, WMO/UNEP, Cambridge University Press, Cambridge, U.K., 364 pp.

IPCC, 1992. *Climate Change 1992: The IPCC Supplementary Report,* WMO/UNEP, J.T. Houghton, B.A. Callander, and S.K. Varney (eds.), Cambridge University Press, Cambridge, U.K., 200 pp.

IPCC, 1995. *Climate Change 1994: Radiative Forcing of Climate Change and an Evaluation of the IPCC IS92 Emission Scenarios,* J.T. Houghton, L.G. Meira Filho, J. Bruce, H. Lee, B.A. Callander, E. Haites, N. Harris, and K. Maskell (eds.), Cambridge University Press, Cambridge, U.K., 339 pp.

IPCC, 1996a. *Climate Change 1995: The Science of Climate Change, Contribution of Working Group 1 to the Second Assessment Report of the Intergovernmental Panel on Climate Change,* J.T. Houghton, L.G. Meira Filho, B.A. Callander, N. Harris, A. Kattenberg, and K. Maskell (eds.), Cambridge University Press, Cambridge, U.K., 572 pp.

IPCC, 1996b. *Climate Change 1995: Impacts, Adaptations and Mitigation of Climate Change: Scientific-Technical Analysis, Contribution of Working Group II to the Second Assessment Report of the Intergovernmental Panel on Climate Change,* R.T. Watson, M.C. Zinyowera, R.H. Moss, and D.J. Dokken (eds.), Cambridge University Press, Cambridge, U.K., 878 pp.

IPCC, 1996c. *Climate Change 1995: Economic and Social Dimensions of Climate Change, Contribution of Working Group III to the Second Assessment Report of the Intergovernmental Panel on Climate Change,* J.P. Bruce, H. Lee, and E.F. Haites (eds.), Cambridge University Press, Cambridge, U.K., 448 pp.

IPCC, 1998. *The Regional Impacts of Climate Change: An Assessment of Vulnerability.* A Special Report of the Intergovernmental Panel on Climate Change Working Group II, R.T. Watson, M.C. Zinyowera, R.H. Moss, and D.J. Dokken (eds.), Cambridge University Press, Cambridge, U.K., 517 pp.

Isdale, P., 1984. Fluorescent bands in massive corals record centuries of coastal rainfall. *Nature* **310**, 578-579.

Iselin, C.D., 1939. Some physical factors which may influence the productivity of New England's coastal waters. *J. Mar. Res.* **2**, 74-85.

Jacobs, G.A., H.E. Hurlburt, J.C. Kindle, E.J. Metzger, J.L. Mitchell, W.J. Teague, and A.J. Wallcraft, 1994. Decade-scale trans-Pacific propagation and warming effects of an El Niño anomaly. *Nature* **370**, 360-363.

Jacoby, G.C., and R.D. D'Arrigo, 1989. Reconstructed Northern Hemisphere annual temperature since 1671 based on high-latitude tree-ring data from North America. *Clim. Change* **14**, 39-59.

Jacoby, G.C., R.D. D'Arrigo, and T. Davaajamts, 1996. Mongolian tree-rings and 20th century warming. *Science* **273**, 771-773.

Jain, S., U. Lall, and M. Mann, in press. Seasonality and interannual variations of Northern Hemisphere temperature: Equator to pole gradient and ocean-land contrast. *J. Climate*.

James, I.N., and P.M. James, 1989. Ultra-low-frequency variability in a simple atmospheric circulation model. *Nature* **342**, 53-55.

Jenkins, W.J., 1998. Studying subtropical thermocline ventilation and circulation using tritium and ^{3}He. *J. Geophys. Res.* **103**, 15817-15831.

Ji, M., A. Kumar, and A. Leetma, 1994. A multi-season climate forecast system at the National Meteorological Center. *Bull. Amer. Meteor. Soc.* **75**, 569-577.

Jiang, N., J.D. Neelin, and M. Ghil, 1995. Quasi-quadrennial and quasi-biennial variability in the equatorial Pacific. *Climate Dyn.* **12**, 101-112.

Jin, F.-F., J.D. Neelin, and M. Ghil, 1994. El Niño on the Devil's Staircase: Annual subharmonic steps to chaos. *Science* **264**, 70-72.

Jin, F.-F., J.D. Neelin, and M. Ghil, 1996. El Niño/Southern Oscillation and the annual cycle: Subharmonic frequency-locking and aperiodicity.

Physica D **98**, 442-465.

Johannessen, O.M., M. Miles, and E. Bjorgo, 1995. Decreases in Arctic sea ice extent and area 1978-1994. *Nature* **376**, 126-127.

Johnson, G.C., and A.H. Orsi, 1997. Southwest Pacific Ocean water-mass changes between 1968/9 and 1990/1. *J. Climate* **10**, 306-316.

Johnston, P.V., R.L. McKenzie, J.G. Keys, and W.A. Matthews, 1992. Observations of depleted stratospheric NO_2 following the Pinatubo volcanic eruption. *Geophys. Res. Lett.* **19**, 211-213.

Jones, P.D., 1988. Hemispheric surface air temperature variations: Recent trends and an update to 1987. *J. Climate* **1**, 654-660.

Jones, P.D., and R.S. Bradley, 1992. Climatic variations over the last 500 years. In *Climate Since AD 1500,* R.S. Bradley and P.D. Jones (eds.), Routledge, London, pp. 649-665.

Jones, P.D., and K.R. Briffa, 1992. Global surface air temperature variations during the twentieth century. *Holocene* **2**, 165-179.

Jones, P.D., S.C.B. Raper, R.S. Bradley, H.F. Diaz, P.M. Kelly, and T.M.L. Wigley, 1986a. Northern Hemisphere surface air temperature variations, 1851-1984. *J. Clim. Appl. Meteor.* **25**, 161-179.

Jones, P.D., T.M.L. Wigley, and P.B. Wright, 1986b. Global temperature variations between 1861 and 1984. *Nature* **322**, 430-434.

Jouzel, J., N.I. Barkov, and J.M. Barnola, 1993. Extending the Vostok ice-core record of paleoclimate to the penultimate glacial period. *Nature* **364**, 407-410.

Joyce, T.M., and P. Robbins, 1996. The long-term hydrographic record at Bermuda. *J. Climate* **9**, 3121-3131.

Kalnay, E., M. Kanamitsu, and R. Kistler, 1996. The NCEP/NCAR 40-year reanaylsis project. *Bull. Amer. Meteor. Soc.* **77**, 437-471.

Kaplan, A., M. Cane, Y. Kushnir, A. Clement, M. Blumenthal, and B. Rajagopalan, in press. Analyses of global sea surface temperature, 1856-1991. *J. Geophys. Res.*

Kapsner, W.R., R.B. Alley, C.A. Shuman, S. Anandakrishnan, and P.M. Grootes, 1995. Dominant influence of atmospheric circulation on snow accumulation in Greenland over the past 18,000 years. *Nature* **373**, 52-54.

Karl, T.R., V.E. Derr, D.R. Easterling, C.K. Folland, D.H. Hofmann, S. Levitus, N. Nicholls, D.E. Parker, and G.W. Withee, 1995. Critical issues for long-term climate monitoring. *Clim. Change* **31**, 185-221.

Karl, T.R., R.W. Knight, D.R. Easterling, and R.G. Quayles, 1996. Indices of climate change for the United States. *Bull. Amer. Meteor. Soc.* **77**, 279-292.

Kaser, G., and B. Noggler, 1991. Observations on Speke Glacier, Ruwenzori Range, Uganda. *J. Glaciol.* **37**, 315-318.

Kates, R.W., B.L. Turner II, and W.C. Clark, 1990. The great transformation. In *The Earth as Transformed by Human Action*, B.L. Turner II, W.C. Clark, R.W. Kates, J.F. Richards, J.T. Mathews, and W.B. Meyer (eds.), Cambridge University Press, Cambridge, U.K., pp. 1-17.

Kawasaki, T., S. Tanaka, Y. Toba, and A. Taniguchi (eds.), 1991. *Long-Term Variability of Pelagic Fish Populations and Their Environment,* Pergamon Press, Tokyo, 402 pp.

Keables, M.J., 1989. A synoptic climatology of the bimodal precipitation distribution in the upper Midwest. *J. Climate* **2**, 1289-1294.

Keeling, R.F., and T.-H. Peng, 1995. Transport of heat, CO_2, and O_2 by the Atlantic's thermohaline circulation. *Phil. Trans. Roy.. Soc. London B* **348**, 133-142.

Keeling, C.D., and T.P. Whorf, 1994. Atmospheric CO_2 records from sites in the SIO air sampling network. In *Trends '93: A Compendium of Data on Global Change,* T.A. Boden, D.P. Kaiser, R.J. Sepanski and F.W. Stoss (eds.). ORNL/CDIAC-65, Carbon Dioxide Information Analysis Center, Oak Ridge National Laboratory, Oak Ridge, Tennessee, pp. 16-26.

Keeling, C.D., R.B. Bacastow, A.E. Bainbridge, C.E. Ekdahl, Jr., P.R. Guenther, L.S. Waterman, and J.F.S. Chin, 1976. Atmospheric carbon dioxide variations at Mauna Loa Observatory, Hawaii. *Tellus* **28**, 538-551.

Keeling, C.D., R.B. Bacastow, A.F. Carter, S.C. Piper, T.P. Whorf, M. Heimann, W.G. Mook, and H. Roeloffzen, 1989. A three-dimensional model of atmospheric CO_2 transport based on observed winds: 1. Analysis of observational data. In *Aspects of Climate Variability in the Pacific and the Western Americas*, D.H. Peterson (ed.), Geophysical Monograph 55, American Geophysical Union, Washington, D.C., pp. 165-236.

Keeling, C.D., T.P. Whorf, M. Wahlen, and J. van der Plicht, 1995. Interannual extremes in the rate of rise of atmospheric carbon dioxide since 1980. *Nature* **375**, 666-670.

Keeling, C.D., J.F.S. Chin and T.P. Whorf, 1996a. Increased activity of northern vegetation inferred from atmospheric CO_2 measurements. *Nature* **382**, 146-149.

Keeling, R.F., S.C. Piper, and M. Heimann, 1996b. Global and hemispheric CO_2 sinks deduced from changes in atmospheric O_2 concentration. *Nature* **381**, 218-221.

Kelly, P.M., and T.M.L. Wigley, 1992. Solar cycle length, greenhouse forcing and global climate. *Nature* **360**, 328-330.

Kimoto, M., 1989. *Multiple Flow Regimes in the Northern Hemisphere Winter.* Ph.D. thesis, University of California, Los Angeles, Calif., 210 pp.

Kinne, S., O.B. Toon, and M.J. Prather, 1992. Buffering of stratospheric circulation by changing amounts of tropical ozone: A Pinatubo case study. *Geophys. Res. Lett.* **19**, 1927-1930.

Knox, F., and M.B. McElroy, 1984. Changes in atmospheric CO_2: Influence of marine biota at high latitudes. *J. Geophys. Res.* **89**, 4629-4637.

Knutson, T., and S. Manabe, 1995. Time-mean response over the tropical Pacific to increased CO_2 in a coupled ocean-atmosphere model. *J. Climate* **8**, 2181-2199.

Knutson, T., and S. Manabe, in press. Model assessment of decadal variability and trends in the tropical Pacific Ocean. *J. Climate.*

Knutson, T. R., S. Manabe, and D. Gu, 1997. Simulated ENSO in a global coupled ocean-atmosphere model: Multidecadal amplitude modulation and CO_2 sensitivity. *J. Climate* **10**, 138-161.

Koike, M., N.B. Jones, W.A. Mathews, P.V. Johnston, R.L. McKenzie, D. Kinnison, and J. Rodriguez, 1994. Impact of Pinatubo aerosols on the partitioning between NO_2 and HNO_3. *Geophys. Res. Lett.* **21**, 597-600.

Komhyr, W.D., R.D. Grass, R.D. Evans, R.K. Leonard, D.M. Quincy, D.J. Hofmann, and G.L. Koenig, 1993. Unprecedented ozone decrease over the United States from Dobson spectrophotometer observations. *Geophys. Res. Lett.* **21**, 201-204.

Krupa, S.V., and J.-J. Jager, 1996. Adverse effects of elevated levels of ultraviolet (UV)-B radiation and ozone (O_3) on crop growth and productivity. In *Global Climate Change and Agricultural Production*, F. Bazzaz and W. Sombroek (eds.), Wiley Publishers, Chichester, U.K., pp. 141-170.

Kumar, A., A. Leetmaa, and M. Ji, 1994. Simulations of atmospheric variability induced by sea surface temperatures and implications for global warming. *Science* **266**, 632-634.

Kunkel, K.E., S.A. Changnon, and J.R. Angel, 1994. Climatic aspects of the 1993 upper Mississippi River Basin flood. *Bull. Amer. Meteor. Soc.* **75**, 811-822.

Kushnir, Y., 1994. Interdecadal variations in North Atlantic sea surface temperature and associated atmospheric conditions. *J. Climate* **7**, 141-157.

Kushnir, Y., and I.M. Held, 1996. Equilibrium atmospheric response to the North Atlantic SST anomalies. *J. Climate* **9**, 1208-1220.

Kushnir, Y., and J.M. Wallace, 1989. Low-frequency variability in the Northern Hemisphere winter: Geographical distribution, structure and time-scale dependence. *J. Atmos. Sci.* **46**, 3122-3142.

Kushnir, Y., V.J. Cardone, J.G. Greenwood, and M. Cane, 1997. On the recent increase in North Atlantic wave heights. *J. Climate* **10**, 2107-2113.

Kushnir, Y., B. Rajagopalan, and Y.M. Tourre, 1998. Decadal climate variability in the Atlantic Basin. Ninth Conference on the Interaction of the Sea and Atmosphere, Phoenix, AZ, 11-16 January 1998, American Meteorological Society, 58-61.

Kusler, J.A., W.J. Mitsch, and J.S. Larson, 1994. Wetlands. *Sci. Amer.* **270**,

64-70.

Kutzbach, J.E., and F.A. Street-Perrott, 1985. Milankovitch forcing of fluctuations in the level of tropical lakes from 18 to 0 kyr BP. *Nature* **317**, 130-134.

Kutzbach, J., R. Gallimore, S. Harrison, P. Behling, R. Selin, and F. Laarif, 1998. Climate and biome simulations for the past 21,000 years. *Quat. Sci. Rev.* **17**, 473-506.

Labitzke, K., and M.P. McCormick, 1992. Stratospheric temperature increases due to Pinatubo aerosols. *Geophys. Res. Lett.* **19**, 207-210.

Labitzke, K., and B. Naujokat, 1983. On the variability and on trends of the temperature in the middle stratosphere. *Contrib. Atmos. Phys.* **56**, 495-507.

Labitzke, K., and H. van Loon, 1993. Some recent studies of probable connections between solar and atmospheric variability. *Ann. Geophys.* **11**, 1084-1094.

Labitzke, K., B. Naujokat, and M.P. McCormick, 1983. Temperature effects on the stratosphere of the April 4, 1982, eruption of El Chichón, Mexico. *Geophys. Res. Lett.* **10**, 24-26.

Lachenbruch, A.H., and B.V. Marshall, 1986. Changing climate: Geothermal evidence from permafrost in the Alaskan Arctic. *Science* **234**, 689-696.

Lacis, A., J. Hansen, and M. Sato, 1992. Climate forcing by stratospheric aerosols. *Geophys. Res. Lett.* **19**, 1607-1610.

Laird, K.R., S.C. Fritz, K.A. Maasch, and B.F. Cumming, 1996. Greater drought intensity and frequency before AD 1200 in the northern Great Plains, USA. *Nature* **384**, 552-554.

Lall, U., and M. Mann, 1995. The Great Salt Lake: A barometer of low frequency climatic variability. *Water Resour. Res.* **31**, 2503-2515.

Lall, U., Sangoyomi, T., and H.D.I. Abarbanel, 1996. Nonlinear dynamics of the Great Salt Lake: Nonparametric short-term forecasting. *Water Resour. Res.* **32**, 975-985.

Lall, U., B. Rajagopalan, and M. Cane, 1998. Nonstationarity in ENSO and its teleconnections: An exploratory analysis and a nonlinear dynamics perspective. In *Proceedings of the PACLIM 98 Workshop on Climate Variability of the Eastern North Pacific and Western North America.*

Lamb, H.H., 1972. *Climate: Present, Past & Future*, vols. I & II, Methuen & Co., London.

Lamb, P.J., 1978. Large-scale tropical Atlantic circulation patterns associated with sub-Saharan weather anomalies. *Tellus* **30**, 240-251.

Lamb, H.H., 1995. *Climate, History and the Modern World*, Routledge, London, 433 pp.

Lamb, P. J., R.A. Peppler, and S. Hastenrath, 1986. Interannual variability in the tropical Atlantic. *Nature* **322**, 238-240.

Landsea, C.W., and W.M. Gray, 1992. The strong association between Western Sahelian monsoon rainfall and intense Atlantic hurricanes. *J. Climate* **5**, 435-453.

Landsea, C.W., N. Nicholls, W.M. Gray, and L.A. Avila, 1996. Downward trends in the frequency of intense Atlantic hurricanes during the past five decades. *Geophys. Res. Lett.* **23**, 1697-1700.

Latif, M., 1998. Dynamics of interdecadal variability in coupled ocean-atmosphere models. *J. Climate* **11**, 602-624.

Latif, M., and T.P. Barnett, 1994. Causes of decadal climate variability over the North Pacific/North American sector. *Science* **266**, 634-637.

Latif, M., and T.P. Barnett, 1996. Decadal climate variability over the North Pacific and North America: Dynamics and predictability. *J. Climate* **9**, 2407-2423.

Latif, M., A. Grötzner, M. Münnich, E. Maier-Reimer, S. Venzke, and T.P. Barnett, 1996. A mechanism for decadal climate variability. In *Decadal Climate Variability: Dynamics and Predictability*. D.L.T. Anderson and J. Willebrand (eds.), NATO ASI Series, Vol. I44, Springer-Verlag, Heidelberg, pp. 263-292.

Latif, M., D. Anderson, T. Barnett, M. Cane, R. Kleeman, A. Leetmaa, J.J. O'Brien, A. Rosati, and E. Schneider, 1998. A review of predictability and prediction of ENSO. *J. Geophys. Res.* **103**, 14375-14393.

Lau, N.-C., 1988. Variability of the observed midlatitude storm tracks in relation to low frequency changes in the circulation pattern. *J. Atmos. Sci.* **45**, 2718-2743.

Lau, N.-C., and M.J. Nath, 1991. Variability of the baroclinic and barotropic transient eddy forcing associated with monthly changes in the midlatitude storm tracks. *J. Atmos. Sci.* **48**, 2589-2613.

Lau, N.-C., and M.J. Nath, 1996. The role of the "atmospheric bridge" in linking tropical Pacific ENSO events to extratropical SST anomalies. *J. Climate* **9**, 2036-2057.

Lazier, J.R., 1988. Temperature and salinity changes in the deep Labrador Sea, 1962-1986. *Deep-Sea Res.* **18**, 1247-1253.

Lazier, J.R., 1995. The salinity decrease in the Labrador Sea over the past thirty years. In *Natural Climate Variability on Decade-to-Century Time Scales*, D.G. Martinson, K. Bryan, M. Ghil, M.M. Hall, T.R. Karl, E.S. Sarachik, S. Sorooshian, and L.D. Talley (eds.), National Academy Press, Washington, D.C., pp. 295-302.

Lea, D.W., E.A. Boyle, and G.T. Shen, 1989. Coralline barium records temporal variability in equatorial Pacific upwelling. *Nature* **340**, 373-376.

Lean, J., 1991. Variations in the sun's radiative output. *Rev. Geophys.* **29**, 505-535.

Lean, J., A. Skumanich, and O. White, 1992. Estimating the sun's radiative output during the Maunder minimum. *Geophys. Res. Lett.* **19**, 1591-1594.

Lean, J., J. Beer, and R. Bradley, 1995. Reconstruction of solar irradiance since 1610: Implications for climate change. *Geophys. Res. Lett.* **22**, 3195-3198.

Legras, B., and M. Ghil, 1985. Persistent anomalies, blocking and variations in atmospheric predictability. *J. Atmos. Sci.* **42**, 433-471.

Leurs, J., and R.E. Eskridge, 1995. Temperature corrections for the VIZ and Vaisala Radiosondes. *J. Appl. Meteor.* **34**, 1241-1253.

Levins, R., T. Awerbuch, U. Brinckmann, I. Eckhardt, P. Epstein, N. Makhoul, C. Albuquerque de Passos, C. Puccia, A. Spielman, and M. Wilson, 1994. The emergence of new diseases. *Am. Sci.* **82**, 52-60.

Levitus, S., 1989. Interpentadal variability of temperature and salinity at intermediate depths of the North Atlantic Ocean, 1970-74 versus 1955-59. *J. Geophys. Res.* **94**, 6091-6131.

Levitus, S., and T.P. Boyer, 1994. World Ocean Atlas 1994, vol. 4: Temperature. NOAA Atlas NESDIS 4. 117 pp.

Levitus, S., R. Burgett, and T.P. Boyer, 1994. World Ocean Atlas 1994, vol. 3: Salinity. NOAA Atlas NESDIS 3. 99 pp.

Lighthill, J., G.J. Holland, W.M. Gray, C. Landsea, K. Emanuel, G. Craig, J. Evans, Y. Kunihara, and C.P. Guard, 1994. Global climate change and tropical cyclones. *Bull. Amer. Meteor. Soc.* **75**, 2147-2157.

Lindzen, R.S., 1996. The importance and nature of the water vapor budget in nature and models. In *Climate Sensitivity to Radiative Perturbations,* H. Le Treut (ed.), NATO ASI Series, Vol. I34, Springer-Verlag, Heidelberg, pp. 51-66.

Linsley, B.K., R.B. Dunbar, G.M. Wellington, and D.A. Mucciarone, 1994. A coral-based reconstruction of Intertropical Convergence Zone variability over Central America since 1707. *J. Geophys. Res.* **99**, 9977-9994.

Liu, A., S.G.H. Philander, and R.C. Pacanowski, 1994. A GCM study of tropical-subtropical upper-ocean water exchange. *J. Phys. Oceanogr.* **24**, 2606-2623.

Loevinsohn, M.E., 1994. Climatic warming and increased malaria incidence in Rwanda. *The Lancet* **343**, 714-718.

Lorenz, E.N., 1963. Deterministic nonperiodic flow. *J. Atmos. Sci.* **20**, 130-141.

Lorenz, E.N., 1967. Nature and theory of the general circulation of the atmosphere. *WMO Bull.* **16**, 74-78.

Lorenz, E.N., 1969. Atmospheric predictability as revealed by naturally occurring analogues. *J. Atmos. Sci.* **26**, 636-646.

Lorenz, E.N., 1982. Atmospheric predictability experiments with a large numerical model. *Tellus* **34**, 505-513.

Lorenz, E.N., 1990. Can chaos and intransitivity lead to interannual variability? *Tellus* **42A**, 378-389.

Lott, N., 1993. *The Big One! A Review of the March 12-14, 1993 "Storm of the Century,"* Special Report TR 93-01, National Oceanic and Atmospheric Administration, Washington, D.C.

Lough, J.M., D.J. Barnes, and R.B. Taylor, 1996. The potential of massive corals for the study of high-resolution climate variation of the past millennium. In *Climatic Variations and Forcing Mechanisms of the Last 2000 Years*, P.D. Jones, R.S. Bradley, and J. Jouzel (eds.), NATO ASI Series, Vol. I41, Springer-Verlag, Heidelberg, pp. 355-372.

Lysne, J., P. Chang, and B. Geise, 1998. Impact of the extratropical Pacific on equatorial variability. *Geophys. Res. Lett.* **24**, 2589-2592.

MacAyeal, D.R., 1994. Binge/purge oscillations of the Laurentide ice sheet, a cause of the North Atlantic's Heinrich Events. *Paleoceanography* **8**, 775-784.

Madole, R.F., 1994. Stratigraphic evidence of desertification in the west-central Great Plains within the past 1000 yr. *Geology* **22,** 483-486.

Madole, R.F., 1995. Spatial and temporal patterns of late Quaternary eolian deposition, eastern Colorado, U.S.A. *Quat. Sci. Rev.* **14**, 155-177.

Magalhaes, A., and P. Magee, 1994. The Brazilian Nordeste. In *Drought Follows the Plough,* M. Glantz (ed.), Cambridge University Press, Cambridge, U.K., pp. 59-76.

Maier-Reimer, E., U. Mikolajewicz, and A. Winguth, 1996. Future ocean uptake of CO_2: Interaction between ocean circulation and biology. *Climate Dyn.* **12**, 711-721.

Manabe, S., and R.J. Stouffer, 1988. Two stable equilibria of a coupled ocean-atmosphere model. *J. Climate* **1**, 841-866.

Manabe, S., and R.J. Stouffer, 1994. Multiple-century response of a coupled ocean-atmosphere model to an increase of atmospheric carbon dioxide. *J. Climate* **7**, 5-23.

Manabe, S., and R. J. Stouffer, 1996. Low-frequency variability of surface air temperature in a 1,000 year integration of a coupled ocean-atmosphere-land model. *J. Climate* **9**, 376-393.

Manabe, S., and R.T. Wetherald, 1975. The effect of doubling CO_2 concentration on the climate of a GCM. *J. Atmos. Sci.* **32**, 3-15.

Mann, M.E., and J. Park, 1996. Greenhouse warming and changes in the seasonal cycle of temperature: Model versus observations. *Geophys. Res. Lett.* **23**, 1111-1114.

Mann, M., U. Lall and B. Saltzman, 1995a. Decadal-to-century scale climate variability: Insights into the rise and fall of the Great Salt Lake. *Geophys. Res. Lett.* **22**, 937-940.

Mann, M.E., J. Park, and R.S. Bradley, 1995b. Global interdecadal and century-scale climate oscillations during the past five centuries. *Nature* **378**, 266-270.

Mann, M.E., R.S. Bradley, and M.K. Hughes, 1998. Global-scale temperature patterns and climate forcing over the past six centuries. *Nature* **392**, 779-787.

Mantua, N.J., and D.S. Battisti, 1994. Evidence for the delayed oscillator mechanism for ENSO: The "observed" oceanic Kelvin mode in the far western Pacific. *J. Phys. Oceanogr.* **24**, 691-699.

Mantua, N.J., S.R. Hare, Y. Zhang, J.M. Wallace, and R.C. Francis, 1997. A Pacific interdecadal climate oscillation with impacts on salmon production. *Bull. Amer. Meteor. Soc.* **78**, 1069-1079.

Mao, J., and A. Robock, 1998. Surface air temperature simulations by AMIP General Circulation Models: Volcanic and ENSO Signals and Systematic Errors. *J. Climate* **11**, 1538-1552.

Markham, C.G., and D.R. McLain, 1977. Sea surface temperature related to rain in Ceara, northeastern Brazil. *Nature* **265**, 320-323.

Marland, G., R.J. Andres, and T.A. Boden, 1994. Global, regional, and national CO_2 emissions. In *Trends '93: A Compendium of Data on Global Change*, T.A. Boden, D.P. Kaiser, R.J. Sepanski, and F.W. Stoss (eds.), ORNL/CDIAC-65, Oak Ridge National Laboratory, Oak Ridge, Tennessee, pp. 505-584.

Maslanik, J.A., M.C. Serreze, and R.G. Barry, 1996. Recent decreases in Arctic summer ice cover and linkages to atmospheric circulation and anomalies. *Geophys. Res. Lett.* **23**, 1677-1680.

Mass, C.F., and D.A. Portman, 1989. Major volcanic eruptions and climate: A critical evaluation. *J. Climate* **2**, 566-593.

Maykut, G.A., and N. Untersteiner, 1971. Some results from a time-dependent thermodynamic model of sea ice. *J. Geophys. Res.* **76**, 1550-1575.

McCartney, M.S., R.G. Curry, and H.F. Bezdek, 1997. The interdecadal warming and cooling of Labrador Sea Water. *ACCP Notes* **IV**, 1-11.

McCormick, M.P., P.-H. Wang, and L.R. Poole, 1993. Stratospheric aerosols and clouds. In *Aerosol-Cloud-Climate Interactions*, P.V. Hobbs (ed.), Int. Geophys. Ser. **54**, Academic Press, San Diego, pp. 205-222.

McCreary, J., and P. Lu, 1994. On the interaction between the subtropical and equatorial ocean circulations: The subtropical cell. *J. Phys. Oceanogr.* **24**, 466-497.

McGuffie, K., and A. Henderson-Sellers, 1988. Is Canadian cloudiness increasing? *Atmos.-Ocean* **26**, 608-633.

McIntyre, A., and B. Molfino, 1996. Forcing of Atlantic equatorial and subpolar millennial cycles by precession. *Science* **274**, 1867-1870.

McKibben, B., 1995. An explosion of green, *Atlantic Monthly*, April, 61-83.

McLaughlin, F.A., E.C. Carmack, R.W. Macdonald, and J.K. Bishop, 1996. Physical and geochemical properties across the Atlantic/Pacific water mass front in the southern Canadian basin. *J. Geophys. Res.* **101**, 1183-1195.

McPhee, M.G., T.P. Stanton, J.H. Morison, and D.G. Martinson, 1998. Freshening of the upper ocean in the Central Arctic: Is perennial sea ice disappearing? *Geophys. Res. Lett.* **25**, 1729-1732

Mechoso, C.R., S.W. Lyons, and J.A. Spahr, 1990. The impact of sea surface temperature anomalies on the rainfall over northeast Brazil. *J. Climate* **3**, 812-826.

Meehl, G.A., and W.M. Washington, 1995. Cloud albedo feedback and the super greenhouse effect in a global coupled GCM. *Climate Dyn.* **11**, 399-411.

Meehl, G.A., and W.M. Washington, 1996. El Niño-like climate change in a model with increased atmospheric CO_2 concentrations. *Nature* **382**, 56-60.

Meese, D.A., A.J. Gow, and G.A. Zielinski, 1994. The accumulation record from the GISP2 core as an indicator of climate change throughout the Holocene. *Science* **266**, 1680-1682.

Mehta, V.M., and T. Delworth, 1995. Decadal variability of the tropical Atlantic ocean surface temperature in shipboard measurements and in a global ocean-atmosphere model. *J. Climate* **8**, 172-190.

Meko, D., E.R. Cook, D.W. Stahle, C.W. Stockton, and M.K. Hughes, 1993. Spatial patterns of tree-growth anomalies in the United States and southeastern Canada. *J. Climate* **6**, 1773-1786.

Melillo, J.M., 1995. Human influences on the global nitrogen budget and their implications for the global carbon budget. In *Toward Global Planning of Sustainable Use of the Earth: Development of Global Eco-Engineering*, S. Murai and M. Kimura (eds.), Elsevier, Amsterdam, pp. 117-134.

Mikolajewicz, U., and E. Maier-Reimer, 1990. Internal secular variability in an ocean general circulation model. *Climate Dyn.* **4**, 145-156.

Mikolajewicz, U., T.J. Crowley, A. Schiller, and R. Voss, 1997. Modelling teleconnections between the North Atlantic and North Pacific during the Younger Dryas. *Nature* **387**, 384-387.

Minnis, P., E.F. Harrison, L.L. Stowe, G.G. Gibson, F.M. Denn, D.R. Doelling, and W.L. Smith, Jr., 1993. Radiative climate forcing by the Mount Pinatubo eruption. *Science* **259**, 1411-1415.

Minobe, S., 1997. Climate variability with periodicity 50-70 years over the North Pacific and North America. *Geophys. Res. Lett.* **24**, 683-686.

Mitchell, J.M., Jr., 1976. An overview of climatic variability and its causal mechanisms. *Quat. Res.* **6**, 481-493.

Mitchell, J.M., Jr., C.W. Stockton, and D.M. Meko, 1979. Evidence of a 22-year rhythm of drought in the western United States related to the Hale solar cycle since the 17th century. In *Solar-Terrestrial Influences on Weather and Climate*, B.M. McCormac and T.A. Seliga (eds.), D. Reidel Publishing Co., Dordrecht, Holland, 125-144.

Mitchell, J.F.B., T.C. Johns, J.M. Gregory, and S.F.B. Tett, 1995. Climate response to increasing levels of greenhouse gases and sulfate aerosols. *Nature* **376**, 501-504.

Mitsuguchi, T., E. Matsumoto, O. Abe, T. Uchida, and P.J. Isdale, 1996.

Mg/Ca thermometry in coral skeletons. *Science* **274**, 961-963.

Moisan, J.R., and P.P. Niiler, 1998. The seasonal heat budget of the North Pacific: Net heat flux and heat storage rates (1950-1990). *J. Phys. Oceanogr.***28**, 401-421.

Molinari, R.B., D. Mayer, J. Festa, and H. Bezdek, 1997. Multi-year variability in the near surface temperature structure of the midlatitude western North Atlantic Ocean. *J. Geophys. Res.* **102**, 3267-3278.

Moon, Y., 1995. *Large-scale Atmospheric Variability and the Great Salt Lake*. Ph.D. dissertation, Utah State University, Logan, 154 pp.

Moon, Y., and U. Lall, 1996. Large scale atmospheric indices and the Great Salt Lake: Interannual and interdecadal variability. *J. Hydrol. Eng.* 1, 55-62.

Mooney, H.A., B.G. Drake, R.J. Luxmoore, W.C. Oechel, and L.F. Pitelka, 1991. Predicting ecosystem responses to elevated CO_2 concentrations. *BioScience* **41**, 96-104.

Montzka, S.A., J.H. Butler, R.C. Myers, T.M. Thompson, T.H. Swanson, A.D. Clarke, L.T. Lock, J.W. Elkins, 1996. Decline in the tropospheric abundance of halogen from halocarbons: Implications for stratospheric ozone depletion. *Science* **272**, 1318-1322.

Morison, J.H., M. Steele, R. Andersen, 1998. Hydrography of the upper Arctic Ocean measured from the nuclear submarine USS Pargo. *Deep-Sea Res. I* **45**, 15-38.

Moritz, R.E., and D.K. Perovich (eds.), 1996. Surface Heat Budget of the Arctic Ocean Science Plan, ARCSS/OAII Report Number 5, University of Washington, Seattle, 64 pp.

Moron, V., R. Vautard, and M. Ghil, 1998. Trends, interdecadal and interannual oscillations in global sea-surface temperatures. *Climate. Dyn.* **14**, 545-569.

Mosley-Thompson, E., J. Dai, L.G. Thompson, P.M. Grootes, J.K. Arbogast, and J.F. Paskievitch, 1991. Glaciological studies at Siple Station (Antarctica): Potential ice-core paleoclimatic record. *J. Glaciol.* **37**, 11-22.

Moulin, C., C.E. Lambert, F. Dulac, and U. Dayan, 1997. Control of atmosphere export of dust from North Africa by the North Atlantic Oscillation. *Nature* **387**, 691-694.

Moura, A.D, 1994. Prospects for seasonal-to-interannual climate prediction and applications for sustainable development. *World Meteor. Soc. Bull.* **43**, 207-215.

Moura, A., and J. Shukla, 1981. On the dynamics of droughts in northeast Brazil: Observations, theory, and numerical experiments with a general circulation model. *J. Atmos. Sci.* **38**, 2653-2675.

Muhs, D.R., and Maat, P.B., 1993. The potential response of eolian sands to greenhouse warming and precipitation reduction on the Great Plains of the United States. *J. Arid Environ.* **25**, 351-361.

Munnich, M., M.A. Cane, and S.E. Zebiak, 1991. A study of self-excited oscillations of the tropical ocean-atmosphere system. II. Nonlinear cases. *J. Atmos. Sci.* **48**, 1238-1248.

Myneni, R.B., C.D. Keeling, C.J. Tucker, G. Asrar, and R.R. Nemani, 1997. Increased plant growth in the northern high latitudes from 1981 to 1991. *Nature* **386**, 698-702.

Mysak, L.A., 1986. El Niño, interannual variability and fisheries in the northeast Pacific Ocean. *Canad. J. Aquat. Sci.* **43**, 464-497.

Mysak, L.A., D.K. Manak, and R.F. Marsden, 1990. Sea ice anomalies in the Greenland and Labrador Seas during 1901-1984 and their relation to an interdecadal Arctic climate cycle. *Climate Dyn.* **5**, 111-133.

Najjar, R., 1992. Marine biogeochemistry. In *Climate System Modeling,* K.E. Trenberth (ed.), Cambridge University Press, Cambridge, U.K., pp. 241-282.

Nakazawa, T., T. Machida, M. Tanaka, Y. Fujii, S. Aoki, and O. Watanabe, 1993. Differences of the atmospheric CH_4 concentration between the Arctic and Antarctic regions in pre-industrial/agricultural era. *Geophys. Res. Lett.* **20**, 943-947.

Namias, J., 1968. Long-range weather forecasting: History, current status, and outlook. *Bull. Amer. Meteor. Soc.* **49**, 438-470.

NAPAP, 1991. *National Acid Precipitation Assessment Program, 1990 Integrated Assessment Report*. NAPAP, Washington, D.C., 520 pp.

Neelin, J.D., and H.A. Dijkstra, 1995. Ocean-atmosphere interaction and the tropical climatology. Part I: The dangers of flux correction. *J. Climate* **8**, 1325-1342.

Nesme-Ribes, E., E.N. Ferreira, R. Sadourny, H. Le Treut, and Z.X. Li, 1993. Solar dynamics and its impact on solar irradiance and the terrestrial climate. *J. Geophys. Res.* **98**, 18923-18935.

Nicholls, N., 1985. Southern Oscillation and Indonesian sea surface temperature. *Mon. Wea. Rev.* **112**, 424-432.

Nicholls, R.J., and S.P. Leatherman, 1994. Global sea-level rise. In *As Climate Changes: Potential Impacts and Implications*, K. Strzepek and J.B. Smith (eds.), Cambridge University Press, Cambridge, U.K., pp. 92-123.

Nicholson, S.E., 1989. Long-term changes in African rainfall. *Weather* **44**, 46-56.

Nitta, T., and S. Yamada, 1989. Recent warming of tropical sea surface temperature and its relationship to the Northern Hemisphere circulation. *J. Meteorol. Soc. Japan* **67**, 1557-1583.

NOAA (National Oceanic and Atmospheric Administration), 1995. *CPC - Stratosphere: Northern Hemisphere Winter Summary - 1994-1995,* National Weather Service, National Centers for Environmental Prediction, Climate Prediction Center.

NOAA, 1996. *Northern Hemisphere Winter Summary - 1995-96. Selected Indicators of Stratospheric Climate,* National Weather Service, National Centers for Environmental Prediction, Climate Prediction Center.

NOAA, 1997. *Northern Hemisphere Winter Summary - 1996-97. Selected Indicators of Stratospheric Climate,* National Weather Service, National Centers for Environmental Prediction, Climate Prediction Center.

Nobre, P., and J. Shukla, 1996. Variations of sea surface temperature, wind stress and rainfall over the tropical Atlantic and South America. *J. Climate* **9**, 2464-2479.

Norris, J., and C. Leovy, 1994. Interannual variability in stratiform cloudiness and sea surface temperature. *J. Climate* **7**, 1915-1925.

Noss, R., and Cooperrider, A., 1994. *Saving Nature's Legacy: Protecting and Restoring Biodiversity,* Island Press, Washington, D.C., 380 pp.

NRC (National Research Council), 1975. *Understanding Climatic Change: A Program for Action*. National Academy Press, Washington, D.C., 239 pp.

NRC, 1994. *Solar Influences on Global Change*. National Academy Press, Washington, D.C., 163 pp.

NRC, 1995. *Natural Climate Variability on Decade-to-Century Time Scales*, D.G. Martinson, K. Bryan, M. Ghil, M.M. Hall, T.R. Karl, E.S. Sarachik, S. Sorooshian, and L.D. Talley (eds.), National Academy Press, Washington, D.C., 630 pp.

NRC, 1996. *Learning to Predict the Climate Variations Associated with El Niño: Accomplishments and Legacies of the TOGA Program*. National Academy Press, Washington, D.C., 171 pp.

NRC, 1997. *The Global Ocean Observing System: Users, Benefits, and Priorities*. National Academy Press, Washington, D.C., 82 pp.

NRC, 1998a. *A Scientific Strategy for U.S. Participation in the GOALS (Global Ocean-Atmosphere-Land System) Component of the CLIVAR (Climate Variability and Predictability) Programme*. National Academy Press, Washington, D.C., 69 pp.

NRC, 1998b. *Global Energy and Water Cycle Experiment (GEWEX) Continental-Scale International Project (GCIP): A Review of Progress and Opportunities*. National Academy Press, Washington, D.C., 93 pp.

O'Brien, S.R., P.A. Mayewski, L.D. Meyer, D.A. Meese, M.S. Twickler, and S.I. Whitlow, 1995. Complexity of Holocene climate as reconstructed from a Greenland ice core. *Science* **270**, 1962-1964.

Oechel, W.C., S.J. Hastings, G. Vourlitis, M. Jenkins, G. Riechers, and N. Grulke, 1993. Recent change of Arctic tundra ecosystems from a net carbon dioxide sink to a source. *Nature* **361**, 520-523.

Ogilvie, A.E.J., 1984. The past climate and sea ice records from Iceland, Part 1: Data to A.D. 1780. *Clim. Change* **6**, 131-152.

Oppo, D.W., J.F. McManus, and J.L. Cullen, 1998. Abrupt climate events 500,000 to 340,000 years ago: Evidence from subpolar North Atlantic sediments. *Science* **279**, 1335-1338.

Oppenheimer, M., 1998. Global warming and the stability of the West Antarctic Ice Sheet. *Nature* **393**, 325-332.

Osborn, M.T., R.J. DeCoursey, and C.R. Trepte, 1995. Evolution of the Pinatubo volcanic cloud over Hampton, Virginia. *Geophys. Res. Lett.* **22**, 1101-1104.

Osterkamp, T.E., T. Zhang, and V.E. Romanovsky, 1994. Evidence for a cyclic variation of permafrost temperatures in Northern Alaska. *Permafrost Periglacial Process.* **5**, 137-144.

Otto-Bliesner, B.L., and G.R. Upchurch, 1997. Vegetation-induced warming of high-latitude regions during the Late Cretaceous Period. *Nature* **385**, 804-807.

Overpeck, J.T., 1996. Varved sediment records of recent seasonal-millennial environmental variability. In *Climatic Variations and Forcing Mechanisms of the Last 2000 Years,* P.D. Jones, R.S. Bradley, and J. Jouzel (eds.), NATO ASI Series, Vol. I41, Springer-Verlag, Heidelberg, pp. 479-498.

Overpeck, J.T., D. Rind, and R. Goldberg, 1990. Climate-induced changes in forest disturbance and vegetation. *Nature* **343**, 51-53.

Overpeck, J.T., P.J. Bartlein, and T. Webb III, 1991. Potential magnitude of future vegetation change in eastern North America: Comparisons with the past. *Science* **254**, 692-695.

Overpeck, J.T., D.M. Anderson, S.E. Trumbore, and W.L. Prell, 1996. The southwest Indian monsoon over the last 18,000 years. *Climate Dyn.* **12**, 213-225.

Overpeck, J.T., K. Hughen, D. Hardy, R. Bradley, R. Case, M. Douglas, B. Finney, K. Gajewski, G. Jacoby, A. Jennings, S. Lamoureux, A. Lasca, G. MacDonald, J. Moore, M. Retelle, S. Smith, A. Wolfe, and G. Zielinski, 1997. Arctic environmental change of the last four centuries. *Science* **278**, 1251-1256.

Owen, J.A., and C.K. Folland, 1988. Modelling the influence of sea-surface temperatures on tropical rainfall. In *Recent Climatic Change, A Regional Approach.* S. Gregory (ed.), Bellhaven Press, London, pp. 141-153.

Palmen, E., and C.W. Newton, 1969. *Atmospheric Circulation Systems, Their Structure and Physical Interpretation.* Academic Press, New York, 603 pp.

Palmer T.N., 1993. Extended-range prediction and the Lorenz model. *Bull. Amer. Meteor. Soc.* **74**, 49-65.

Palmer, T.N., and D.L.T. Anderson, 1993. *Scientific Assessment of the Prospects for Seasonal Forecasting: A European Perspective.* European Centre for Medium-Range Weather Forecasting Berkshire, U.K., 34 pp.

Palmer, T.N., C. Brankovic, P. Viterbo, and M.J. Miller, 1992. Modeling interannual variations of summer monsoons. *J. Climate* **5**, 399-417.

Parilla, G., A. Lavin, H. Bryden, M. Garcia, and R. Millard, 1994. Rising temperatures in the subtropical North Atlantic Ocean over the past 35 years. *Nature* **369**, 48-51.

Paterson, W.S.B., 1994. *The Physics of Glaciers*, 3rd ed. Pergamon (Elsevier Science Inc.), Tarrytown, New York, 480 pp.

Pedlosky, J., 1987. An inertial theory for the equatorial undercurrent. *J. Phys. Oceanogr.* **18**, 880-886.

Penland, C., and P.D. Sardeshmukh, 1995. The optimal growth of tropical sea surface temperature anomalies. *J. Climate* **6**, 1067-1076.

Pielke, R.A., Jr., and C.W. Landsea, 1998. Normalized hurricane damages in the United States: 1925-1995. *Weather and Forecasting* **13**, 351-361.

Pierrehumbert, R.T., 1995. Thermostats, radiator fins, and the local runaway greenhouse. *J. Atmos. Sci.* **52**, 1784-1806.

Pittock, A.B., 1978. A critical look at long-term sun-weather relationships. *Rev. Geophys. Space Physics* **16**, 400-420.

Pittock, A.B., 1983. Solar variability, weather, and climate: An update. *Quart. J. Roy. Meteor. Soc.* **109**, 23-55.

Plaut, G., M. Ghil, and R. Vautard, 1995. Interannual and interdecadal variability in 335 years of central England temperatures. *Science* **268**, 710-713.

Pohjola, V.A., and J.C. Rogers, 1997. Atmospheric circulation and variations in the Scandinavian glacier mass balance. *Quat. Res.* **47**, 29-36.

Polzin, K.L., J.M. Toole, J.R. Ledwell, and R.W. Schmitt, 1997. Spatial variability of turbulent mixing in the Abyssal Ocean. *Science* **276**, 93-96.

Poorter, H., 1993. Interspecific variation in the growth response of plants to an elevated ambient CO_2 concentration. *Vegetation* **104/105**, 77-97.

Postel, S.L., G.C. Daily, and P.R. Ehrlich, 1996. Human appropriation of renewable fresh water. *Science* **271**, 785-788.

Prather, M.J., 1992. Catastrophic loss of stratospheric ozone in dense volcanic clouds. *J. Geophys. Res.* **97**, 10187-10191.

Prather, M.J., 1994. Lifetimes and eigenstates in atmospheric chemistry. *Geophys. Res. Lett.* **21**, 801-804.

Prentice, I.C, and M.T. Sykes, 1995. Vegetation geography and global carbon storage changes. In *Biotic Feedbacks in the Global Climate System,* G.M. Woodwell and F.T. Mackenzie (eds.), Oxford University Press, New York, pp. 304-312.

Prentice, I.C., W. Cramer, S.P. Harrison, R. Leemans, R.A. Monserud, and A.M. Solomon, 1992. A global biome model based on plant physiology and dominance, soil properties and climate. *J. Biogeogr.* **19**, 117-134.

Pringle, H., 1997. Death in Norse Greenland. *Science* **275**, 924-926.

Prinn, R.G., R.F. Weiss, B.R. Miller, J. Huang, F.N. Alyea, D.M. Cunnold, P.J. Fraser, D.E. Hartley, and P.G. Simmonds, 1995. Atmospheric trends and lifetime of CH_3CCl_3 and global OH concentrations. *Science* **269**, 187-192.

Qiu, B., and R.X. Huang, 1994. Subduction/obduction in the North Atlantic and North Pacific. In *The Atlantic Climate Change Program. Proceedings from the Principal Investigators Meeting* (May 9-11, 1994), A.-M. Wilburn (ed.), NOAA Geophysical Fluid Dynamics Laboratory, Princeton, New Jersey, pp. 97-99.

Qiu, B., and F.-F. Jin, 1997. Antarctic circumpolar waves: An indication of ocean-atmosphere coupling in the extratropics. *Geophys. Res. Lett.* **24**, 2585-2588.

Qiu, B., and T.M. Joyce, 1992. Interannual variability in the mid- and low-latitude western North Pacific. *J. Phys. Oceanogr.* **22**, 1062-1079.

Quadfasel, D., 1991. Warming in the Arctic. *Nature* **350**, 385.

Quinn, W.H., 1993. A study of Southern Oscillation-related climatic activity for A.D. 622-1900 incorporating Nile River flood data. In *El Niño: Historical and Paleoclimatic Aspects of the Southern Oscillation*, Cambridge University Press, Cambridge, U.K., pp. 119-149.

Quinn, W.H., and V.T. Neal, 1984. Long-term variations in the Southern Oscillation, El Niño, and Chilean subtropical rainfall. U.S. National Marine Fisheries Service, *Fishery Bulletin* **81**, 363-374.

Quinn, T.M., T.J. Crowley, and F.W. Taylor, 1996. New stable isotope results from a 173-year coral from Espiritu Santo, Vanuatu. *Geophys. Res. Lett.* **23**, 3413-3416.

Rajagopalan, B., and U. Lall, 1995. Seasonality of precipitation along a meridian in the Western United States. *Geophys. Res. Lett.* **22**, 1081-1084.

Rajagopalan, B., U. Lall, and M.A. Cane, 1997. Anomalous ENSO occurrences: An alternate view. *J. Climate* **10**, 2351-2357.

Randall, D., J. Curry, D. Battisti, G. Flato, R. Grumbine, S. Hakkinen, D. Martinson, R. Preller, J. Walsh, and J. Weatherly, 1998. Status of and outlook for large-scale modeling of atmosphere-ice-ocean interactions in the Arctic. *Bull. Amer. Meteor. Soc.* **79**, 197-219.

Rao, V.B., M.C. De Lima, and S.H. Franchito, 1993. Seasonal and interannual variations of rainfall over eastern northeast Brazil. *J. Climate* **6**, 1754-1763.

Rasmusson, E.M., and P.A. Arkin, 1993. A global view of large-scale precipitation variability. *J. Climate* **6**, 1495-1522.

Rasmusson, E.M., and J.M. Wallace, 1983. Meteorological aspects of the El Niño/Southern Oscillation. *Science* **222**, 1195-1202.

Rastetter, E.B., R.B. McKane, G.R. Shaver, and J.M. Melillo, 1992. Changes in C storage by terrestrial ecosystems: How C-N interactions restrict responses to CO_2 and temperature. *Water Air Soil Pollut.* **64**, 327-344.

Raynaud, D.M., J. Jouzel, J.M. Barnola, J. Chapellaz, R.J. Delmas, and C. Lorius, 1993. The ice record of greenhouse gases. *Science* **259**, 926-

934.

Reason, C.J.C., R.J. Allan, and J.A. Lindesay, 1996a. Evidence for the influence of remote forcing on interdecadal variability in the southern Indian Ocean. *J. Geophys. Res.* **101**, 11867-11882.

Reason, C.J.C., R.J. Allan, and J.A. Lindesay, 1996b. Dynamical response of the oceanic circulation and temperature to interdecadal variability in the surface winds over the Indian Ocean. *J. Climate* **9**, 97-114.

Reich, P.B., and R.G. Amundson, 1985. Ambient levels of ozone reduce net photosynthesis in tree and crop species. *Science* **230**, 566-570.

Reid, G.C., 1991. Solar total irradiance variations and the global sea surface temperature record. *J. Geophys. Res.* **96**, 2835-2844.

Reinhold, B.B., and R.P. Pierrehumbert, 1982. Dynamics of weather regimes: Quasi-stationary waves and blocking. *Mon. Wea. Rev.* **110**, 1105-1145.

Reynolds, R.W., 1988. A real-time global sea surface temperature analysis. *J. Climate* **1**, 75-86.

Rind, D., 1998. Latitudinal temperature gradients and climate change. *J. Geophys. Res.* **103**, 5943-5971.

Rind, D., and J.T. Overpeck, 1993. Hypothesized causes of decade-to-century-scale climate variability: Climate model results. *Quat. Sci. Rev.* **12**, 357-374.

Rind, D., R. Healy, C. Parkinson, and D. Martinson, 1995. The role of sea ice in 2xCO_2 climate model sensitivity. Part I: The total influence of sea-ice thickness and extent. *J. Climate* **8**, 449-463.

Rind, D., C. Rosenzweig, and M. Stieglitz, 1997. The role of moisture transport between ground and atmosphere in global change. *Ann. Rev. Energy Environ.* **22**, 47-74.

Rinsland, C.P., M.R. Gunson, M.C. Abrams, L.L. Lowes, R. Zander, E. Mathieu, A. Goldman, M.K.W. Ko, J.M. Rodriguez, and N.D. Sze, 1994. N_2O_5 to NHO_3 in the post-Mount Pinatubo eruption stratosphere. *J. Geophys. Res.* **99**, 8213-8219.

Ritchie, J.C., C.H. Eyles, and C.V. Haynes, 1985. Sediment and pollen evidence for an early- to mid-Holocene humid period in the eastern Sahara. *Nature* **314**, 352-355.

Robinson, D.A., K.F. Dewey, and R.R. Heim, Jr., 1993. Global snow cover monitoring: An update. *Bull. Amer. Meteor. Soc.* **74**, 1689-1696.

Robock, A., 1991. The volcanic contribution to climate change of the past century. In *Greenhouse Gas-Induced Climatic Change: A Critical Appraisal of Simulations and Observations*, M.E. Schlesinger (ed.), Elsevier, Amsterdam, pp. 429-444.

Robock, A., and Mao, J., 1995. The volcanic signal in surface temperature observations. *J. Climate* **8**, 1086-1103.

Roemmich, D., and C. Wunsch, 1984. Apparent changes in the climatic state of the deep North Atlantic Ocean. *Nature* **307**, 447-450.

Rogers, J.C., 1981. The North-Pacific Oscillation. *J. Climatol.* **1**, 39-57.

Rogers, J.C., 1984. The association between the North Atlantic Oscillation and the Southern Oscillation in the Northern Hemisphere. *Mon. Wea. Rev.* **112**, 1999-2015.

Rogers, J.C., 1990. Patterns of low-frequency monthly sea level pressure variability (1899-1986) and associated wave cyclone frequencies. *J. Climate* **3**, 1364-1379.

Rogers, H.H., G.B. Runion, and S.V. Krupa, 1994. Plant responses to atmospheric CO_2 enrichment with emphasis on roots and the rhizosphere. *Environ. Pollut.* **83**, 155-189.

Ropelewski, C.F., and M.S. Halpert, 1987. Global and regional scale precipitation and temperature patterns associated with El Niño/Southern Oscillation. *Mon. Wea. Rev.* **115**, 1606-1626.

Ropelewski, C.F., P.J. Lamb, and D.H. Portis, 1993. The global climate for June to August 1990: Drought returns to sub-Saharan West Africa and warm Southern Oscillation episode conditions develop in the central Pacific. *J. Climate* **6**, 2188-2212.

Rossow, W.B., and B. Cairns, 1995. Monitoring changes in clouds. *Clim. Change* **31**, 175-217.

Rowell, D.P., C.K. Folland, K. Maskell, J.A. Owen, and M.N. Ward, 1992. Modelling the influence of global sea surface temperatures on the variability and predictability of seasonal Sahel rainfall. *Geophys. Res. Lett.* **19**, 905-908.

Rudels, B., E.P. Jones, L.G. Anderson, and G. Kattner, 1994. On the intermediate depth waters of the Arctic ocean. In *The Polar Oceans and Their Role in Shaping the Global Environment*, O.M. Johannessen, R.D. Muench, and J.E. Overland (eds.), American Geophysical Union, Washington, D.C., pp. 33-46.

Russell, J.M., M.-Z. Luo, R.J. Cicerone, and L.E. Deaver, 1996. Satellite confirmation of the dominance of chlorofluorocarbons in the global stratospheric chlorine budget. *Nature* **379**, 526-529.

Sandweiss, D.H., J.B. Richardson, E.J. Reitz, H.B. Rollins, and K.A. Maasch, 1996. Geoarchaeological evidence from Peru for a 5000 years B.P. onset of El Niño. *Science* **273**, 1531-1533.

Santer, B.D., K.E. Taylor, T.M.L. Wigley, J.E. Penner, P.D. Jones, and U. Cubasch, 1995. Towards the detection and attribution of an anthropogenic effect on climate. *Climate Dyn.* **12**, 77-100.

Santer, B.D., K.E. Taylor, T.M.L. Wigley, T.C. Johns, P.D. Jones, D.J. Karoly, J.F.B. Mitchell, A.H. Oort, J.E. Penner, V. Ramaswamy, M.D. Schwartzkopf, R.J. Stouffer, and S. Tett, 1996. A search for human influences on the thermal structure of the atmosphere. *Nature* **382**, 39-46.

Sarachik, E.S., M. Winton, and F.L. Yin, 1996. Mechanisms for decadal-to-centennial climate variability. In *Decadal Climate Variability: Dynamics and Predictability*. D.L.T. Anderson and J. Willebrand (eds.), NATO ASI Series, Vol. I44, Springer-Verlag, Heidelberg, pp.157-210.

Sarmiento, J.L., and J.R. Toggweiler, 1984. A new model for the role of the oceans in determining atmospheric CO_2. *Nature* **308**, 621-624.

Sarmiento, J.L., T.M.C. Hughes, R.J. Stouffer, and S. Manabe, 1998. Simulated response of the ocean carbon cycle to anthropogenic climate warming. *Nature* **393**, 245-249.

Sato, M., J.E. Hanson, M.P. McCormick, and J.B. Pollack, 1993. Stratospheric aerosol optical depths, 1850-1990. *J. Geophys. Res.* **98**, 22987-22994.

Saunders, M.A., and A.R. Harris, 1997. Statistical evidence links exceptional 1995 Atlantic hurricane season to record sea warming. *Geophys. Res. Lett.* **24**, 1255-1258.

Schindler, D.W., and S.E. Bailey, 1993. The biosphere as an increasing sink for atmospheric carbon: Estimates from increased nitrogen deposition. *Global Biogeochem. Cycles* **7**, 717-734.

Schlesinger, M.E., and N. Ramankutty, 1992. Implications for global warming of intercycle solar irradiance variations. *Nature* **360**, 330-333.

Schlosser, P., G. Bonisch, M. Rhein, and R. Bayer, 1991. Reduction of deep water formation in the Greenland Sea during the 1980s: Evidence from tracer data. *Science* **251**, 1054-1056.

Schmitt, R.W., and S.E. Wijffels, 1992. The role of the ocean in the global water cycle. In *Interactions Between Global Climate Subsystems*, G.A. McBean and M. Hantel (eds.), Geophysical Monograph 75. American Geophysical Union, Washington, D.C., 77-84.

Schmitz, W.J. 1996. *On the World Ocean Circulation: The Pacific and Indian Oceans/A Global Update, Vol. 2*. WHOI Technical Report, WHOI-96-08, 237 pp.

Schneider, D., 1997. The rising seas. *Sci. Amer.* **276**, 112-117.

Schwartz, M.L., and E.C.F. Bird, 1990. Artificial beaches. *J. Coastal Res.* **6**.

Seager, R., and R. Murtugudde, 1997. Ocean dynamics, thermocline adjustment and regulation of tropical SST. *J. Climate* **10**, 521-534.

Sear, C.B., P.M. Kelly, P.D. Jones, and C.M. Goodess, 1987. Global surface temperature response to major volcanic eruptions. *Nature* **330**, 365-367.

Sellers, P.J., L. Bounoun, G.J. Collatz, D.A. Randall, D.A. Duzlich, S.O. Los, J.A. Berry, I.Y. Fung, C.J. Tucker, C.B. Field, and T.G. Jensen, 1996. Comparison of radiative and physiological effects of doubled atmospheric CO_2 on climate. *Science* **271**, 1402-1406.

Semazzi, F.H.M., L.-H. Neng, L.-L. Yuh, and F. Giorgi., 1993. A nested model study of the Sahelian climate response to sea-surface temperature anomalies. *Geophys. Res. Lett.* **20**, 2897-2900.

Serreze, M.C., J.A. Maslanik, J.R. Key, and R.F. Kokaly, 1995. Diagnosis of the record minimum in Arctic sea ice area during 1990 and associated snow cover. *Geophys. Res. Lett.* **22**, 2183-2186.

Severinghaus, J.P., T. Sowers, E.J. Brook, R.B. Alley, and M.L. Bender, 1998. Timing of abrupt climate change at the end of the Younger Dryas interval from the thermally fractionated gases in polar ice. *Nature* **391**, 141-146.

Sevruk, B., 1982. *Methods of Correcting for Systematic Error in Point Precipitation Measurements for Operational Use.* Operational Hydrology Report 21, World Meteorological Organization, Geneva, 91 pp.

Shackleton, N.J., and N.D. Opdyke, 1973. Oxygen isotope and paleomagnetic stratigraphy of equatorial Pacific core V28-238: Oxygen isotope temperatures and ice volumes on a 10^5 and 10^6 year scale. *Quat. Res.* **3**, 39-55.

Shapiro, L.J., 1982. Hurricane climatic fluctuations, Part 2: Relation to large-scale circulation. *Mon. Wea. Rev.* **110**, 1014-1023.

Shen, G.T., and R.B. Dunbar, 1996. Environmental controls on uranium in reef corals. *Geochim. Cosmochim. Acta* **59**, 2009-2024.

Shen, G.T., E.A. Boyle, and D.W. Lea, 1987. Cadmium in corals as a tracer of historical upwelling and industrial fallout. *Nature* **328**, 794-796.

Shen, G.T., L.J. Linn, T.M. Campbell, J.E. Cole, and R.G. Fairbanks, 1992. A chemical indicator of trade wind reversal in corals from the western tropical Pacific. *J. Geophys. Res.* **97**, 12689-12698.

Shen, C.-C., T. Lee, C.-Y. Chen, C.-H. Wang, C.-P. Dai, and L.-A. Li, 1996. The calibration of D[Sr/Ca] versus sea surface temperature relationship for Porites corals. *Geochim. Cosmochim. Acta* **60**, 3849-3858.

Shinoda, M., 1990. Long-term Sahelian drought from the late 1960's to the mid 1980's and its relation to the atmospheric circulation. *J. Meteor. Soc. Japan* **68**, 613-624.

Shukla, J.M., C. Nobre, and P. Sellers, 1990. Amazon deforestation and climate change. *Science* **247,** 1322-1324.

Simmons, A.J., J.M. Wallace, and G.W. Branstator, 1983. Barotropic wave propagation and instability, and atmospheric teleconnection patterns. *J. Atmos. Sci.* **40**, 1363-1392.

Sirocko, F., M. Sarnthein, H. Erlenkeuser, H. Lange, M. Arnold, and J.C. Duplessy, 1993. Century-scale events in monsoonal climate over the past 24,000 years. *Nature* **364**, 322-324.

Slonosky, V.C., L.A. Mysak, and J. Derome, 1997. Linking Arctic sea-ice and atmospheric circulation anomalies on interannual and decadal timescales. *Atmos.-Ocean* **35**, 333-366.

Smethie, W.M., Jr., 1993. Tracing the thermohaline circulation in the western North Atlantic using chlorofluorocarbons. *Prog. Oceanogr.* **31**, 51-99.

Smith, R.C., B. Prézelin, K.S. Baker, R.R. Bidigare, N.P. Boucher, T. Coley, D. Karentz, S. MacIntyre, H.A. Matlick, D. Menzies, M. Ondrusek, Z. Wan, and K.J. Waters, 1992. Ozone depletion: Ultraviolet radiation and phytoplankton biology in Antarctic waters. *Science* **255**, 952-959.

Smith, R.C., 1995. Implications of increased solar UV-B for aquatic ecosystems. In *Biotic Feedbacks in the Global Climatic System*, G.M. Woodwell and F.T. Mackenzie (eds.), Oxford University Press, New York, pp. 263-277.

Solomon, S., R.W. Portman, R.R. Garcia, L.W. Thomason, L.R. Poole, and M.P. McCormick, 1996. The role of aerosol variability in anthropogenic ozone depletion at northern mid-latitudes. *J. Geophys. Res.* **101**, 6713-6727.

Spall, M.A., 1996. Dynamics of the Gulf Stream/Deep Western Boundary current crossover, Part II: Low-frequency internal oscillations. *J. Phys. Oceanogr.* **26**, 2169-2182.

Speich, S., H. Dijkstra, and M. Ghil, 1995. Successive bifurcations in a shallow-water model, applied to the wind-driven ocean circulation. *Nonlin. Proc. Geophys.* **2**, 241-268.

Spencer, R.W., 1993. Global oceanic precipitation from the MSU during 1979-91 and comparisons to other climatologies. *J. Climate* **6**, 1301-1326.

Spencer, R.W., and J.R. Christy, 1992a. Precision and radiosonde validation of satellite gridpoint temperature anomalies, Part I: MSU channel 2. *J. Climate* **5**, 847-857.

Spencer, R.W., and J.R. Christy, 1992b. Precision and radiosonde validation of satellite gridpoint temperature anomalies, Part II: A tropospheric retrieval and trends during 1979-90. *J. Climate* **5**, 858-866.

Spencer, N.E., and P.L. Woodworth, 1993. *Data Holdings of the Permanent Service for Mean Sea Level (November 1993).* Permanent Service for Mean Sea Level, Bidston, Birkenhead, U.K., 81 pp.

Staehelin, J., J. Thudium, R. Buhler, A. Volz-Thomas, and W. Graber, 1994. Trends in surface ozone at Arosa (Switzerland). *Atmos. Environ.* **28**, 75-87.

Staehelin, J., R. Kegel, and N.R.P. Harris, 1998. Trend analysis of the homogenized total ozone series of Arosa (Switzerland), 1926-1996. *J. Geophys. Res.* **103**, 8389-8399.

Stager, J.C., and P.A. Mayewski, 1997. Abrupt early to mid-Holocene climatic transition registered at the equator and the poles. *Science* **276**, 1834-1836.

Steele, M., and T. Boyd, in press. Retreat of the cold halocline layer in the Arctic Ocean. *J. Geophys. Res.*

Stephens, B.B., R.F. Keeling, M. Heimann, K.D. Six, R. Murnane, and K. Caldeira, 1998. Testing global ocean carbon cycle models using measurements of atmospheric O_2 and CO_2 concentrations. *Global Biogeochem. Cycles* **12**, 213-230.

Stine, S., 1994. Extreme and persistent drought in California and Patagonia during mediaeval time. *Nature* **369**, 546-549.

Stockdale, T.N., D.L.T. Anderson, J.O.S. Alves, and M.A. Balmaseda, 1998. Global seasonal rainfall forecasts using a coupled ocean-atmosphere model. *Nature* **392**, 370-373.

Stocker, T.F., and A. Schmittner, 1997. Influence of CO_2 emission rates on the stability of the thermohaline circulation. *Nature* **388**, 862-865.

Stommel, H., 1979. Determination of water mass properties of water pumped down from the Ekman layer to the geostrophic flow below. *Proc. Nat. Acad. Sci. USA* **76**, 3051-3055.

Stouffer, R.J., S. Manabe, and K.Ya. Vinnikov, 1994. Model assessment of natural variability in recent global warming. *Nature* **367**, 634-636.

Sturges, W., 1987. Large-scale coherence of sea level at very low frequencies. *J. Phys. Oceanogr.* **17**, 2084-2094.

Sturges, W.E., and B.G. Hong, 1995. Wind forcing of the Atlantic thermocline along 32 degrees north at low frequencies. *J. Phys. Oceanogr.* **25**, 1706-1715.

Sun, D., 1997. El Niño: A coupled response to radiative heating? *Geophys. Res. Lett.* **24**, 2031-2034.

Sun, D.-Z., and Z. Liu, 1996. Dynamic ocean-atmosphere coupling: A thermostat for the tropics. *Science* **272**, 1148-1150.

Sun, D.-Z., and K.E. Trenberth, 1998. Coordinated heat removal from the equatorial Pacific during the 1986-87 El Niño. *Geophys. Res. Lett.***25**, 2659-2662.

Sutton, R.T., and Allen, M.R., 1997. Decadal predictability of North Atlantic temperature and climate. *Nature* **388**, 563-567.

Takahashi, T., R.A. Feely, R.F. Weiss, R.H. Wanninkhof, D.W. Chipman, S.L. Sutherland, and T.T. Takahashi, 1997. Global air-sea flux of CO_2: An estimate based on measurements of sea-air pCO_2 difference. *Proc. Nat. Acad. Sci. USA.* **94**, 8292-8299.

Talley, L.D., 1991. An Okhotsk Sea water anomaly: Implications for subthermocline ventilation in the North Pacific. *Deep-Sea Res.* **38**, S171-190.

Tanimoto, Y., N. Iwasaka, K. Hanawa, and Y. Toba, 1993. Characteristic variations of sea surface temperature with multiple time scales in the North Pacific. *J. Climate* **6**, 1153-1160.

Tans, P.P., 1998. Why carbon dioxide from fossil fuel burning won't go away. In *Perspectives in Environmental Chemistry,* D.L. Macalady, (ed.), Oxford University Press, New York, pp. 271-291.

Tans P.P., and White, J.W.C., 1998. In balance, with a little help from the plants. *Science* **281**, 183-184.

Tans, P.P., I.Y. Fung, and T. Takahashi, 1990. Observational constraints on the global atmospheric CO_2 budget. *Science* **247**, 1431-1438.

Tans, P.P., J.A. Berry, and R.F. Keeling, 1993. Oceanic $^{13}C/^{12}C$ observations: A new window on ocean CO_2 uptake. *Global Biogeochem. Cycles*

7, 353-368.

Thomason, L.W., G.S. Kent, C.R. Trepte, and L.R. Poole, 1997a. A comparison of the stratospheric aerosol background periods of 1979 and 1989-1991. *J. Geophys. Res.* **102**, 3611-3616.

Thomason, L.W., L.R. Poole, and T.R. Deshler, 1997b. A global climatology of stratospheric aerosol surface area density as deduced from SAGE II: 1984-1994. *J. Geophys. Res.* **102**, 8967-8976.

Thompson, D.W.J., and J.M. Wallace, 1998. The Arctic Oscillation signature in the wintertime geopotential height and temperature fields. *Geophys. Res. Lett.* **25**, 1297-1300.

Thompson, L.G., E. Mosley-Thompson, W. Dansgaard, and P.M. Grootes, 1986. The Little Ice Age as recorded in the stratigraphy of the tropical Quelccaya Ice Cap. *Science* **234**, 361-364.

Thompson, L.G., E. Mosley-Thompson, M.E. Davis, J.F. Bolzan, J. Dai, T. Yao, N. Gundestrup, X. Wu, L. Klein, and Z. Xie, 1989. Holocene-Late Pleistocene climatic ice core records from Qinghai-Tibetan Plateau. *Science* **246**, 474-477.

Thompson, A.M., J.A. Chapellaz, I.Y. Fung, and T.L. Kuscera, 1993a. The atmospheric CH_4 increase since the last glacial maximum: 2. Interaction with oxidants. *Tellus* **45B**, 242-257.

Thompson, L.G., E. Mosley-Thompson, M.E. Davis, P.N. Lin, T. Yao, M. Dyurgerov, and J. Dai, 1993b. Recent warming: Ice core evidence from tropical ice cores with emphasis on Central Asia. *Global Planet. Change* **7**, 145-156.

Thompson, L.G., E. Mosley-Thompson, M.E. Davis, P.-N. Lin, K.A. Henderson, J. Cole-Dai, J.F. Bolzan, and K.-B. Liu, 1995. Late Glacial stage and Holocene tropical ice core records from Huascarán, Peru. *Science* **269**, 46-50.

Thompson, L.G., T. Yao, M.E. Davis, K.A. Henderson, E. Mosley-Thompson, P.-N. Lin, J. Beer, H.-A. Synal, J. Cole-Dai, and J.F. Bolzan, 1997. Tropical climate instability: The last glacial cycle from a Qinghai-Tibetan ice core. *Science* **276**, 1821-1827.

Timmerman A., J. Oberhuber, A. Bacher, M. Esch, M. Latif, and R. Roeckner, 1998. ENSO response to greenhouse warming. Max-Planck Institut für Meteorologie, Report No. 251, 13 pp.

Ting, M., and Wang, H., 1997. Summertime United States precipitation variability and its relation to Pacific SST. *J. Climate* **10**, 1853-1873.

Tissue, D.T., and W.C. Oechel, 1987. Response of *Eriophorum vaginatum* to elevated CO_2 and temperature in the Alaskan Arctic tundra. *Ecology* **68**, 401-410.

Todling, R., S.E. Cohn, and N.S. Sivakumaran, 1998: Suboptimal schemes for retrospective data assimilation based on the fixed-lag Kalman smoother. *Mon. Wea. Rev.* **126**, 2274-2286.

Tourre, Y.M., and W.B. White, 1997. Evolution of the ENSO signal over the Indo-Pacific domain. *J. Phys. Oceanogr.* **27**, 683-696.

Tourre, Y.M., Y. Kushnir, and W.B. White, in press. Evolution of interdecadal variability in sea level pressure, sea surface temperature and upper ocean temperature over the Pacific Ocean. *J. Phys. Oceanogr.*

Townsend, A.R., B.H. Braswell, E.A. Holland, and J.E. Penner, 1996. Spatial and temporal patterns in terrestrial carbon storage due to deposition of atmospheric nitrogen. *Ecol. Appl.* **6**, 806-814.

Tremblay, L.-B., L.A. Mysak, and A.S. Dyke, 1997. Evidence from driftwood records for century-to-millennial scale variations of the high latitude atmospheric circulation during the Holocene. *Geophys. Res. Lett.* **24**, 2027-2030.

Trenberth, K., 1984. Signal versus noise in the Southern Oscillation. *Mon. Wea. Rev.* **112**, 326-332.

Trenberth, K., 1990. Recent observed interdecadal climate changes in the Northern Hemisphere. *Bull. Amer. Meteor. Soc.* **71**, 988-993.

Trenberth, K.E., 1995. Atmospheric circulation climate changes. *Clim. Change* **31**, 427-453.

Trenberth, K.E., in press. Atmospheric moisture residence times and cycling: Implications for rainfall rates with climate change. *Clim. Change* **34**.

Trenberth, K.E., and G.W. Branstator, 1992. Issues in establishing causes of the 1988 drought over North America. *J. Climate* **5**, 159-172.

Trenberth, K.E., and C. J. Guillemot, 1998: Evaluation of the atmospheric moisture and hydrogical cycle in the NCEP/NCAR reanalyses. *Climate Dyn.* **14**, 213-231.

Trenberth, K.E., and T.J. Hoar, 1995. The 1990-1995 El Niño-Southern Oscillation event: Longest on record. *Geophys. Res. Lett.* **23**, 57-60.

Trenberth, K.E., and T.J. Hoar, 1997. El Niño and climate change. *Geophys. Res. Lett.* **24**, 3057-3060.

Trenberth, K.E., and J.W. Hurrell, 1994. Decadal atmosphere-ocean variations in the Pacific. *Climate Dyn.* **9**, 303-319.

Trenberth, K.E., and A. Solomon, 1994. The global heat balance: Heat transports in the atmosphere and ocean. *Climate Dyn.* **10**, 107-134.

Trepte, C.R., and M.H. Hitchman, 1992. Tropical stratospheric circulation diagnosed in satellite aerosol data. *Nature* **355**, 626-628.

Trepte, C.R., R.E. Veiga, and M.P. McCormick, 1993. The poleward dispersal of Mount Pinatubo volcanic aerosols. *J. Geophys. Res.* **98**, 18563-18573.

Tucker, C.J., H.E. Dregne, and W.W. Newcomb, 1991. Expansion and contraction of the Sahara Desert from 1980 to 1990. *Science* **253**, 299-301.

Turner, B.L. II, R.H. Moss, and D.L. Skole (eds.), 1993. *Relating Land Use and Global Landcover Change: A Proposal for an IGBP-HDP Core Project.* IGBP report 24/HDP report 5, International Geosphere-Biosphere Programme, Stockholm.

Tziperman, E., 1997. Inherently unstable climate behaviour due to weak thermohaline ocean circulation. *Nature* **386**, 592-595.

Unal, Y.S., and M. Ghil, 1995. Interannual and interdecadal oscillation patterns in sea level. *Climate Dyn.* **11**, 255-278.

Uvo, C.B., C.A. Repelli, S.E. Zebiak, and Y. Kushnir, 1997. The relationships between tropical Pacific and Atlantic SST and northeast Brazil monthly precipitation. *J. Climate* **11**, 551-562.

van Loon, H., and R.L. Jenne, 1972. The zonal harmonic standing waves in the Southern Hemisphere. *J. Geophys. Res.* **77**, 992-1003.

van Loon, H., and J.C. Rogers, 1978. The seesaw in winter temperatures between Greenland and Northern Europe, Part I: General description. *Mon. Wea. Rev.* **106**, 296-310.

van Loon, H., R.L. Jenne, and K. Labitzke, 1973. Zonal harmonic standing waves. *J. Geophys. Res.* **78**, 4463-4471.

VEMAP (Vegetation/Ecosystem Modeling and Analysis Project), 1995. Vegetation/ecosystem modeling and analysis project: Comparing biogeography and biogeochemistry models in a continental-scale study of terrestrial ecosystem responses to climate change and CO_2 doubling. *Global Biogeochem. Cycles* **9**, 407-437.

Vinnikov, K.Ya., P.Ya. Groisman, and K.M. Lugina, 1990. Empirical data on contemporary global climate changes (temperature and precipitation). *J. Climate* **3**, 662-667.

Vitousek, P.M., P.R. Ehrlich, A.H. Ehrlich, and P.A. Matson, 1986. Human appropriation of the products of photosynthesis. *BioScience* **36**, 368-373.

Vitousek, P.M., H.A. Mooney, J. Lubchenco, and J.M. Melillo, 1997. Human domination of Earth's ecosystems. *Science* **277**, 494-499.

Wagner, R.G., and A.M. da Silva, 1994. Surface conditions associated with anomalous rainfall in the Guinea coastal region. *Int. J. Climatol.* **14**, 179-199.

Walker, G.T., and E. Bliss, 1932. World Weather V. *Mem. Roy. Meteor. Soc.* **4**, 53-84.

Wallace, J.M., 1996. Observed climatic variability: Spatial structure. In *Decadal Climate Variability*, D.L.T. Anderson and J. Willebrand (eds.), NATO ASI Series, Vol. I44, Springer-Verlag, Heidelberg, pp. 31-81.

Wallace, J.M., and D.S. Gutzler, 1981. Teleconnections in the geopotential height field during the Northern Hemisphere winter. *Mon. Wea. Rev.* **109**, 784-812.

Wallace, J.M., and P. Hobbs, 1977. *Atmospheric Science: An Introductory Survey.* Academic Press, New York, 467 pp.

Wallace, J.M., C. Smith, and C.S. Bretherton, 1992. Singular value decomposition of wintertime sea surface temperature and 500-mb height anomalies. *J. Climate* **5**, 561-576.

Wallace, J.M., Y. Zhang, and K.-H. Lau, 1993. Structure and seasonality of interannual and interdecadal variability of the geopotential height and temperature fields in the Northern Hemisphere troposphere. *J. Climate* **6**, 2063-2082.

Wallace, J.M, Y. Zhang, and J.A. Renwick, 1995. Dynamic contribution to hemispheric mean temperature trends. *Science* **270,** 780-783.

Walland, D.J., and I. Simmonds, 1997. North American and Eurasian snow cover co-variability. *Tellus* **49A**, 503-512.

Walsh, J.E., W.L. Chapman, and T.L. Shy, 1996. Recent decrease of sea level pressure in the central Arctic. *J. Climate* **9**, 480-486.

Ward, M.N., 1992. Provisionally corrected surface wind data, worldwide ocean-atmosphere surface fields, and Sahelian rainfall variability. *J. Climate* **5**, 454-475.

Ward, M.N., and C.K. Folland, 1991. Prediction of seasonal rainfall in the north Nordeste of Brazil using eigenvectors of sea surface temperature. *Int. J. Climatol.* **11**, 711-743.

WASA Group, 1998. Changing waves and storms in the northeast Atlantic. *Bull. Amer. Meteor. Soc.* **79**, 741-760.

Washington, W.M., and G.A. Meehl, 1995. High-latitude climate change in a global coupled ocean-atmosphere-sea ice model with increased atmospheric CO_2. *J. Geophy. Res.* **101**, 12795-12801.

Watanabe, T., and K. Mizuno, 1994. Decadal changes of the thermal structure in the North Pacific. *Int. WOCE Newsletter* **15**, 10-13.

WCRP (World Climate Research Programme), 1995. *CLIVAR Science Plan: A Study of Climate Variability and Predictability*. Report No. 89, World Climate Research Programme, Geneva, 157 pp.

Weaver, A.J., and T.M.C. Hughes, 1994. Rapid interglacial climate fluctuations driven by North Atlantic ocean circulation. *Nature* **367**, 447-450.

Weaver, A.J., and E.S. Sarachik, 1991. Evidence for decadal variability in an ocean general circulation model: An advective mechanism. *Atmos.-Ocean* **29**, 197-231.

Weaver, A.J., J. Marotzke, P.F. Cummins, and E.S. Sarachik, 1993. Stability and variability of the thermohaline circulation. *J. Phys. Oceanogr.* **23**, 39-60.

Wedin, D.A., and D. Tilman, 1996. Influence of nitrogen loading and species composition on the carbon balance of grasslands. *Science* **274**, 1720-1723.

Weiss, H., H.-A. Courty, W. Wetterstrom, L. Senior, R. Meadow, F. Guichard, and A. Curnow, 1993. The genesis and collapse of third millennium north Mesopotamian civilization. *Science* **261**, 995-1004.

Wellington, G.M., G. Merlen, and R.B. Dunbar, 1996. Calibration of stable oxygen isotope signatures in Galápagos corals. *Paleoceanography* **11**, 467-480.

White, W.B., and R. Peterson, 1996. An Antarctic Circumpolar Wave in surface pressure, wind, temperature, and sea ice extent. *Nature* **380**, 699-702.

White, J.W.C., L.K. Barlow, D. Fischer, P. Grootes, J. Jouzel, S.J. Johnson, M. Stuiver, and H. Clausen, 1997a. The climate signal in the stable isotopes of snow from Summit, Greenland: Results of comparisons with modern climate observations. *J. Geophys. Res.* **102**, 26425-26439.

White, W.B., J. Lean, D.R. Cayan, and M.D. Dettinger, 1997b. Response of global upper ocean temperature to changing solar irradiance. *J. Geophys. Res.* **102**, 3255-3266.

Whitlock, C., and P.J. Bartlein, 1997. Vegetation and climate change in northwest America during the past 125 kyr. *Nature* **388**, 57-61.

Whung, P.-Y., E.S. Saltzmann, M.J. Spencer, P.A. Mayewski, and N. S. Gundestrup, 1994. A two hundred year record of biogenic sulfur in a South Greenland ice core (20D). *J. Geophys. Res.* **99**, 1147-1156.

Wigley, T.M.L., and T.P. Barnett, 1990. Detection of the greenhouse effect in the observations. In *Climate Change: The IPCC Scientific Assessment,* J.T. Houghton, G.J. Jenkins, and J.J. Ephraums (eds.), Cambridge University Press, Cambridge, U.K., pp. 239-256.

Wigley, T.M.L., and D. Schimel (eds.), 1998. *The Carbon Cycle*. Cambridge University Press, Stanford, California. 420 pp.

Willson, R.C., and H.S. Hudson, 1991. A solar cycle of measured and modeled total irradiance. *Nature* **351**, 42-44.

Wilson, J.C., M.R. Stolzenburg, W.E. Clark, M. Lowenstein, G.V. Ferry, K.R. Chan, and K.K. Kelly, 1992. Stratospheric sulfate aerosol in and near the Northern Hemisphere polar vortex: The morphology of the sulfate layer, multimodal size distributions, and the effect of denitrification. *J. Geophys. Res.* **97**, 7997-8013.

Winton, M., and E.S. Sarachik, 1993. Thermohaline oscillations of an oceanic general circulation model induced by strong steady salinity forcing. *J. Phys. Oceanogr.* **23**, 1389-1410.

WMO (World Meteorological Organization), 1984. *Summary Report on the Status of the WMO Background Air Pollution Monitoring Network as at May 1984.* World Meteorological Organization Technical Document WMO/TD 13, Geneva, 21 pp.

WMO, 1995. *Scientific Assessment of Ozone Depletion: 1994. Global Ozone Research and Monitoring,* Report No. **37,** World Meteorological Organization, Geneva.

Wofsy, S.C., M.L. Goulden, J.W. Munger, S.-M. Fan, P.S. Bakwin, B.C. Daube, S.L. Bassow, and F.A. Bazzaz, 1993. Net exchange of CO_2 in a mid-latitude forest. *Science* **260**, 1314-1317.

Woodwell, G.M., and F.T. Mackenzie (eds.), 1995. *Biotic Feedbacks in the Global Climate System*, Oxford University Press, Oxford, U.K., 411 pp.

WRI (World Resources Institute), 1996. *World Resources 1996-1997, A Guide to the Global Environment.* Oxford University Press, New York, 365 pp.

Xie, S.-P., and Y. Tanimoto, 1998. A pan-Atlantic decadal climate oscillation. *Geophys. Res. Lett.* **25**, 2185-2188.

Yao, T., L.G. Thompson, E. Mosley-Thompson, Y. Zhihong, Z. Xingping, and P.-N. Lin, 1996. Climatological significance of the ^{18}O in north Tibetan ice cores. *J. Geophys. Res.* **101**, 29531-29537.

Yin, F.L., and E.S. Sarachik, 1994. An efficient convective adjustment scheme for ocean general circulation models. *J. Phys. Oceanogr.* **24**, 1425-1430.

Yuan, X., M.A. Cane, and D.G. Martinson, 1996. Climate variations: Cycling around the South Pole. *Nature* **380**, 673-674.

Zakharov, 1997. *Sea Ice in the Climate System.* Arctic Climate System Study, World Climate Research Programme, WMO/TD-No.782, World Meteorological Organization, Geneva, 80 pp.

Zebiak, S.E., 1993. Air-sea interaction in the equatorial Atlantic region. *J. Climate* **6**, 1567-1586.

Zenk, W., and N. Hogg, 1996. Warming trend in Antarctic Bottom Water flowing into the Brazil Basin. *Deep-Sea Res.* **43**, 1461-1473.

Zhang, R.-H., and S. Levitus, 1996. Structure and evolution of interannual variability of the tropical Pacific upper ocean temperature. *J. Geophys. Res.* **101**, 20501-20524.

Zhang, A., W.H. Soon, S.L. Baliunas, G.W. Lockwood, B.A. Skiff, and R.R. Radick, 1994. A method of determining possible brightness variations of the sun in past centuries from observations of solar-type stars. *Astrophys. J.* **427**, L111-L114.

Zhang, Y., J.M. Wallace, and D.S. Battisti, 1997. ENSO-like decade-to-century scale variability: 1900-93. *J. Climate* **10**, 1004-1020.

Zhang, R.-H., L.M. Rothstein, and A.J. Busalacchi, 1998. Origin of upper-ocean warming and El Niño change on decadal scales in the tropical Pacific Ocean. *Nature* **391**, 879-883.

Zhou, Z., 1996. *The Cause of Sea Surface Temperature Variability in the Tropical Atlantic Ocean,* Ph.D. thesis, University of Maryland, College Park, Maryland.

Zielinski, G.A., P.A. Mayewski, and L.D. Meeker, 1994. Record of volcanism since 7000 BC from the GISP2 Greenland ice core and implications for the volcano-climate system. *Science* **264**, 948-950.

Zorita, E., J.P. Hughes, D.P. Lettenmaier, and H. von Storch, 1995. Stochastic characterization of regional circulation patterns for climate model diagnosis and estimation of local precipitation. *J. Climate* **8**, 1023-1042.

Zuidema, G., G.J. van den Born, J. Alcamo, and G.J.J. Kreileman, 1994. Simulating changes in global land cover as affected by economic and climatic factors. *Water Air Soil Pollut.* **76**, 163-198.

Acronyms and Abbreviations

ACSYS	Arctic Climate System Study
ACW	Antarctic Circumpolar Wave
CFC	Chlorofluorocarbon
CGCM	Coupled general-circulation model
CLIMAP	Climate Mapping Project
CLIVAR	Climate Variability and Predictability Programme (WCRP)
CMDL	Climate Monitoring and Diagnostics Laboratory (NOAA)
COWL	Cool ocean-warm land pattern
CTD	Conductivity-Temperature-Depth instrument
Dec-cen	Decade-to-century-scale
DU	Dobson unit (of ozone amount)
EOF	Empirical orthogonal function
ENSO	El Niño/Southern Oscillation
GCM	General-circulation model
GCOS	Global Climate Observing System
GEWEX	Global Energy and Water Cycle Experiment
GLOSS	Global Sea Level Observing System
GOALS	Global Ocean-Atmosphere-Land System Program
GOOS	Global Ocean Observing System
GSA	Great Salinity Anomaly
GSDW	Greenland Sea Deep Water
Gt	Gigaton (10^{15} grams)
GTOS	Global Terrestrial Observing System
IGY	International Geophysical Year
IPCC	Intergovernmental Panel on Climate Change
ITCZ	Intertropical Convergence Zone
LSW	Labrador Sea Water
MARVOR	A multi-cycle RAFOS float
masl	meters above sea level
NAO	North Atlantic Oscillation
NOAA	National Oceanic and Atmospheric Administration
NPP	Net primary productivity
OCS	Carbonyl sulfide
PAGES	Past Global Changes core project
PALACE	Profiling Autonomous Lagrangian Circulation Explorer
PDO	Pacific (inter)Decadal Oscillation
PDSI	Palmer drought severity index
PNA	Pacific-North American pattern
RAFOS	Floating measurement device utilizing the Sound Fixing and Ranging (SOFAR) channel
SAGE	Stratospheric Aerosol and Gas Experiment
SAM I & II	Stratospheric Aerosol Measurement experiments I & II
SPMW	Subpolar Mode Water
SLP	Sea-level pressure
SMMR	Satellite Multifrequency Microwave Radiometer
SOI	Southern Oscillation Index
SPARC	Stratospheric Processes and their Role in Climate
SSM/I	Special Sensor Microwave/Imager
SST	Sea surface temperature
Sv	Sverdrup (10^6 m^3/s)
THC	Thermohaline circulation
TOGA	Tropical Oceans and Global Atmosphere program
TOMS	Total Ozone Mapping Spectrometer

UV-B	Ultraviolet B radiation
XBT	Expendable bathythermograph
XCTD	Expendable conductivity-temperature-depth instrument
WCRP	World Climate Research Programme
WOCE	World Ocean Circulation Experiment